JESSIE LUTHER AT THE GRENFELL MISSION

MCGILL-QUEEN'S ASSOCIATED MEDICAL SERVICES (HANNAH INSTITUTE)

Studies in the History of Medicine, Health, and Society Series
Editors: S.O. Freedman and J.T.H. Connor

Volumes in this series have financial support from Associated Medical Services, Inc., through the Hannah Institute for the History of Medicine Program.

1 Home Medicine
The Newfoundland Experience
John K. Crellin

2 A Long Way From Home
The Tuberculosis Epidemic among the Inuit
Pat Sandiford Grygier

3 Labrador Odyssey
The Journal and Photographs of Eliot Curwen on the Second Voyage of Wilfred Grenfell, 1899
Edited by Ronald Rompkey

4 Architecture in the Family Way
Doctors, Houses, and Women, 1870–1900
Annmarie Adams

5 Local Hospitals in Ancien Régime France
Rationalization, Resistance, Renewal, 1530–1789
Daniel Hickey

6 Foisted upon the Government?
State Responsibilities, Family Obligations, and the Care of the Dependant Aged in Nineteenth-Century Ontario
Edgar-André Montigny

7 A Young Man's Benefit
The Independent Order of Odd Fellows and Sickness Insurance in the United States and Canada, 1860–1929
George Emery and J.C. Herbert Emery

8 The Weariness, the Fever, and the Fret
The Campaign against Tuberculosis in Canada, 1900–1950
Katherine McCraig

9 The War Diary of Clare Gass, 1916–1918
Edited by Susan Mann

10 Jessie Luther at the Grenfell Mission
Edited by Ronald Rompkey

Jessie Luther at the Grenfell Mission

Edited by Ronald Rompkey

McGill-Queen's University Press
Montreal & Kingston · London · Ithaca

ISBN 0-7735-2176-3
Legal deposit first quarter 2001
Bibliothèque nationale du Québec

Printed in Canada on acid-free paper

This book has been published with the help of a grant from the
Humanities and Social Sciences Federation of Canada,
using funds provided by the Social Sciences
and Humanities Research Council of Canada.

McGill-Queen's University Press acknowledges the financial support
of the Government of Canada through the Book Publishing
Industry Development Program (BPIDP) for its activities.
It also acknowledges the support of the Canada Council for
the Arts for its publishing program.

Canadian Cataloguing in Publication Data

Luther, Jessie, 1860-1952
Jessie Luther at the Grenfell Mission
(McGill-Queen's/Associated Medical Services (Hannah Institute)
studies in the history of medicine)
Includes bibliographical references and index.
ISBN 0-7735-2176-3
1. Luther, Jessie, 1860–1952 – Diaries. 2. Occupational therapists – Newfoundland – St. Anthony – Diaries. 3. Grenfell Labrador Medical Mission – History. I. Rompkey, Ronald. II. Title. III. Series.
RM699.7.L88A3 2001 615.8'515'092 C00-901042-4

This book was typeset by Typo Litho Composition Inc.
in 10/12 Baskerville.

IN MEMORIAM

Sandra Gwyn (1935–2000)

Advocate for the Arts

Contents

Thus they discoursed together till late at night; and after they had committed themselves to their Lord for protection, they betook themselves to rest. The pilgrim they laid in a large upper chamber, whose window opened towards the sun rising; and the name of the chamber was Peace, where he slept to break of day.

Bunyan, *The Pilgrim's Progress*

Acknowledgments

Jessie Luther's experiences as a crafts teacher at the Grenfell Mission were first recorded in notebooks compiled during four years at St Anthony, Newfoundland, and in letters Miss Luther wrote to her mother at Providence, Rhode Island. Following a busy career as an occupational therapist, she combined the two in the memoir "When the Grenfell Mission Was Young" and submitted it to Houghton Mifflin in Boston in 1948. When the manuscript was dismissed as too "specialized" for a general trade publisher, she appealed to her friend S. Foster Damon, the William Blake scholar at Brown University, for editorial advice, but Damon was busy with his own publishing projects and recommended offering an extract to the *Atlantic Monthly* to generate public interest. Miss Luther chose instead to submit the whole manuscript to Atlantic Monthly Press, now bearing the fresh title "The Uncharted Coast," and this too was rejected, in May 1949. Finally, she turned to Peggy Hitchcock, a professional editor who had worked on *The Making of Yesterday* (1947), the diaries of the French correspondent Raoul de Roussy de Sales, and the ensuing correspondence between the two reveals that once Mrs Hitchcock agreed to assist, the manuscript took on a new shape. Its gradual transformation over a period of two years provides a very good example of how the ordering of personal events in diaries and collections of letters engages in what Hayden White calls the fictions of factual representation.

After reading the manuscript during Christmas 1949, Mrs Hitchcock recognized its value as a "charming and unpretentious postscript" to the saga of Wilfred Grenfell, the mission's founder. But it was too long and disorganized. "I would be inclined to suggest attempting to use the diary form universally, making excerpts from letters in such a way that they did not clash with this, and linking up where necessary with paragraphs written from the viewpoint of today," she advised.

Thus, it was Peggy Hitchcock who imposed the diary form throughout and called for a reduction to 350 pages of typescript with the elimination of extraneous matter and with new explanatory passages that could be added with the benefit of hindsight. That spring, having read the original notebooks, she recognized also their potential appeal as an intimate, unheroic portrait of Grenfell himself. However, she also introduced a fundamental difference of opinion about whether the work should be offered to the publisher simply as a record of events or as self-representation.

Mrs Hitchcock now began to wrestle with editorial questions and by April 1951 had produced a typescript of the early chapters, polishing the phrasing and adjusting the punctuation while Miss Luther pruned the chapters to follow. In May she concluded, "For the average reader I believe that Part I will be the important part of the book. This introduces him to the setting and the day to day life of the Mission. A relatively small caste [sic] of characters is involved, and all their activities are interesting. I do not think that in Part II we can attempt to follow events in such detail, and should try as far as possible to pick out entries that add a definitely new element to the picture." By abandoning the diary format at the end of the first part, she wanted to make the second part more episodic, bearing in mind especially the change of tone introduced by the loss of what she called the "happy family" atmosphere of the first part.

At this point, Miss Luther rebelled. Responding in May, she resisted the change of format and restored the edited sections to their original state. She was also troubled by the extensive stylistic changes. "What troubles me most," she said, "is so much re-phrasing and substitution of words that I have carefully chosen to give the impression I want to convey. I think there is often added interest in the use of inference and suggestion instead of merely factual statement for, after all, the average reader is supposed to have a little imagination, even in the 'realistic' world of today." Miss Luther had anticipated some judicious cutting and some correction of grammatical and rhetorical errors, but it had not occurred to her that her style would be called into question. "It is my book," she maintained, "which I want to recognize as my own – however faulty – and not feel that it is, partly, a changeling." Mrs Hitchcock therefore acquiesced and for the remainder of the process merely suggested alterations, leaving it to Miss Luther to accept or reject them.

By January 1952, Mrs Hitchcock appeared satisfied with the structure of the manuscript, although less satisfied with the phraseology and with Miss Luther's rather haphazard spelling and punctuation. "I have done what I could with punctuation and in a few places (on the carbon copies) suggested rewording," she reported. "There is unfortunately nothing I *can* do about the paragraphing, which is quite wild and woolly." She had by now abandoned all hope of a clean manuscript and knew that the finished product would lack the professional appearance a publisher would expect, but it was more important for Miss Luther, now ninety-one, to be content. So once again Miss Luther sent the manuscript to Houghton Mifflin – bearing yet another title, "Mission to Labrador" – and once again it was sent back. In his letter of August 1952, the publisher allowed that she had done an "excellent job" but rejected it for much the same reason as before – that the public's memory of Grenfell had faded and that her own experiences would not excite immediate interest.

Two months later Miss Luther died, leaving the manuscript, photographs, and associated documents in the hands of her nephew, Carleton Goff. In his efforts to publish the manuscript, Goff ran into exactly the same response, for in January 1953 Scribner's wrote him to say that "the general public would not be interested in such a detailed account of the Grenfell Mission." Goff subsequently released the documents to his niece, Martha Gendron, who was equally unsuccessful. Houghton Mifflin found the typescript "intriguing" in 1988 but lacking sufficient interest either as history or adventure.

I became aware of the manuscript's qualities as a representation of northern life from a woman's perspective while I was writing a biography of Wilfred Grenfell, and I am most grateful to Martha Gendron for permission to edit and publish it. While Grenfell's name does not command the recognition it once did, the manuscript has acquired importance as a window on the rural culture existing in northern Newfoundland and Labrador in the early years of the twentieth century, as perceived by a cultivated and educated visitor. In addition to their preoccupation with the Grenfell Mission, these recollections trace the beginnings of modernism in a small fishing village, where a community of British and American volunteers with a shared moral vision had assembled to participate in Grenfell's enterprise. When a switch is thrown on 27 July 1908, introducing electricity into a traditional way of life, it does not simply improve the medical service. It brings to an end what Jessie Luther calls "our simple life and cherished remoteness from urban glare, confusion and artificiality." Change followed rapidly. Her account of this experience offers a unique interpretation of "the quiet and beauty of this interesting place before the heavy hand of 'technology' is laid upon it to diminish its natural and unusual charm."

This edition presents the final typescript of "Mission to Labrador" in full. Inconsistencies of spelling and punctuation still exist in the original, and errors of fact, unexplained events and other difficulties remain. Even now, the material appears at times haphazard. I have silently corrected technical faults as far as possible without disturbing Miss Luther's characteristic tone and style. Her abbreviations and her spelling practices, especially her devotion to the hyphen and her appropriation of the exclamation mark to signify tone, remain as they were. Where variations of spelling occur, I have followed the preponderant one, and where local expressions cannot be interpreted in context I have glossed them in a note. The greatest challenges occurred with casual references to individuals and to local language and lore: even though Miss Luther aimed to present what she called "daily life," she sometimes left factual details to stand on their own, as though intrinsically interesting or self-evident. These details I have amplified in the notes.

In the process of identifying individuals, events, and phenomena, I have relied on the knowledge and good will of others. I thank Ki Adams, Joan Andersen, the late Edward Andrews, Linda Anstey, Colin Banfield, Sandra Beardsall, David Bell, the late Wesley Biles, the Reverend Frank Cluett, Norm Cohen, John Crellin, Carleton Goff, the late Rosamond Grenfell, Gordon Handcock, James K. Hiller, Gillian Hillyard, Jane Jacobs, Murray Lankester, Paula Laverty, John Mannion, Dorothy McNeill, the late Horace McNeill, William Kirwin, Stephen Loring, the Most Reverend Stewart Payne, Andrew Parsons, Ted Patey, Victor Penny, David Peters, Mabel Peterson, Geoffrey Place, Peter Roberts, Senator William Rompkey, Doris Saunders, Christina (Ashdown) Sidey, Thomas M. Smith, William L. Soule Jr, Anthony Spencer, Michael Taft, Robert W. Wakefield, and Bridget Williams.

I thank Manuscripts and Archives, Yale University Library, for permission to publish material in the Wilfred Thomason Grenfell Papers, and I thank the Grenfell Historical Society, St Anthony, for the use of photographs in its collection. The following institutions have provided guidance and direction: the Centre for Newfoundland Studies and the Maritime History Archive (Memorial University), the Providence Art Club, the Handicraft Club (Providence), the Rhode Island Historical Society Library, the Providence Public Library, and the Wheeler School. At Memorial University, I thank Gary McManus and Charles Conway, Cartographic Lab, for map reproduction, and Sharon Merils, Photographic Services, for the reprinting of archival photos.

I am grateful to those who have contributed archival information: Elizabeth Adams (Oxford University Archives), Mary Ann Bamberger (University of Illinois Library), Laura Beardsley (Historical Society of Pennsylvania), Odile Bourbigot (McGill University Archives), Mary Brinson (Program Development, Butler Hospital), Heather Colombo (American Occupational Therapy Foundation), Roberta Copp (South

Caroliniana Library), Kate Currie (Beaton Institute, Cape Breton), Barbara DeWolfe (Waltham Public Library), Larry Dohey (Archives of the Roman Catholic Archdiocese, St John's), Ruthann Gildea (Butler Hospital Library), George F. Henderson (Queen's University Archives), Philip Hiscock (Memorial University of Newfoundland Folklore Archive), Margot Karp (Pratt Institute Library), Edward L. Lach Jr (Johns Hopkins University Medical Archives), Evelyn Lannon (Boston Public Library), Elaine LeBlanc (Centre acadien, Université Sainte-Anne), Andrew Martinez (Rhode Island School of Design Archives), Lynne K. Fonteneau McCann (Williams College Library), Lucretia McClure (Harvard Medical Library), Martha Mitchell (Brown University Library), Regina Monteith (Historic Columbia Foundation, South Carolina), Anne Morton (Hudson's Bay Company Archives), Sharman Prouty (Deerfield Memorial Libraries), Susan Ravdin (Bowdoin College Library), Judith Schiff (Yale University Library), Jonathan Smith (Trinity College Library, Cambridge University), Brian Sullivan (Harvard University Library), Julie Thomas (Chicago Historical Society), Claire Troughton (Mildmay Mission Hospital, London), Lorraine Weiss (Rensselaer County Historical Society, Troy, New York), Susan White (Grenfell Regional Health Services, St Anthony), Laurie Whitehill (Rhode Island School of Design Library), Harold F. Worthley (Congregational Library, Boston), Nanci A. Young (Princeton University Archives), and Philip Worthington (National Library of New Zealand).

I am particularly grateful to Joan Rusted and the late Ivan Curson for research assistance carried out in the United Kingdom, to Jacob Larkin for research assistance at St John's, and to Carlotta Lemieux for her superb copy editing.

Research grants for this project were awarded by the Vice-President's Fund, Memorial University, and by the Hannah Institute for the History of Medicine.

R.G.R.
Department of English
Memorial University

Introduction

ST ANTHONY

Near the top of the northern peninsula of Newfoundland, within an inlet of St Anthony Bight, St Anthony harbour offers the mariner two advantages. As an anchorage, it provides deep water and good holding ground with plenty of room to swing; as a port, it gives ample protection from wind and swell and from floating ice borne south by the Labrador Current. St Anthony and its adjacent ports attracted European fishermen as early as the sixteenth century – Jacques Cartier mentioned it in 1534 on his way north from Cap Rouge.[1] By that time, it was already a seasonal fishing station.

The neighbouring harbours of Baie St-Antoine and St-Méen (later St Anthony Bight) appear in Lieutenant de Courcelle's chart of 1675,[2] when the two formed part of the system of fishing ports centred upon Croque (*l'havre du petit maître*), twenty miles to the south, in accordance with the rule of 1640 contrived at the request of the outfitters of St Malo. This rule not only named hospitable harbours but set forth the number of men to be accommodated in them, and it was further extended and elaborated by Colbert in the *Ordonnance de la Marine* of 1681.[3] The French migratory fishery in Newfoundland reached its peak in the 1780s and collapsed altogether in 1793, with the onset of war, but the French returned after the defeat of Napoleon. A census of 1828 shows over 9,000 men fishing on the northeast coast.[4] But the restriction placed on permanent French settlement by the Treaty of Utrecht (1713) left St Anthony one of the last harbours to be populated. Instead, northern harbours such as St Anthony were occupied by Newfoundlanders trading with the French in local items such as timber and bait. More permanent settlement followed the resolution of difficulties presented by the interpretation of French fishing rights on the peninsula, known to the French as Le Petit Nord (the near north).

The differences existing between England and France remained unresolved even though the French had abandoned their claims to the island of Newfoundland, for they had retained the right to a seasonal fishery on the coast between Cape Bonavista and Pointe Riche – the so-called French Shore or Treaty Shore. In 1783 the limits of the French Shore were adjusted to the section between Cape St John and Cape Ray (effectively White Bay and all of the west coast), provoking a series of disputes over the niceties of language governing the use of the shoreline. Were the French rights exclusive or concurrent with those of other fishermen? Could the French take all species or just cod? Were they permitted to build permanent structures and dwell in them throughout the year? Were the English damaging the property of French fishermen? Such inevitable questions arising from the joint exploitation of the fishing grounds did not encourage new settlement. Indeed, as Captain A.H. Hoskins, RN, wrote in his report of 1872, "On the French Shore I was invariably met with the objection that the inhabitants feel their position with regard to the French as so precarious and uncertain, that they do not trouble themselves to improve the land, but prefer to depend solely on the fishery as their means of subsistence."[5]

When the English migratory fishery declined and permanent settlement of the island began in the 1820s, Newfoundland's emerging sense of colonial identity created further tensions. "The history of the question from 1783 until about 1830 consists of a record of quarrels between the rival fishermen," wrote the Newfoundland journalist P.T. McGrath, "interspersed with proclamations by the Governors of the colony against the native fishers; for by this time the base of the industry had been transferred from the Mother Country to the colonial ports and the settlement of the coast-line was proceeding apace, for this was the heyday of the cod-fishing."[6] Newfoundland elected its first assembly in 1832, encouraging permanent settlement on the French Shore and forcing the French into awkward and sometimes absurd contrivances in order to preserve the equipment they were required to abandon for the winter. Thus, French fishermen resorted to leaving their *échaffauds* (stages) in the hands of *gardiens* (custodians), English-speaking settlers to whom they could entrust their effects. "Aux termes des traités," wrote Joseph Arthur de Gobineau in 1859, "nos équipages ne sauraient hiverner sur la côte. Cependant force leur est, en partant à l'automne, de laisser derrière eux, avec leurs habitations, des barques en grand nombre qu'il serait fort embarrassant et très-coûteux de rapporter en France à chaque voyage; puis, enfin, les meubles, les ustensiles, les filets, des amas de sel. Afin que tout ce butin ne soit pas dilapidé et pillé pendant leur absence, ils sont, depuis un temps immémorial, dans l'habitude de le confier à la surveillance de gardiens sujets anglais, dont l'hivernage ne peut donner lieu à aucune contestation."[7]

Under such conditions, the French presence continued seasonally in harbours such as St Anthony, where the Church of England bishop Edward Feild took the trouble to record the state of French "rooms," or stations, in 1849. "There are three in this harbour," he noted, "and four large vessels, one of them a barque. The tricolour was hoisted on the flag-staff when we came in."[8] These three rooms are shown in Lieutenant Pierre's survey of 1857: a headquarters (*amirauté*) on the north side of the entrance, inside Lamage Point (Le Dos de Cheval); a second installation at Marguerite Point (Point à la Marguerite); and a third on the south side, where the government wharf stands today. Commander William Chimmo, RN, visiting in 1867 aboard HMS *Gannet* in search of new fishing grounds and harbours of refuge, observed that there were still "several fishing boats & stages on the shore" and offered the opinion that "the Frenchmen looked more substantial and comfortable than those of the N.fdLand; There were at anchor a Bark, Brig and several Schooners. Although full of French people and french vessels we did not see a French Flag – they were either ashamed of them or had none to hoist, so I imagine!"[9] Dr Frederic H. Crowdy recorded further details of the rivalry during his visit in 1873:

In this arm there are only one or two British families, but in the summer 2 or 3 vessels full of men come from France, and set up their stages here: they seem to be well enough in their way, but the men, who are all paid wages and dieted, not having any share in the profits, annoy the English settlers by stealing salmon out of their nets, and Mr. [Alfred] Simms told me he frequently had to fire his gun at them to frighten them away from his nets; on the whole however he seems to lead a quiet life with them and dines with the Captain of the crew every Sunday. He takes care of their boats during the winter when they are away in France, and is paid for that purpose: Mr. Simms is a very substantial planter having some money in the Bank at St John's, however his money is no good when he can get nothing to buy with it, & this year the ice was so late on the coast that the traders could not get there[;] hence they were all very nearly starved. Simms would have had ample for his own family wants, but he had to share with his neighbours, till at last they were almost at the last biscuit when the French vessels arrived and they borrowed some provisions from them.[10]

As late as 1886, Julien Thoulet mentioned a *bureau de poste* at Cape St Anthony, where French ships could pick up their mail in bulk, and a cemetery at Croque, where erstwhile colleagues and competitors lay side by side: "Matelots et officiers, Français et Anglais, catholiques et protestants, un Villaret de Joyeuse entre un quartier-maître français et un novice anglais, dorment côte à côte, dans la suprême égalité."[11] The two great fisheries proceeded independently, so independently that there is hardly any overlap between the fishing vocabularies of the two language groups.[12]

As the bank fishery developed in the nineteenth century, the French presence on the northern coast declined, and France eventually renounced its privileges in an *entente cordiale* signed in 1904. By then, St Anthony had grown into something of a regional centre. The census of 1857 (wherein St Anthony still appears as St-Antoine) shows only eleven houses and seventy-one inhabitants, but that number had doubled within a generation, and there were now four resident merchants. When Wilfred Grenfell established a small clinic there in 1900, the population stood at 233, of whom 194 were Methodists.

The Great Northern Peninsula was not the most hospitable part of the island for settlement. It featured a climate closer to that of Labrador and parts of northern Canada, determined largely by the cold water of the Labrador Current that bathed both sides of the peninsula until July. In summer, the northern tip offered an ambiance quite unlike what was experienced even on the southern portion of the peninsula, with a cooler season accompanied by low cloud and sometimes fog. Nowhere else on earth did arctic conditions such as these intrude so far south into the middle latitudes.[13] Frost began in early November, and harbours were normally frozen solid before Christmas. Thaw might not occur until early May. "Poorer and more desolate-looking places than St Anthony and Griquet, near the straits, could not be imagined," observed the Canadian author Agnes Laut in 1899. "Winter closes in earlier here than at the other settlements and cuts the inhabitants off from all possible aid."[14] At St Anthony, the mean annual temperature hovered just above freezing, making the growing season correspondingly short.

The surrounding countryside presented a backdrop of wild, subarctic grandeur. Beneath skies of aquamarine, turning at dusk into a phantasmagoria of pinks, yellows, and mauves, acres of rolling moss and shrub barrens, punctuated with patches of stunted spruce and tamarack, abruptly came to a halt at a rocky cove or a sequence of cliffs. In places, forest gave way to lichen woodland with black spruce growing in open areas, tundra covering all but the lowest ground and making possible the quarrying of peat. Geologists have concluded that the area was overrun by glaciers at various times, producing a variety of deposits and erosional effects:[15] visible rock twisted into knobs and basins, and erratic boulders littering the surface. A second type of terrain consisted of bedrock barely supporting vegetation, although it was possible to raise crops in selected areas. James P. Howley, in his geological survey of 1902, found the soil in the vicinity of Conche to be "of superior quality," sufficient to produce "very fine" root crops and hay.[16] But these were not the features that attracted Wilfred Grenfell. Grenfell required a harbour with access to the Strait of Belle Isle for half the year. At the invitation of a growing population in need of a permanent medical service, he set about building a hospital there in 1899.

WILFRED GRENFELL

More than any other individual, Wilfred Grenfell (1865–1940) transformed St Anthony from a small fishing community into a centre of some substance. When he arrived for his first winter there, he was thirty-five and already experienced in northern medicine. The son of the headmaster and proprietor of Mostyn House School at Parkgate, Cheshire, he had been educated at his father's school and at Marlborough before entering the London Hospital Medical College in 1883. A vigorous, idealistic young man of athletic disposition, he played rugby and football for both the hospital and Richmond, and in his spare time displayed a taste for utilitarian social improvement, running a summer camp for working boys and opening a Sunday school and boxing club at his house in Hackney. Awarded the conjoint diploma LRCP, MRCS in 1888, he then failed the MB examination at the University of London and for the Michaelmas term took up residence at Queen's College, Oxford, where he won blues for rowing and rugby and a half blue for athletics.

Throughout his medical training, Grenfell was susceptible to new ideas for social improvement and was eager to adapt them. Moved by the American evangelists Dwight L. Moody and Ira D. Sankey to dedicate his life to Christian social action, he soon put his beliefs into practice as a physician aboard the first hospital ship fitted out by the Royal National Mission to Deep Sea Fishermen for work among the North Sea fishing fleets. An opportunity for more independent action followed in 1892. At the invitation of Newfoundland mercantile interests, he inaugurated a floating medical service for the migratory fishing communities of coastal Labrador and the Quebec north shore, and a year later he returned to build small hospitals at Battle Harbour and Indian Harbour. He had found his life's work.

A student of visionary and unconventional ideas of the kind sometimes associated with communalist movements, Grenfell was seized with the desire to reform. For the next forty years, he built not only hospitals and nursing stations but schools and orphanages, and started farms, mills, and other small industries. He became in his early years an admirer of such worthies as H.H. Almond, the hygienic optimist of Loretto School in Scotland, and later embraced prohibition, vegetarianism, eugenics, Emmanualism, self-help, cooperatives, the settlement movement, and the psychic research conducted by Sir Oliver Lodge. In the United States, he was attracted to the Social Gospel and admired the liberal American Baptist minister Harry Emerson Fosdick. He was also an enthusiastic supporter of Dr John Harvey Kellogg, the health propagandist and promoter of flaked cereals. In 1899, well established in Labrador, he decided to remain in northern Newfoundland over the winter as an itinerant.

Grenfell had planned to remain in Labrador for the first part of the winter and then walk to Quebec. Finding this impossible, he decided to cross the ice in the Strait of Belle Isle in February and walk to the nearest point of the Newfoundland railway. But finding this equally impossible, he chose to install himself at St Anthony and visit the settlements within a fifty-mile radius by dog team. In September 1899 he left England aboard a tramp steamer bound for Tilt Cove with A.J. Beattie, a Scot he had met at Oxford, intending to return in the spring for the launch of his new steam-driven hospital vessel *Strathcona.* Arriving at St Anthony, the two set themselves up in the house of the merchant Frederick Moore and made plans for building a house of their own. "Mr. Moore was the real factor in determining my coming here," Grenfell wrote to the Mission supporters at home, "as he most generously put at my disposal half his house: and while he arranges for all our wants, food, housing, firing, &c., and so leaves us an absolutely free hand for our work, with no domestic cares whatever, yet we can be as private as we like, having even a door of our own by which people can come directly to our room without in any way passing through other parts of the house."[17] But the Mission would not support overseas expansion, and Grenfell was required to take the responsibility himself, successfully raising money through lecture tours of Canada, the United States, and Great Britain.

During the winter of 1900–1, he left on a tour of the eastern United States, leaving the St Anthony medical service in the hands of Dr George Simpson, who reported as early as November that the foundation for a new wooden hospital had been laid with local labour and that the ground around it was cleared. Grenfell's tour was organized by Emma White of the Congregationalist Library in Boston, and by the following summer, assisted by a grant from the Government of Newfoundland, Grenfell was roughing out plans for the interior of the hospital. At New Year's, however, the building had still not been opened, because the heating system had to be installed below the ground floor, and the rock beneath it needed to be blasted away with dynamite. Construction was further delayed to allow the local lumber to dry, and thus Grenfell was not able to get inside until December 1903. Meanwhile, eleven in-patients were treated that year and thirteen operations carried out under anaesthetic. With the signing of the *entente cordiale* between France and Britain in 1904, Grenfell predicted, "The settlement of the French Shore question should mean much for this coast. It will bring more southerners down to settle, and will open up the coast to enterprise that it has hitherto been closed to."[18] During the first half of 1904, 618 patients were seen in the new hospital, which was officially opened in March. St Anthony had quickly become the medical hub of the French Shore.

In the spring of 1905, Grenfell made a successful tour of New York and Massachusetts, recruiting young volunteers attracted by his idealism. In his absence, the number of patients rose to 790. A wharf was built, together with an industrial building and a sawmill. On his return, he wrote, "I found a road had been built up to this, and the corner of St Anthony around the hospital is beginning to look quite like a settlement."[19] In 1906, the first year of Jessie Luther's residence, the number of patients increased to 807, and for his efforts Grenfell was made a companion of the Order of St Michael and St George, the first of many public honours to be bestowed upon him. The size of the hospital had doubled by 1909: there were now 1,667 outpatients and 153 treated within. Grenfell also added an orphanage with 28 children under the superintendency of Eleanor Storr, a British volunteer. In 1910, the last year recorded by Jessie Luther, 195 patients were treated in the hospital and 1,707 outpatients, and the fortnightly admissions brought by the mail steamer were taxing the staff to the limit. In one instance, no fewer than 91 people were being fed there.

In the spring of 1908, Grenfell was suddenly transformed from a physician and reformer into a popular hero. After going through the ice in Hare Bay with a team of dogs, he spent a night adrift on a floating fragment and suffered severe frostbite before a small crew of fishermen arrived the next day to pluck him off. The details were recorded in his widely read booklet *Adrift on an Icepan* (1909). Regarded by now as one of the colony's chief benefactors, he next built a house near the hospital and brought there his bride of twenty-three, Anna Elizabeth Caldwell MacClanahan of Lake Forest, Illinois. The couple took up permanent residence, and two sons were born: Wilfred Thomason Jr (1910) and Kinloch Pascoe (1912). (A daughter, Rosamond Loveday, was born in 1917.) But Grenfell's new status as husband and father changed the tone of the small community of workers he had recruited, for Anne Grenfell, a strong, well-educated woman, exerted considerable influence on her husband's life. He dressed more carefully. He paid attention to social conventions.

Following Jessie Luther's departure in 1910 – the end of the pioneer period – Grenfell expanded his operations with the aid of supporting associations in Canada, the United States, Britain, and Ireland, all incorporated under the aegis of the International Grenfell Association, founded in 1913. Moreover, he formally detached himself from the RNMDSF. Not content merely to improve the medical service, he concentrated on eliminating social and economic deficiencies and earned further public recognition for his benevolence and self-sacrifice. Such a wide range of activities amounted to cultural intervention, however, and the new ventures inevitably brought him into conflict with the elected authorities, particularly when he revealed through his many books, articles, and public lectures the chronic poverty and starvation among the population. Nevertheless, the Grenfell Mission, as it came to be known, continued as a benevolent force in northern Newfoundland and Labrador.

After the First World War, Grenfell resumed his activities as full-time lecturer, writer, and publicist, aided in large measure by his wife in the role of secretary and editor. His autobiography, *A Labrador Doctor* (1919), and his absorbing lectures on life in the North brought him a host of fresh admirers and volunteers. In 1927, as he opened a new concrete hospital, he was created a knight of the Order of St Michael and St George. The running of the Grenfell Mission passed to younger, more practical hands, notably those of Dr Harry Paddon at North West River, Labrador, and Dr Charles Curtis at St Anthony. Grenfell resigned from the direct management in 1936 but remained active as a figurehead and fundraiser, dividing his time between his residence at Kinloch House, near Charlotte, Vermont, and the southern United States, where he journeyed periodically to recuperate from the effects of chronic arhythmia. Lady Grenfell predeceased him in 1938 after a struggle with cancer. Grenfell himself died of a heart attack at Kinloch House on 9 October 1940, and his ashes were brought to St Anthony, where they were placed inside a rock face overlooking the harbour, near those of his wife and his close colleagues.

JESSIE LUTHER

By the time she met Grenfell in 1905, Jessie Luther was a mature artist and teacher with considerable experience in craft production. She was born at Providence, Rhode Island, on 3 November 1860, the daughter of Joseph J. Luther, a watchmaker and clockmaker, "ingenious and

Jessie Luther, age seventeen
(Jessie Luther Papers [JLP])

clever," as she put it, in the use of tools.[20] Her mother, Sarah Godfrey Luther, sprang from the seagoing Godfrey family, claiming descent from Roger Williams, the founder of Providence in 1636. There were two sisters: Alice Luther (later Goff) and Mabel L. Luther, also a craftswoman. Jessie herself was born at home, 12 Williams Street, and attended small private schools until the age of ten; but showing an interest in handmade objects and an aptitude for visual art, she was sent for private instruction to a school run by Miss J.L. Abbott until 1879. At nineteen, she seemed destined for a life in the art world of New England.

First she entered the Rhode Island School of Design for drawing and painting but left a year later to study watercolour with the distinguished Sidney R. Burleigh, who had just returned from Paris after three years of study with Jean-Paul Laurens. Burleigh's studio on Thomas Street, the first studio building erected at Providence, was the right place for a young artist to start. For the next four years, Jessie Luther studied at the Studio, a small art school run by Mary C. Wheeler, who had been raised in the heady transcendentalist and communitarian environment of Concord, Massachusetts. Devoted to French art, Miss Wheeler saw herself as a pioneer, having returned after six years in France to establish herself at 12 Cabot Street. "The art in this city is pitiable," she wrote a colleague in 1883, "hardly an artist here who deserves the name of one. I find a desire to learn, and shall have a good working class from the 'first families.'"[21] The Studio, a large room in the house which Mary Wheeler later built at 26 Cabot Street – decorated with ancient Normandy chests, tapestries, velvet and brocade hangings, antique Persian rugs, and her own paintings – presented one of the formative influences of Jessie Luther's early life. Three more years of private studio work followed, and in 1890 Jessie left with the Wheeler students for Paris, where she painted for several months with the American prodigy Paul Bartlett and the French artist Raphaël Collin.

After returning to the United States in 1894 to continue her studio work and teach art classes, she took her first institutional appointment at the College for Women in Columbia, South Carolina, part of the women's education movement gathering force in the South after the end of the Civil War. Run as a private business, its shares bought and sold on the open market, it was a small college situated in the Hampton-Preston Mansion purchased from the Ursuline Convent. It had opened its doors to its first class in the fall of 1890 and was run on tight Presbyterian principles. Jessie Luther was featured prominently in the college catalogue as an artist trained in "the best art schools in New England" and in "eminent artists' studios in Europe."[22] The art department offered instruction in painting and drawing, including oil, watercolour, and mineral painting. The course in decorative art included painting on satin, silk, plush, china, plaques, panels, and tiles.[23] But a year later, Jessie was back in Providence. She was thirty-five with no marriage prospects and no permanent position. Returning to her private studio in an outbuilding of the family home, she began again to weave and paint, and to make pottery for firing in the kiln at the Wheeler School.

By 1899, Jessie Luther was a disciple of arts and crafts, the ideology of nostalgic aestheticism imported from Britain and well established in Boston and Chicago. Arts and crafts ideology, incorporating the ideas and example of John Ruskin and William Morris as a response to the perceived ugliness and repetitiveness of industrialism, had produced a profound impact on American culture and continued to do so up to the beginning of the First World War. Ruskin and Morris inspired a generation dedicated to craft production, workmanship, and artistic integrity, a way of reuniting art and labour. For Jessie Luther, the appeal of crafts, especially weaving, sprang from what she considered a "primitive" instinct to create practical objects from materials within reach. In a rare theoretical statement published later in *House Beautiful*, a forum for arts and crafts ideas, she wrote,

It is a far cry from the work of primitive man to the modern mill, with its hundreds of looms, its whirring machinery, and its many toilers, each performing the allotted task without knowledge of or interest in the whole structure of which their own work is a part. And now, in the midst of this mechanical age, when methods are constantly invented for substituting the machine for the hand, thereby forcing the hand simply to become a part of the machine itself, without thought, will, or ingenuity, there has come a reaction, a reversion to hand industry, to the work of the individual, with its interest in the worker, its opportunity for originality in color and design, and for the development and co-ordination of hand and brain.[24]

In the United States, arts and crafts activity took hold among élite women and art professionals, and by the turn of the century it had spread deep into the middle class, producing not a political revolution but an alternative within the existing social order, based on the cultivation of art in everyday life. It found expression in the household furniture, rugs, pottery, metalwork, jewellery, bookbinding, woodwork, textiles, architecture, and other products of individual workshops and craft societies.

What had begun in Britain as an alternative to the art and labour associated with industrial capitalism was transformed into a style of art and a middle-class leisure activity as well as a means of personal, social, and occupational therapy. The history of arts and crafts is part of the history of the middle class and its concept of the role of the craftsman. "The idealistic, uplifting, optimistic yet paternalistic spirit of the movement reflects the class that turned to arts and crafts as solution to and escape from the industrial world it did so much to forge," writes Eileen Boris.[25] Although some proponents developed art manufacturing, the movement thrived on the participation of individual women through the revival of traditional "women's work" in rural folk societies and communities of urban immigrants.

The arts and crafts "movement" found expression in the network of organized groups formed in the 1890s, particularly in Boston and Chicago, among professional artists such as Jessie Luther, and also among architects, art workers, and ultimately amateurs and patrons. The Society of Arts and Crafts, founded in Boston in 1897, provided a model for these associations and gave Jessie Luther her first opportunity to have her work assessed. (She was elected a craftsman in 1905 and promoted to master craftsman in 1923.)[26] In Boston, the focus of the British movement narrowed to a preoccupation with good taste, with the production of unique, precious, and decorative objects. The society began to cater to the consumer, and the salesroom replaced the cooperative workshop as the defining space. Not so in Chicago.

There, the residents of Hull House, the social settlement founded by Jane Addams, promoted the craftsman ideal. At Hull House, where the Chicago Society of Arts and Crafts was founded, the society openly admired Ruskin and Morris and took strength from what Jane Addams called "collective living," the give and take of colleagues.[27]

Jessie Luther took full advantage of the craft training available at Providence. She studied modelling and woodcarving with J. Liberty Tadd. She enrolled for courses in metalwork, enamelling, and weaving at the Rhode Island School of Design. She pursued woodcarving and carpentry with Allen Weeks of the Moses Brown School. In Massachusetts, she studied leather tooling with Bessie Cram at Cambridge and wove baskets at Deerfield Industries. She did bookbinding with Clara Buffum and pottery with Arthur E. Baggs. In 1901 she responded to an invitation from Jane Addams to reside at Hull House, where residents pledged to remain for at least six months and paid their own board and lodging. For the next two years she worked at Hull House, one of the pre-eminent arts and crafts centres. Early in her tenure, she made her mark as a craftswoman at an exhibition of the Chicago Arts and Crafts Society, where a reviewer observed, "Miss Jessie Luther (a new name, and a welcome one in this Society) reveals a clever use of her hammer in the copper bowls and boxes, some of which are enriched with transparent enamels."[28]

At Hull House, Jessie taught woodwork, woodcarving, pottery, basket weaving, and metalwork.[29] Jane Addams made her director of the Labor Museum, a new program established as a working demonstration of old, discarded craft processes in textiles, metalwork, woodwork, and bookbinding as they were practised by Chicago's immigrant population. The museum had been opened in November 1900 with five departments: woods, bookbinding, textiles, grains, metals, and pottery. The workers paid for the materials themselves, and the finished products were sold by the workers or the shop directors.[30] Strictly speaking, the Labor Museum was not a museum at all but a space where immigrant women were encouraged to feel confident in their traditional accomplishments. The demonstrations were supplemented with a lecture series by speakers influenced by the ideology of Ruskin and Morris, and by this means the Labor Museum showed how the revival of domestic crafts could be integrated into the new industrial order. It was a challenging and demanding job that Jessie Luther sustained until 1903.

The making of pottery and its decorative forms was perhaps the most popular of the museum's activities, followed closely by instruction in the use of grains and the operation of kitchen equipment. Cooking classes were held every day. Hull House was also the meeting place of the Chicago Arts and Crafts Society, and its members assisted workers in the shops. The museum therefore concentrated on the labour rather than the commercial features. "It was further believed," wrote Jessie Luther in 1902, "although it is perhaps difficult to demonstrate, that when the materials of daily life and contact remind the student of the subject of his lesson and its connections, it would hold his interest and feed his thought as abstract and unconnected study utterly fails to do."[31] Hull House offered a living example of arts and crafts ideals. In 1904, when the *Craftsman*, the arts and crafts journal founded in 1901 by Gustav Stickley, reported on Hull House's progress, it sounded lyrical. "As Ruskin and Morris have shrieked in our ears, he who tends a machine all his life and is treated like a machine ... tends to become himself a machine, grows less and less human," it began. "Here, at any rate, in these few rooms, is an attempt to substitute Renascence [sic] and mediaeval ideals of industry, or the better part of them, for that Puritan utilitarianism which crushed them out, and which seem to have retained, without holding on to its compensating religiousness."[32] However, running such a program taxed body and soul, and in 1903 Jessie found herself a patient at St Mary's Hospital, suffering from exhaustion. That summer, as she recovered in rural New Hampshire, she entered the next phase of her life, occupational therapy.

The Ark was a large family house built by Joseph Cutter early in the nineteenth century in Jaffrey, at the foot of Mount Monadnock.[33] By the end of the century, the whole estate had become a popular resort. Here Jessie Luther met Herbert Hall (1870–1923), a general practitioner from Manchester, New Hampshire, with a large practice at Marblehead, Massachusetts. Although not a psychiatrist, Hall was preoccupied with the enervating effects of neurasthenia, a condition of ennui and depression, which was then regarded as one of the severe consequences of modern urban living. "Neurasthenia" was a general term encompassing what might now be regarded as a variety of neurotic symptoms. "Tortured by indecision and doubt," writes T.J. Jackson Lears, "the neurasthenic seemed a pathetic descendant of the iron-willed Americans who had cleared forests, drained swamps, and subdued a continent."[34] Hall challenged the rest cure favoured almost universally for its treatment. Inspired by the ideas of Elwood Worcester and Emmanuelism, a popular self-help movement, he advocated occupation and activity to engage hands and mind, and fixed upon the practice of crafts to do it. "The modern Arts and Crafts idea appealed very strongly," he wrote later, "because of the growing interest in the movement and because of the clean, wholesome atmosphere which surrounds such work, and because of the many-sided appeal which such a work as the making of pottery, for

Jessie Luther, age forty (JLP)

instance, has to most educated minds."[35] Through the fall of 1903, Hall and Jessie Luther worked out a plan for occupational therapy in a small sanatorium known as the Handicraft Shop, and by May 1904 they had established a collection of workshops in the Devereux Mansion, at the entrance to the harbour. The *Craftsman* wrote enthusiastically, "Although this enterprise is now in its infancy, and doubtless many changes and improvements will suggest themselves with its growth, its immediate success would seem to indicate that it has met a need which may exist in many towns and cities throughout the country."[36]

The new venture attracted a succession of physicians from the Boston area in search of results, and in March 1905 Wilfred Grenfell was brought to observe their activities during one of his fundraising tours. What struck Grenfell at once was the non-medical application of such training, and he invited Jessie to come north and teach basic crafts in Newfoundland, but she was too involved with the new project. In January 1906 she moved into full-time occupational therapy, a profession still in its infancy, by taking on the direction of a program at the Butler Hospital in Providence.[37] There she spent the rest of her career, leaving Hall to make a commercial success of his small industries, especially pottery, by hiring Arthur Eugene Baggs, Jessie Luther's former instructor. The Marblehead Pottery, one of his occupational therapy "handicraft shops," was especially productive, not only in pottery but handweaving, woodcarving, and metalwork. By 1908, it was turning out nearly two hundred pieces a week, using brick clay from Massachusetts and stoneware clay from New Jersey. Hall next opened the Devereux Mansion Sanatorium in 1912 with facilities for forty workers. The pottery eventually moved from the Devereux Mansion to a house on Front Street, and Baggs took over ownership in 1916.

At the Butler Hospital, Jessie Luther met her second mentor, the psychiatrist Alder Blumer (1857–1940). A native of Sunderland, County Durham, Blumer had received his early education in England, France, and Germany but came to the United States in 1877, receiving the MD from the University of Pennsylvania in 1879. He was appointed superintendent of the Utica State Hospital, New York, in 1886 and in 1899 superintendent of the Butler Hospital, where he remained until 1922. Blumer is significant for his leadership in abolishing mechanical restraint, for the development of occupational therapy, and for championing the humanitarian and scientific study of the treatment of mental illness. For him, the most hopeful direction for psychotherapy lay in securing a proper environment for the patient's restoration, especially in the provision of "systematic and diversified occupation."[38] Jessie Luther was instrumental in creating such an environment, beginning with a class of eight. The next summer, when her patients moved on to other activities outside the hospital, she responded to Grenfell's persistent requests, and with her departure from Boston her travel account begins.

The slow development of the occupational therapy program at the Butler Hospital made it possible for her to spend two winters at St Anthony. Otherwise, she made the annual trek in late June to check on the standard of work, look for markets, and extend her network to other communities, returning to Providence in October or November. Using new looms constructed with local labour, she founded weaving centres at Griquet, Cook's Harbour, and Flower's Cove, Newfoundland; at Forteau and Red Bay, Labrador; and at Harrington Harbour, Quebec. In Labrador, she encouraged native embroidery on deerskin and found a market for baskets made from native grass. Her most successful initiative, however, was the extensive hooked-mat industry, the largest and best known of the Grenfell crafts, later developed by Anne Grenfell and marketed through shops in Boston, New York, and Philadelphia. Years later, Jessie Luther declared, "I know of no work of the kind found anywhere to compare with the excellence of workmanship of these mats, many of which resemble tapestry. It is difficult to make purchasers believe they are hooked, the work is so fine and even."[39]

The mat industry began when Miss Luther discovered women making mats in their homes with basic designs of blocks, triangles, or floral patterns. At about the same time, she became aware of the mats designed and dyed by Helen Albee in rural New Hampshire, and it occurred to her that a similar industry might be developed. She wrote, "To her my thanks are due for a correspondence full of interest and encouragement and her little book [*Abnàkee Rugs*] has been a most helpful guide in my own attempt to develop a hooked mat industry."[40] Convinced that the existing designs would not be marketed easily outside Newfoundland and that the colours would not match the interiors of contemporary households, she offered new designs, consisting of a plain centre and a border featuring images of seals, walruses, deer, rabbits, komatiks, dogs, and people in winter dress. (Grenfell later added other local images.) Most of the material was dyed in the loom room at St Anthony using vegetable dyes such as madder and indigo, fast colours, and experimental hues produced from spruce twigs mordanted with alum so as to yield a fast beige colour. The women were also given copperas and lime to produce iron-rust shades, and the patent bark powder used by fishermen to dye sails. With the patterns marked or stencilled on burlap, the mats were given to women to work with in their homes.

Although Miss Luther regarded herself as the founder of the Grenfell crafts, she did not continue her close association after her return to Providence. After the first summer, the industry was not funded by the Grenfell Mission at all but relied on donations, appeals, and public lectures given by Jessie Luther and Grenfell. All the teachers were volunteers. Yet the weavers and mat makers needed to be paid for their work, and materials had to be provided in advance to fill orders during the winter.[41] Jessie Luther revealed something of her frustration with the financial arrangements in a statement written in December 1913 for the mission's magazine, *Among the Deep Sea Fishers*:

> There is no definite Industrial Fund to depend upon and it has been difficult to extend the work under the circumstances as the necessary expenditure of extension would be a financial risk. I am glad to say that we are not in debt at present, but the payment of the workers in the new weaving centers during the winter, before the product can be sold, will tax the fund heavily and possibly deplete it. I have hoped that an endowment fund of $1,000 for the work might be possible. If the income derived from such a sum could be depended upon annually it would solve many problems and enable us to carry the work to many localities where it is much needed.[42]

Sensing the end of her direct involvement, she forwarded a copy to Jane Addams, who replied, "I was enormously interested in the article you sent me, and I hope very much that you will always feel that you are part of Hull-House. We certainly think so. It would be very nice indeed if you could make us a visit, some time when your work among the deep-sea fisheries is impossible."[43] But there was another dimension. Offstage, Miss Luther was engaged in a major disagreement with Grenfell about the management of the industry, and by the fall of 1915 she was ready to resign.

Her correspondence with Grenfell does not reveal the source of her anxiety, but the draft of one letter shows her growing impatience with the way the industry proceeded. Miss Luther was disappointed by the lack of capital investment in the Industrial Department and perhaps felt exploited. The industry she had founded was being managed by Anne Grenfell, and the direction of the work was falling to new craft teachers whom Grenfell recruited in the United States and Great Britain. She wrote,

Your attitude in regards to the Industrial work & your frank statement that your written arguments are to be disregarded make it utterly impossible for me to longer work in connection with one whose aims are so at variance with mine and on whose word I cannot rely. I am therefore sending my resignation to Mr. Sheard to be laid before the International Board. I have been notified that at the August meeting of the Board it was decided you and I should have a conference some time during your stay in New England in October, at which time the Industrial situation should be thoroughly discussed. As the decision in regard to my resignation is absolutely final, it seems to me that such a meeting is neither necessary nor advisable, as after the first of January [1916] my efforts and interests in connection with the Industrial Department and the Mission generally will come to an end and the interval will be devoted merely to settling all the business affairs connected with my ten years of effort. ~~You will of course understand that after that date the proceeds from any of my lectures or writings concerning the Mission and my work in connection with it will be devoted to some other cause.~~[44]

I am extremely sorry that after all these years, in which I have given my utmost of time, strength, ability and the fruit of other years of training and experience, the end should come like this, ~~but I feel that the seed I have sown cannot die, but I have the satisfaction of knowing that I have established & stimulated industrial work among the people of the coast, also that the work has been accomplished, as everyone knows, from a truly altruistic spirit, and with no thought of self-interest in any way – you and Mrs. Grenfell know this in your hearts no matter what impression you may give to others. But the saddest thought I have to take with me in severing my connection with the Mission is the thought of what it once was and of what it is now in spirit. Thus I remember my belief in its high ideals, its Christian charity, its consideration, courtesy~~

~~and kindness, its freedom from malice, unkind criticism and ridicule – its manly courage, its simple womanly helpfulness & its lack of self-seeking & self-aggrandizement, & then I think of it as it is now become during the past few years, the reverse of all those Christian qualities in the vy air one breathed, it makes me sick at heart. The idol's feet of clay are vy ugly to behold.~~

In my future memories of the Mission, I shall try to remember only the earlier years of my connection with it when the Christian spirit of high ideals were still alive, the Mission affairs less complicated and the life at St Anthony exemplified by what you wrote under a photograph of the Guest House, "The Abode of Peace."[45]

Grenfell did not take long to respond, suggesting that Miss Luther's health was to blame for her attitude (Dr Little had been treating her for an unspecified illness for several years)[46] and that unstated differences existed between her and the permanent staff on the coast:

I am sorry that you misinterpreted my letter and my attitude to your department. The increasing differences between yourself and the workers, the Grieves, Wakefields, Paddons, Mrs. Blackburn, Miss Lesley, Miss Bedford, etc., made me appeal to you to interpret our actions in the spirit of camaraderie and believe that we could still be loyal to you. You cannot expect anyone living right alongside hungry people not to find work for them to earn bread at once because of any rules. I could not work with any colleague who would expect it. What you have looked upon as opposition was simply apposition. Our aim, as well as yours, was to help the unfortunate.

I have read your letter over carefully with Dr Little and I am familiar with your physical troubles of late years, and I see no other way now but to advise the Committee to accept your resignation. We must all resign sometime. My great sorrow is that your long connection with the work in the North should end in misunderstanding.[47]

Miss Luther also had made her objections known to William R. Stirling, Grenfell's financial adviser in Chicago. After Stirling forwarded a copy to Anne Grenfell for comment, Mrs. Grenfell quickly responded to correct what she regarded as distortions, explaining,

For one thing she says she was entirely responsible for collecting the money for the Industrial and that it was a hand to mouth existence. I know for a fact that Wilf helped that department out of his own pocket, and also by supplying it with raw materials to be made up by the people. It was he who started all of the local industries on the Coast long before Miss Luther came. Of course she started the homespun, which is good; she also started pottery making, which never amounted to anything; and at one time she tried to start polishing stone and making Arts and Crafts jewelry.

There was more to it than that. According to Anne Grenfell, Miss Luther was demanding a very high standard at a time when people were in dire need of cash. She continued,

The real crux of that situation was that Miss Luther was not willing to sell except at a very high profit for her department, and also, and chiefly, that she would not employ but a very few women, and never any but those who could do extra good work. Naturally in many cases this left out the needy ones. She implies that she only stopped giving out her mats in large numbers to be hooked when we came into the field. I do not think this is accurate. Mrs. Blackburn had charge of Miss Luther's work here in the winter, and she will agree with me that Miss Luther never gave out mats in any quantities, and certainly did not begin to withdraw when we began.[48]

She also believed Miss Luther had made the mistake of "trading on the name of the Mission" to justify the prices she charged and not on the merits of the goods themselves. Thus, with Miss Luther's departure, Anne Grenfell added the mat industry to her own responsibilities. In a further letter to Stirling, she reported, "It takes me three and a half hours every day to get ready the necessary mats; and with them and Wilf's correspondence I can refute the statements of certain ones who declare that I do nothing all day long! I am really not nearly so vindictive as I sound, but I have worked now for years very hard on this mat industry, and I am interested to see it go ahead."[49]

Relations between Miss Luther and the Grenfells improved in later years, and Miss Luther continued to advise the Mission on marketing in the United States, but they were never quite as warm again. Anne Grenfell managed the sale of crafts at the beginning of the First World War, when even less cash circulated in northern Newfoundland, and Miss Luther concentrated on her primary responsibilities as director of occupational therapy at the Butler Hospital. There, with the exception of fourteen months in 1918–19 when she directed the occupational therapy program for convalescing soldiers at the Walter Reed Hospital in Washington, she remained until her retirement in 1938, living at home until her mother's death in 1928, when the family house was sold to the Providence Country Day School. That year, she moved to an apartment at 50 Olive Street, and in 1933 she built a small house at Westport Harbor, Massachusetts, where she resided during the summer, travelling to and fro by bus even after she turned ninety.

Her retirement left her free to engage in her many pursuits. Devoted to the outdoors, she hiked with the Appalachian Club. She supported the Audubon Society, the League of Women Voters, and the Animal Rescue League. Her greatest devotion, however, continued to be her art. A talented painter in watercolours and pastels, she exhibited at the Providence Art Club and attended numerous exhibitions and lectures there. She was a subscriber to the Boston Symphony Orchestra. She remained a member of the Society of Arts and Crafts in Boston and arranged sales for Grenfell crafts at the Handicraft Club in Providence, where Grenfell came to lecture. Each time this occurred, a succession of large wooden crates would arrive bearing an assortment of northern treasures: sealskins, bearskin rugs, decorated Inuit garments, laminated whips, soapstone carvings, sleds and kayaks, caribou antlers, hooked and woven rugs, and dolls. Many similar objects decorated the studio in which she kept her looms. In an adjacent building she kept a kerosene-fired kiln for baking clay pots but seldom used it. Her life remained full and active, except for the last few weeks before her death on 17 October 1952 at a nursing home on Thayer Street. She was buried at the North Burial Ground in Providence.

JESSIE LUTHER AT ST ANTHONY

As a literary form, the travel narrative is never simply a report of day-to-day events. It follows a set of discursive practices involving the persona of the narrator, the choice of incidents, and the description of significant places and objects. Broadly speaking, it offers two possibilities for disclosure: a detached, impersonal investigation of manners and customs or a more sentimental account of an unfolding drama in which the narrator is a part.[50] While Jessie Luther advertises her observations as a survey of daily life at St Anthony – its customs, speech patterns, dress, habits, ways of thinking, and religious beliefs – she assumes at length the voice of the sentimental observer engaged in an important enterprise with her associates and the very community she seeks to reveal. As narrator, she presents herself in what has been called "womanly" fashion,[51] partaking in the conventions of decency and propriety that preoccupied itinerant women writers a century ago. Accordingly, female vulnerability is assumed as a rhetorical posture, one by which men are valued for manifesting the gallant and honorable behaviour associated with the ideal of the gentleman or faulted for other unacceptable behaviour. The feminine perspective is confirmed by the foregrounding of episodes in which men are judged by standards of propriety, by the giving or withholding of approval, or by a preoccupation with domestic matters not normally discussed in men's travel accounts. Occasionally, Miss Luther adopts a patronizing tone, characteristic of similar accounts by educated, middle-class travellers from Britain and the United States, who have been charged with the duty to improve, educate, or convert in their chosen fields.

Part 1 begins with her journey to St Anthony for the winter of 1906–7, and much is made of the physical challenges presented to women travellers. She arrives to find that with Grenfell absent, no special arrangements have been made for her, and the staff are in disarray. A week later, she comes to realize that her work will involve much more than weaving, and on 20 July she declares, "This is more like camping than anything else, except that in camping as I have known it one has no responsibilities and a guide or someone to cook, not usually for so many." She is impressed more than ever by Grenfell at close quarters, his "strength" and "manliness," and by the time she returns to Rhode Island in September, having started weaving classes and introduced a variety of other crafts, she has started to consider coming back on a more permanent basis after Grenfell suggests he will build her a house. That winter, Grenfell convinces Dr Blumer that the crafts program will not advance without her help, and by October 1907 she is on her way again aboard the narrow-gauge railroad built in Newfoundland at the turn of the century. In keeping with travel-writing conventions, her comments on the railway and the system of coastal boats serving the northern communities are more than simply descriptions. They are signs of the peculiarity of foreign ways and projections of her own fortitude in tolerating discomfort.[52]

During an interlude at St John's, she plans for the winter and frets excessively about the character of her new co-workers, especially William George Lindsay, and the sleeping arrangements. But on her arrival, she discovers that Grenfell has bought a house for the resident staff and enlarged it to accommodate them all. The winter's events are therefore played out in what came to be called the Guest House, where she evokes a regime of genteel decorum prevailing among her American and British co-workers. Lindsay is characterized approvingly as the gentlemanly Irishman, even though, according to his military record, he was born in Liverpool of Irish parents in 1865 and had fought as a lieutenant of the King's (Liverpool) Regiment in the Boer War during 1901–2.[53] Lindsay's activities in the wider world are difficult to trace, but in St Anthony he enjoyed a certain status as manager of the peat bog and the reindeer herd. Cuthbert Lee, aged seventeen, who had volunteered for a year of manual work and teaching before entering Harvard, emerges as the helpful young man. Dr John Mason Little is the compleat Bostonian, admired by all. Little joined the mission staff in 1907 and remained at St Anthony for another ten years as chief surgeon before returning to Boston as the husband of Ruth Keese, who is presented here as the spirited and innovative teacher. Miss Luther herself is artist-in-residence, superintending mother, and adviser on household affairs.

From January 1908, the group are preoccupied by the arrival of a herd of reindeer, which Grenfell has procured in Norway with the aid of private fundraisers in Boston and a grant from the Government of Canada. The reindeer are destined eventually for the Canadian mainland. Once they arrive, the members of the household are consumed by the animals' daily movements and those of the Saami herders (Lapps) brought to care for them and train local men. The delivery of the reindeer was one of the most spectacular events to occur on the French Shore, and Miss Luther is absorbed by the treatment of the animals as well as by the appearance and qualities of the Saami – members of a minority ethnic group concentrated in northern Norway, Sweden, Finland, and the Kola peninsula of Russia. Miss Luther's remarks, therefore, typify the Saami as Other: their "odd" physical appearance, exotic dress, and strange behaviour.[54] Intended as a solution to the social problems presented by sled dogs, the reindeer remained at St Anthony until 1918, but the noble experiment did not meet with the success anticipated. In the course of time, the herd was diminished by poachers, and the remaining animals were transported by the Canadian government to the north shore of the Gulf of St Lawrence.

The winter activities selected for comment by Miss Luther leave the impression that she enjoyed a close rapport with Grenfell, whom she idealizes. Indeed, her descriptions of daily events suggest even deeper feelings. For example, she observes on 19 January 1907, "It is interesting to have him consult me about household matters, and he seems to enjoy the whole atmosphere of the house. I see on his face what I call the 'purring' expression which means content." On 21 January she adds, "When Dr. Grenfell first came he often forgot about coming to meals or took them at the hospital without notifying us. I never waited for him, for we are prompt and Dr. Grenfell begged me not to because he hated to feel hurried. Lately, however, he has been on time – sometimes even ahead." In numerous other ways, Grenfell is shown paying her close attention, attention she may have interpreted as personal interest. When in early February he and Dr Little leave for two weeks on a round of medical visits, she writes, "Miss Keese and I feel quite widowed, and the house does not seem a house at all, only a shelter." She adds on 23 March 1908, "Mr. Lindsay is still here, for which we are thankful, for it is lonely without our doctors." By this time, the group has drawn together into what she refers to as a "family," a small enclave rife with personal jokes and intimacies. She has never experienced a close-knit household like this outside her own family and will not experience it again.

That is perhaps why what happens next is dramatized in such detail. On Easter Sunday, en route to attend a patient, Grenfell goes through the ice with komatik and dog team as he cuts across Hare Bay; he drags himself onto a pan of ice and drifts helplessly towards open water throughout the night. After taking extraordinary measures to conserve his body heat, he is rescued twenty-four hours later and brought home exhausted from hunger, exposure, and frostbite. The retelling of these events and the monitoring of his slow recovery preoccupy Miss Luther for several weeks, and much of her commentary is taken up with her own part in his convalescence.

By mid-May, she realizes that the intimate relations of the group have reached their term and will not be repeated. On a photo of the Guest House, Grenfell inscribes "The Abode of Peace," an allusion to the episode in *The Pilgrim's Progress* where Christian enjoys the hospitality of the House Beautiful and reposes in a large upper chamber of the same name. She writes on 20 May, "We expect to hear the steamer's whistle at any moment, and I do not like to think of the ensuing changes." But even as she awaits the influx of summer volunteers, she does not realize how profound those changes will be. In August the electric lights are switched on. "They seem like intruders," she writes. "The indoor lights are most welcome, but no one wants the whole country illuminated, competing with the starlight and moonlight – even the aurora. We hope they can be turned off on clear nights." She journeys briefly to Battle Harbour, Labrador, and returns to Providence in November.

Part 2 introduces a more detached tone as Miss Luther absorbs further dramatic change. In June 1909, Grenfell brings his mother to the United States to witness the conferring of an honorary degree by Harvard University and introduces her to Miss Luther. "I fell in love with Mrs. Grenfell at once: a charming, gracious lady, and the doctor's affectionate devotion to her was beautiful to see," she writes, not realizing that Grenfell has met Anne MacClanahan on the return voyage and had proposed to her before the ship reached New York. Grenfell has decided to keep his plans to himself until August, after his mother has visited his fiancée's family at Lake Forest, Illinois. When he does make the announcement, Miss Luther writes with some misgiving, "Everyone rejoiced with the Doctor in his happiness, but it was difficult to think of him as a domestic man … for his absorption in his work and its precedence in all his thoughts and actions was so great … that any thought of his connection with the possible restraints of married life had seemed remote." Granted a second leave from the Butler Hospital, she returns to St Anthony in October 1909 for quite a different experience.

Not only are the members of the Guest House family dispersed, but a "lack of harmony" exists following the arrival of the new manager of local affairs, Mr. Waldron, whom Miss Luther refers to disapprovingly as Mr. Manager. "I foresee adjustments and changes," she notes, "some of them difficult, but discouragement is not a part of my philosophy, and this is only the beginning of a new era." The hospital and orphanage are being extended again, and work has begun on Grenfell's house, which she describes as "palatial." The autumn is charged with expectation as the community awaits its first glimpse of Grenfell and his bride, who are married on 18 November. Delays in the construction and decoration of their house raise the level of anxiety. Christmas passes, then New Year's. A reception worthy of a vice-regal visit is planned and delayed repeatedly until finally the couple arrive on 12 January aboard the *Prospero.* "She has a lovely voice, is very pretty, very tall and beautifully dressed," Miss Luther reports, but two days later, following the staff dinner, she adds, "I cannot yet realize that Dr. Grenfell is married." Grenfell decides he wants the staff to gather at his house on Sunday afternoons for tea, but by the end of February Anne Grenfell, now pregnant, is not up to entertaining, and the practice is suspended.

With the arrival of spring, Miss Luther gets her kiln working and begins to make pottery. St Anthony is overrun by young volunteers from schools and colleges. On 18 July she embarks for her second trip to the Labrador coast to expand her network of weavers, calling at Battle Harbour, Batteau, Indian Harbour, Rigolet, and Turnavik. A month later she arrives at Hopedale aboard the Reid coastal vessel *Invermore* and spends two weeks engrossed in the life of the Moravians, the German sect that has been working among the Inuit as missionaries and traders since the eighteenth century. She is interested not only in the language of the Moravians but in their worship, customs, music, dress, food, gardens, furnishings, and of course their crafts. Always conscious of the visual, she is impressed by the terrain and transfixed by the northern sky. "The colors danced and swirled like living things, green, pink, crimson, yellow and violet in great waves, constantly changing and darting from one part of the heavens to another," she writes. "From a crown overhead, long fingers of light fell like a fringed curtain, then suddenly changed, and a swirl of wonderful color intensified in another place. I never saw so much movement."

She departs with Grenfell aboard the *Strathcona* on 2 September, passing at Long Tickle, Turnavik, West Turnavik, Cartwright, Grady Harbour, and Indian Tickle, impressed by his multiple role as physician, lay minister, and magistrate. "Dr. Grenfell is at his best in this kind of life," she decides, "and it is a privilege to see him in this role, so like what is often written about him. In some ways he is a different Dr. Grenfell from the one most people meet in a more sophisticated environment." *Strathcona* bounces across the Strait of Belle Isle and on 10 September arrives at St Anthony. Miss Luther realizes as she prepares to go home that she will not return again for an extended stay. "Something tells me that the past winter was my last on the coast," she reflects. "I shall never again see the country half submerged under its beautiful white blanket of snow, enjoy the thrill of travelling behind a dog-team or know the comfortable intimacy of a small group, isolated far from the outside world."

The St Anthony papers end here except for a brief retrospective written in 1916, during her break with Grenfell. During the summer of 1914, her last working tour of the industrial centres, she finds herself at Harrington Harbour when news arrives of the outbreak of war. The modern age has reached the coast of Labrador, and the Grenfell Mission will be changed forever. Men she has worked with side by side will perish in Europe. Grenfell himself will depart for six months in France to stitch up the wounded. She sails to St Anthony aboard the *Prospero,* which has darkened ship for fear of attracting the German submarines rumoured to be lurking offshore, and three weeks later she leaves aboard the *Prospero* again. She returns only once, for a visit in 1930 at the age of seventy. In her afterword of 1952, as she reads of these events again, she is struck by fresh insights into "certain psychological aspects and reactions of which I was unconscious when events and contacts with the people and environment were actually taking place." At ninety-one, it is as if she sees them for the first time.

Ronald Rompkey

Map 1 Newfoundland and Labrador

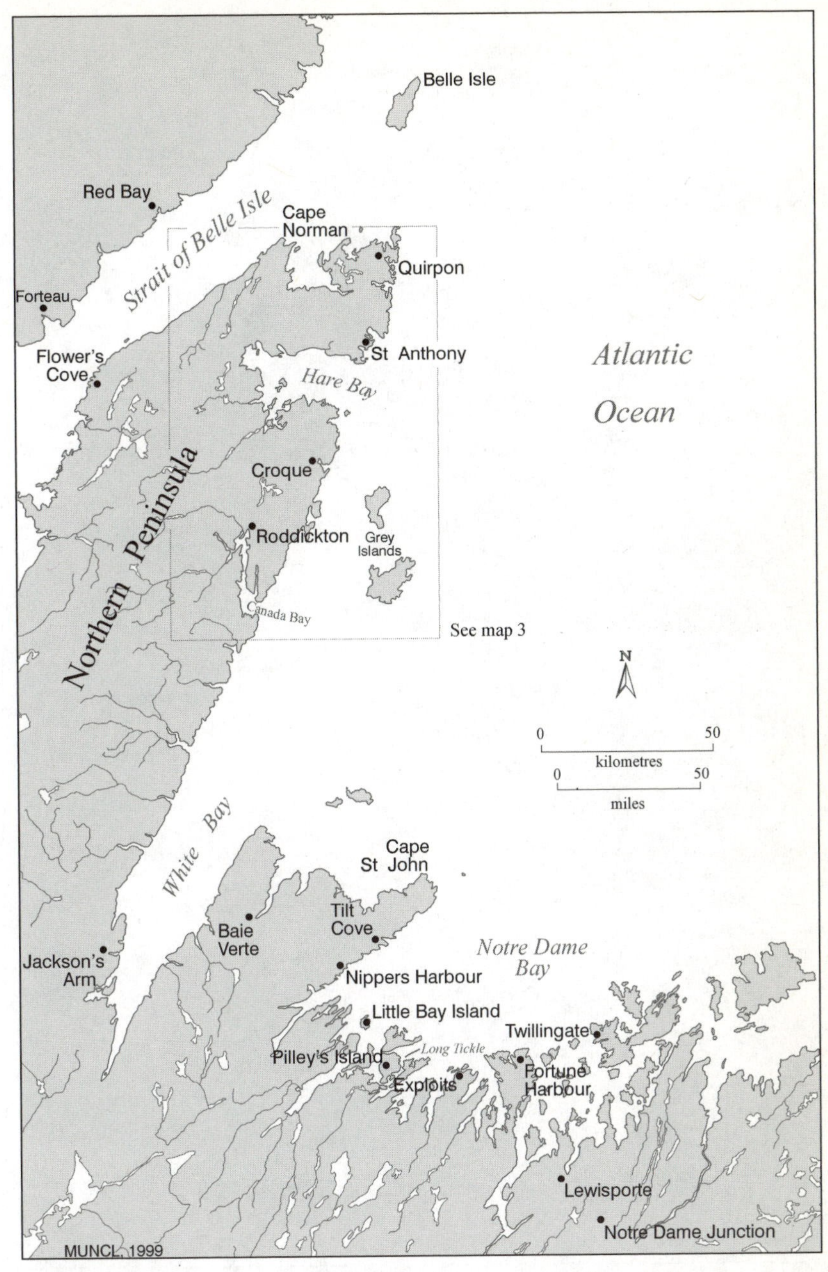

Map 2 Northern Peninsula with part of the French Shore

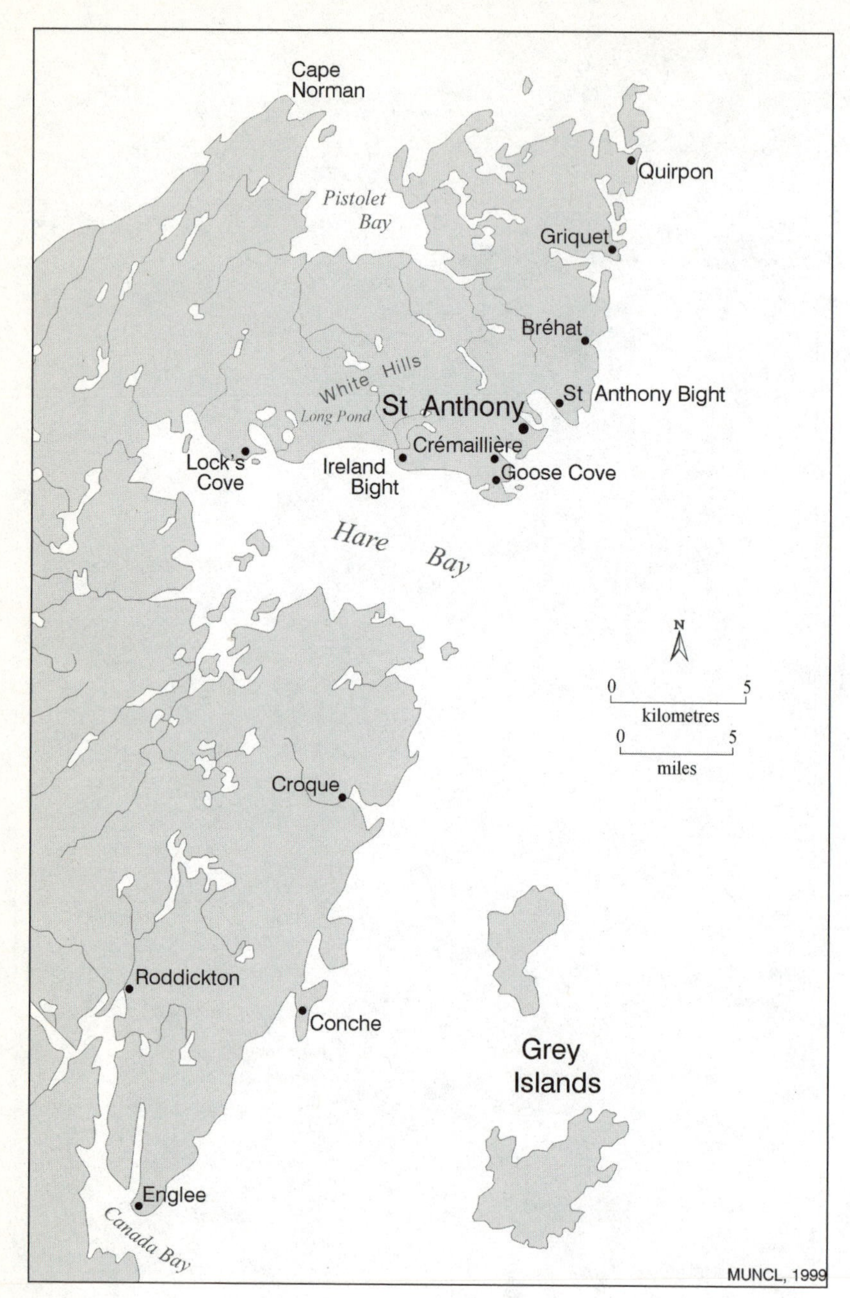

Map 3 St Anthony and environs

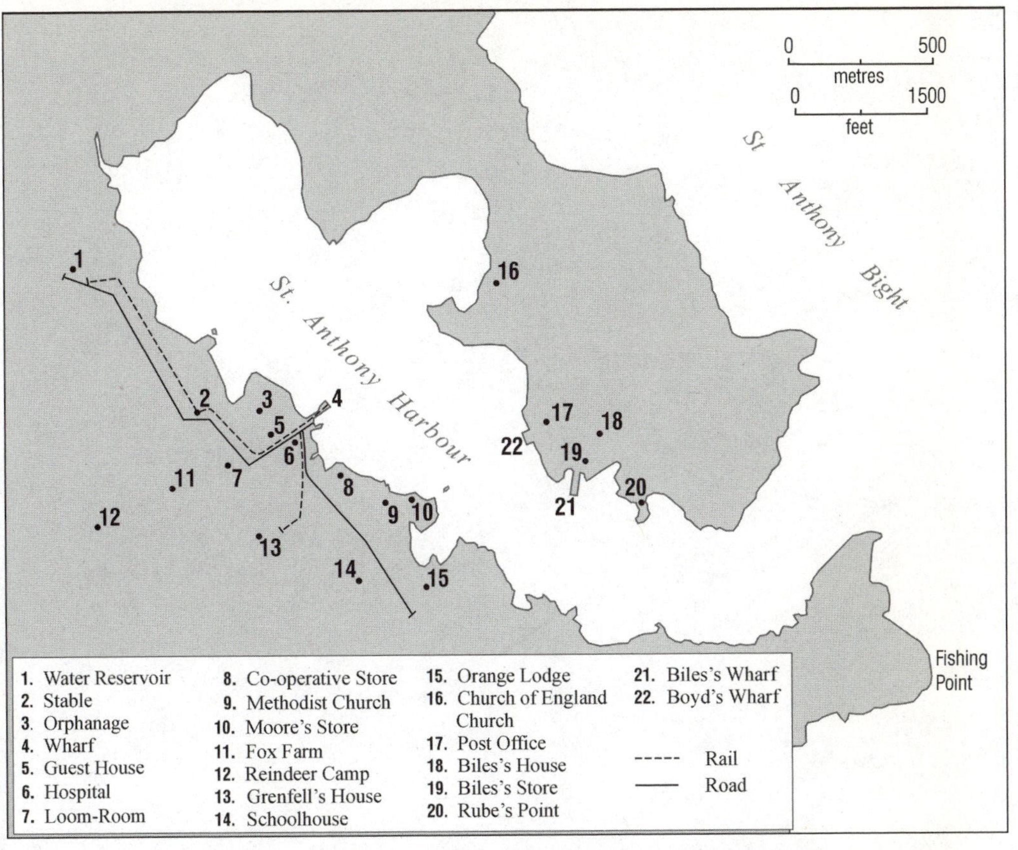

Map 4 St Anthony Harbour and community, c. 1908

JESSIE LUTHER AT THE GRENFELL MISSION

Foreword

Off the southern tip of Labrador, a small island, rocky, bleak and barren like the nearby coast, rises precipitously from the sea – Belle Isle. Like a sentinel it guards the way to the greater island of Newfoundland, held in the jaws of the Gulf of St. Lawrence, its northern boundary the Strait of Belle Isle and Labrador, Cabot Strait and Cape Breton Island to the south. In my early school days that irregular island, shaped like a mutton chop and colored pink in my geography book, stimulated my curiosity. It seemed remote and inaccessible. I wondered how one reached it, what the people were like, how they lived. The wild country, as I pictured it, with its possibilities of hazardous existence, appealed to my imagination and may have been the first stirring of an interest that later led me to adventure on unbeaten paths.

To this barren coast a young English doctor, Wilfred Thomason Grenfell, came in 1892 to aid its people. This young man, strong, athletic and optimistic, fond of adventure and the sea, was later to become internationally known as Sir Wilfred Grenfell, famous for his humanitarian and constructive work among the fishermen of northern Newfoundland and Labrador. His initial experience, however, was with fishermen of the North Sea, where for two years, under the sponsorship of the Royal National Mission to Deep Sea Fishermen, he cruised with them on their boats and shared their hazardous lives that, but for him, would have been without medical help or spiritual guidance. He often referred to the incident that led to his life interest: a chance attendance at a Moody and Sankey meeting in London when, as a medical student on his way home from a clinic, he entered the evangelists' brightly lighted hall through curiosity and left it with the determination to dedicate his youthful energies and his medical profession to the service of his fellow men.

As the years passed, Dr. Grenfell's name, associated with the greatly extended field of his altruistic, adventurous work, became widely known in Canada and the United States, resulting in many speaking engagements. He lectured in New England during the winter of 1905, and I met him when he visited the small occupational therapy sanitarium at Marblehead, Massachusetts, of which I was then in charge. A life-long, treasured friendship developed from that chance meeting, which led to my privilege of contributing in a small measure to the early development of the Grenfell Mission. In response to Dr. Grenfell's earnest request, I went to St. Anthony, Newfoundland, in the summer of 1906 to introduce weaving as a native industry, and thus after many years my curiosity concerning this northern island and its inhabitants was satisfied.

Much has been written by Sir Wilfred and others of that austere country, of the development and scope of the Mission's work, now full grown and bearing fruit, not only as material benefit to those for whom it was initiated but in the enduring spiritual influence of the man whose vision, faith and optimism made the beginning possible. Little has been written in detail, however, of daily life during that early formative period, for very few people had found their way to Newfoundland, and it is of those days of primitive conditions, unexpected happenings and thrilling effort to solve difficult problems that this book is written.

PART ONE

1

The Story Begins

29 June – 6 July 1906

This story began a long time ago. I might use the old fairy story phrase "once upon a time" – but it is a story of realities, not make believe. In the fall of 1903, Dr. Herbert J. Hall, a general practitioner of Marblehead, Massachusetts, and I were trying an experiment. We had met at the "Ark" in Jaffrey, New Hampshire,[1] as convalescents from illness, and during long, inactive days, while friends climbed Mount Monadnock and roamed the country, we discussed innumerable subjects, among them the treatment of those recovering from, or on the verge of, nervous disorders. We agreed that the current treatment known as a "rest cure" was in many cases a mistake, as a patient should be encouraged to look outward, not inward, to forget real or imaginary ills instead of being given the opportunity, through inaction, for brooding on them. The basic idea was not new. Records have shown that from time to time through centuries, certain individuals or groups have realized the beneficial results of occupation, but as far as we knew, no one had acted upon that realization to the extent of advocating it as a definite means of treatment and cure.

As the days passed, what began as an interesting discussion actually took concrete form, and on parting we decided to try an experiment with a group of patients by offering them varied occupations in an effort to stimulate interest and restore self-confidence through accomplishment. The character of our program was influenced by the fact that as director of the Labor Museum at Hull-House, Chicago, prior to my illness, I was not only familiar with but had taught and directed several crafts which might serve our purpose. In the spring of 1904 we found a house on the rocky shore of old Marblehead, Massachusetts. We called it the "Handcraft Shop" that it might not suggest a sanitarium, and from a very small beginning our experiment – afterward known as "occupational therapy" – took shape and increased in scope and development. The details of this effort from its inception to establishment on a working basis would alone provide an additional chapter to this book but would have no direct bearing on its objective.

It was in the second year of our venture that Sir Wilfred (then Doctor) Grenfell was in Boston and came to Salem for a lecture engagement. A sister of one of our patients was his hostess and in the morning brought "The Doctor" to see what we were doing.[2] The reason for his visit was hand-loom weaving – an effective form of occupational therapy treatment – but Dr. Grenfell's interest in looms was not as a treatment. He had in mind a possible industry for the people of northern Newfoundland and Labrador, not only to augment the uncertain income of a fisherman's family but to provide interesting employment during the long winter months between fishing seasons, when idleness, especially among the younger people, sometimes meant disaster.

I have a clear mental picture of Dr. Grenfell as he came up the short flight of steps leading to the loom-room – alert, vigorous, his direct, grey-blue eyes noticing everything. He was at that time about forty years old, of medium height, erect and muscular; his curly brown hair waved above a rather high, broad forehead. His short moustache was brown and his skin firm and ruddy, suggesting fresh air, cold water and healthy out-of-door living. He wore a dark blue, double-breasted serge suit and no hat. He wandered about the little work-shop, talking with the patients, admiring their work and asking sudden questions. Once he glanced through the window as a white sail flashed by on the blue sea and asked, "Is that an iceberg?" [and] then laughed as he realized how far we were from such icy visitors. When he reached the looms, he stopped. "That is exactly what we want in the North," he said. "It would be something for the women to do in the long winter; they could make things for use; they could sell things to help out when the catch is poor. When can you come down to get us started?" I explained that our little Handcraft Shop was my present responsibility; we were just getting on our feet and beginning to make a reputation. I could not possibly leave then, but if he could find anyone able to go I would gladly teach her at Marblehead, and if at some future time I were free to go myself, I would do so.

It was evident the weaving idea made a strong appeal to Dr. Grenfell's vision and imagination, and he spoke with enthusiastic optimism of its possibilities when lecturing in Boston and New York. As a result, interested Boston friends donated four looms and a varied assortment of material. In New York two ladies, after hearing him lecture, volunteered to go north to establish the industry; but then difficulties began.[3] The ladies were strong in zeal and personal interest but weak in the knowledge of weaving. One held a municipal position in New York in connection with street cleaning, I was told, and being an executive told Dr. Grenfell she would take charge. He, in ignorance of weaving requirements, consented. She said she had no time to learn the process. The other volunteer, connected with some welfare work, had lived in the south in her youth and having seen the Negro women weave on plantation looms was sure she could qualify as instructress. Both were put in touch with me, and the ensuing correspondence was both interesting and confusing.

In vain I argued it would be useless to attempt to start an industry without technical knowledge and experience. The executive woman compromised by sending some material more or less useful and trying to interest others in doing so. The southern woman consented to spend four days at Marblehead to prepare her for teaching and establishing a major craft industry in a place where no one had ever seen a hand loom or had any knowledge of, or interest in, its use. The natural result was failure. My pupil came to me for four days. I worked her from 8 a.m. till 10 p.m. each day, and she went away with a note-book and confused idea of the process. She was not an apt pupil. The women went north and so did the looms and material. The looms reached St. Anthony, some with parts missing, and by that time my pupil had forgotten how to assemble them properly. On reaching St. Anthony a year later I found the ladies had quarrelled violently, the one in charge having locked the other out of the loom-room, and after a short stay both had gone farther north on the coast steamer and collected furs. The weaving industry was yet to be established.

The following year our Handcraft Shop was well established and running so smoothly my continued guidance did not seem essential, and in January 1906 I entered a broader field of the same work at Butler Hospital in Providence, Rhode Island, when the superintendent, Dr. G. Alder Blumer, having visited our shop at Marblehead, wished to establish a department of occupational treatment. As at Marblehead, the beginning was small and tentative, but craft occupation was confined to the colder indoor months, out-of-door interests being provided for the summer. With this free time in prospect, I wrote Dr. Grenfell that if he were still interested in the weaving project, I would go north for two or three months and do my best. As a result of his enthusiastic reply, I started for St. Anthony and a summer of adventure on June 29th, 1906.

At that time few were familiar with routes to Newfoundland. I knew only one person who had ventured north: Professor Edmund B. Delabarre,[4] who with a small group of Harvard scientists took the supposedly hazardous journey to Labrador in 1900 and assumed, in my imagination, the status of an arctic explorer. When he left Providence we wondered if he would ever come back. The local travel agencies were of little help. Finally, through the Grenfell Mission office at Boston, I found there were several possible routes, and Miss Emma E. White, secretary and devoted friend of the Mission, helped plan the journey.[5] Like many in those days, Miss White shared the popular belief that ladies disliked or feared travelling alone, and when a writer for the *Ladies' Home Journal* in need of material for an article came to her office to ask about Newfoundland, she suggested we should join forces.[6] The thought of an unknown companion on what might be a difficult and uncertain journey was not reassuring, but as it seemed inevitable, Miss G. and I started on our way together.

To the novice, all routes to St. Anthony seemed complicated, but with familiarity (as I afterward discovered) one became resigned to the many delays with their inconveniences and enjoyed the uncertainties of the journey as one accepts the idiosyncrasies of a friend. After lengthy correspondence, Miss G. and I decided on one of the most complicated routes, starting from Boston on the night boat for Yarmouth, Nova Scotia; and to make the journey a true "fancy trip" (that is, longer and more involved than necessary) we planned to include a side trip to Halifax and a two-night stop at Baddeck, on the beautiful Bras d'Or Lakes en route to North Sydney, just for variety, without realizing to what extent "variety" lay before us in our regular itinerary. On one of the hottest days of late June, 1906, I left Providence laden with heavy clothing and rugs with the mistaken idea that I might encounter ice and snow at my journey's end. Miss G. met me at the pier, and we sailed on the night boat for Yarmouth.

Almost inevitably, one forms a mental picture of an expected acquaintance, but the plump, rather colorless woman bore little resemblance to what was in my mind. She wore flat shoes and a wide leghorn hat trimmed with black velvet ribbon and yellow roses; her drab cloth suit had shapeless lines. Besides her suitcase she carried a camera and a large lunch-box. She greeted me with a friendly, ever-ready smile. Friends who came to see me off left the ship when the whistle blew, and I found a stool on the cool deck where I relaxed as we headed out toward the soft blue sea. It was very still. Across violet-grey and pale-green water the setting sun threw a red pathway to the ship. Only low swells stirred the calm surface of the sea.

It was early morning when I went on deck, and my first sight of Nova Scotia was a long line of land partly veiled in light mist. It seemed to have just risen from the sea. The train was waiting when we landed at Yarmouth, but it took a long time to start. I gazed at the boats left stranded on the mud-flats by the tide, lying so helpless on their slanted sides. No wonder affairs move slowly in this region, when nature herself enforces so many waits. A light mist which I am told is characteristic of Nova Scotia softened outlines, intensified the lovely coloring and magnified distances. Hills on the other side of Digby Gut looked like mountains in the soft haze. The abnormal tide in this region was out. Where the train crossed an arm of the gut, it was so narrow one could step over it. We stopped at Digby for the usual ham sandwich and coffee, but Miss G. still used the contents of her lunch-box. It was beginning to look messy, but barring some accident it might last a day longer.

Beyond wide green meadows at Wolfville, the Bay of Fundy gleamed blue. The incoming tide, flooding red mud-flats, turned lurid as though subterranean fires were burning beneath them – a strange effect against dark purple rain clouds. We passed the fertile "Evangeline Country," the farm houses built on sites of those from which the Acadians were driven long ago, the ancient elms and willows – even the old well – still of interest, although a pump now stands over it. Our "fancy trip" to Halifax proved to be wasted time, but it had one cheering aspect: Miss G's lunch-box was left on the train.

The most interesting part of our journey began the next morning when we started on our way across Cape Breton Island. The train ran within sight of the Bras d'Or Lakes, often following the shore. At Grand Narrows, connecting the two lakes, we left the train and boarded the tiny steamer *Blue Hill*[7] for Baddeck, a Scottish village where nearly everyone's name began with "Mac" and many spoke Gaelic. It was still spring at Baddeck. Lilacs and spring flowers bloomed in gardens; frogs were peeping in swamps nearby. At the comfortable little hotel, someone told us a Gaelic service was to be held the next morning in the country a few miles away, and one of the villagers offered to drive us there; so after breakfast we went careening over miles of rocky, rutty roads, clattering over rude log bridges, reminding me of M.W. Morley's book *Down North and Up Along*.[8]

The little church stood on the bank of a talkative stream, but in fair weather services were held in an adjoining field where logs, laid in two sections on the sloping ground, seated the congregation. Between the sections, a long, weather-worn table extended to within a few feet of the pulpit, a tiny cubicle with a roof, a door at the side and an opening in front, very much like a Punch and Judy booth. Beneath the opening was a bench for the four cantors. The Gaelic service was to us unintelligible but impressive because of the simple, devout people, the lovely, quiet countryside and the little river murmuring an undertone for the cantors' chants, with their strange intervals and cadences. Two posts marked the entrance to the field, and on one of them a tin box suggested offerings. We dropped in our coins and started on our return ride over the hazardous bridges.

The next morning [1 July] we boarded the *Blue Hill* on its way to North Sydney, quite unprepared to find that it was Dominion Day, a Canadian holiday, and most of Baddeck and the surrounding country were junketing. The lake is beautiful. At its northern end, high on a steep, heavily wooded shore, stood the imposing home of Dr. Alexander Graham Bell.[9] All day we sailed over the blue lake, keeping a watchful eye on the loaded deck so near the water's level, and in the early evening the *Blue Hill* left us on the dock at North Sydney and went on with its crowd to the larger town of Sydney. The dock was deserted: no vehicles, no signs of life. We knew nothing of the place. Not daring to leave our possessions on the dock unguarded, we sat on our bags, forlornly waiting for something to happen as the twilight deepened. At last I was starting to explore, when a man appeared driving slowly towards us in a dilapidated buggy. I asked where everyone was.

"Oh, there's great doin's goin' on tonight," said he. "There's a magician come to town, an' everybody's gone to th' hall to see him do his tricks" – and added, "There's been a fire, too. Think th' rest has gone to that."

I asked him to take us to the best hotel.

"There ain't but one," he answered, "an' it's pretty full – men from Montreal here on business."

He looked at our luggage.

"You want all that?" he asked.

We told him we wanted some of it, and he could return for the rest.

The vehicle was a "box buggy," the square body projecting behind the one seat. The man drove negligently. Sitting on the edge of the small portion of the seat left him by two ladies, he held the reins loosely, leaning forward with his foot on the step outside.

To enter the small hotel we passed through a lobby with large windows overlooking the street; a row of comfortable chairs faced the windows, a large spittoon beside each chair. The men looked at us curiously as we passed to the desk, where the clerk assigned to us the one vacant room containing a double bed with a thin cotton mattress and woven-wire spring, sagging in the middle. It was not a comfortable night. My companion was a heavy woman, and in her sleep she slid into the mid-valley of the bed and stayed there. I was continually climbing the up-grade of the mattress and clinging to its outer edge to escape contact with her warm body. And she snored.

North Sydney is not an attractive town. It was a relief to board the steamer for the night crossing of Cabot Strait to Newfoundland, and on the North Sydney dock before sailing I first became aware of the Newfoundlander's kindly spirit. There was confusion on the dock. Passengers – many with children – pushed among piles of freight; the donkey engine clanked loudly. While asking the purser about connections with the *Portia*,[10] I heard a kindly voice ask if he could help and saw a big, burly man at my elbow. From that time, Patrick Finlay took charge. I afterward learned that he was an Irish-Canadian, an agent for contract laborers in the copper mines on the Newfoundland coast, just returning with twenty-two men, and he knew all the ropes. Moreover, he was to leave the train, as we were, at Notre Dame Junction, half way to St. John's, and all the details of a rather complicated journey were made simple by his thoughtfulness.

The little steamer *Bruce*[11] that sailed three times weekly was crowded. Four women were packed into the tiny stateroom. The ports were closed and the air was stifling. At 5 a.m. I rose and dressed, balancing with difficulty on the luggage-covered floor, and went on deck. The air was cold, and a thick fog blotted out everything. We were nearing the Newfoundland coast and moving slowly, feeling our way. Then we stopped, waiting, listening in the muffling fog. Suddenly, just ahead, a rock rose through the mist, surf showing white teeth against its jagged side. Our steamer veered sharply, then went ahead at full speed into the little harbor of Port aux Basques. The captain had found out where we were.

The shore of Newfoundland was quite as I had imagined it: rocky, wild and barren, a few little houses among the rocks, a few fishing boats and bones of wrecks. There was a custom-house and building for freight at the terminal of the one and only railroad in Newfoundland, and the little train waited on its single narrow-gauge track. There were four cars: a first-class, second-class, diner and sleeping-car. Part of the diner was used for luggage. The sleeping-car accommodated sixteen passengers, four narrow sections on each side. We had not expected to find a sleeper and would not have reserved berths anyway, as our train was supposed to leave at 8 a.m. and arrive at Notre Dame Junction at 9 p.m. The sleeping car was full, and we found seats in the "first-class."

Our good friend Patrick Finlay helped us get our luggage through the custom-house and safely aboard the train. Then we waited for no apparent reason. Suddenly, after several false starts, we were on our way, clattering and swaying on the rough road. Often the road-bed seemed to follow the natural contour of the land, which accounted for our variable rate of speed as we crept, groaning, on the up-grade and gained momentum going down. At one place there was difficulty in making the grade: the engine puffed and sputtered, its wheels whirring on the track, and stopped to get breath before jerking on again.

Sometimes the conductor ran along beside the train as if to lighten the load or lend encouragement. We passed low mountains where snow still lay in hollows. The air was chilly, and heavy winds rocked our little train on its uncertain rails. Native passengers told of a dangerous gap where a train was sometimes blown off the track.[12] I took this story with a grain of salt but confess to a sense of relief after safely passing the ominous spot.

The barren countryside was untouched by industrial enterprise or development. Rough country, the few communities merely small groups of simple houses. We stopped at most of them with long waits, giving the trainmen and villagers a chance to "pass the time o' day." We were scheduled to reach Notre Dame Junction at 9 p.m., but as the hours passed, the meaning of the foot-note on the time-table became clear: "The Company will not be responsible for loss or inconvenience caused by delays. The time-table may be changed at any time without notice."

July twilight is late in Newfoundland, not beginning until nine o'clock, but before then the poor children on the train grew very tired and restless in their cramped quarters. There were seventeen babies under three years of age in the car, many of them wailing. Some tired mothers put them to bed on the seats and sat or lay on the floor. In the middle of the night the conditions in the car were nauseating: sleeping people, dirty clothes, remains of food and litter of every sort. My companion slumbered peacefully, her head on the back of the seat. The air was stifling, and when darkness obscured everything but my surroundings, I struggled from my seat, stepping carefully over sleeping people who now filled the aisle, and reached the open platform, where I sat on the top step, my arm around the brake to prevent falling off, and filled my lungs with clean, cool air. It was raining, and the odor of damp earth was very sweet though mixed with cinders.

There are times when annoyance and inconveniences are so great they cease to irritate and become ridiculous. In this case my sense of humor came to the rescue, and I laughed aloud as I clung to the brake in my precarious seat, and I sat there until, just as the dawn was breaking at 1:30 a.m., we left the train at Notre Dame Junction (pronounced locally *not'-er-dam*). A primitive engine attached to a train known as the "Accommodation Mixed" (two freight cars, a caboose and one rough passenger car that was half smoker) waited on a spur track. Patrick Finlay, who also left the train, again helped us aboard. We started for Lewisporte, ten miles away, and an hour later reached the unattractive little lumber village.

Before leaving Boston, Miss White, the Mission secretary, had told us of a possible place to stay while waiting for the boat for Exploits, and

Patrick Finlay at Lewisporte, July 1906 (JLP)

Patrick Finlay knew the family, so we followed him down the road toward a group of small houses. At that hour the village seemed deserted. We passed high piles of lumber stacked by the roadway. A wandering pig and a woman milking her goat were the only signs of life. Mr. Finlay stopped under a window of a small house and called. A moment later a head appeared at an upper window, and he explained the situation to the kindly women, who opened the door, led us to a room with two large feather beds, and made us comfortable with quilts, water and towels. But even as she left the room I sank into one of the beds and pulled the quilt over me without washing my face or removing my shoes.

Lewisporte is on a bay with many small islands. Every other day a small boat, the *Clyde*,[13] meanders through the narrow channels with passengers for Exploits, a fishing settlement on Notre Dame Bay where the coast steamer *Portia* from St. John's makes a brief bi-weekly call on its way north. We went on board the *Clyde* in the afternoon, and Patrick Finlay saw us off, helpful to the last. We said good-by with regret, for he was a kindly, courteous gentleman. I have only grateful memories of him, for I never saw him again.

It was a quiet sail among the little islands of Notre Dame Bay. The shore was rocky and wooded. As we approached the bay's wide mouth the air freshened, and far off to the north we saw our first iceberg, gleaming white in the distance against the blue sea. Once, long ago, on my way to Europe, a steward called passengers to look through their port-holes at ice, a small piece nearby, and later a large berg was seen on the distant horizon. It was interesting, but my reaction on sighting the berg on Notre Dame Bay was different. We were on our way toward the region of many bergs, and as a forerunner of what lay ahead it stimulated my imagination.

We were so late the captain was not sure the *Portia* would wait for us. It was therefore a relief to find, on reaching Exploits, when the little boats put out from shore, that the steamer had not arrived. Exploits is a beautiful place, a deep harbor of crystal clear water, the tiny settlement of small houses and fish stages on its edge clinging "like limpets to the rocks" (to quote Dr. Grenfell). It looked very foreign, and so did the boats that came out to the ship. They are quite different from ours, larger and heavier, and have two short, adjustable masts and tiny, triangular or square sails stained dark yellow, red or brown. When I asked the captain if he knew of any place on shore where we could find food, a passenger came forward and told the captain he would look after us. The captain turned to me. "This is Mr. Manuel,"[14] said he. "He will see that you are all right," an introduction that led to one of my most charming experiences of hospitality.

During all my following years on the coast I found the helpfulness offered us that day, although sometimes in lesser degree, was characteristic of the Newfoundland people. It may be that in addition to a naturally kindly spirit, it is due to the fact that help and hospitality are often a matter of life or death, at least of comfort or suffering, as winter or summer travellers face emergencies, and it is taken for granted that when assistance is given a similar service would be received in the same spirit. Mr. Manuel did everything for us: introduced us to his family, gave us a delicious dinner, and put us comfortably aboard the steamer when it arrived. I shall not forget him, his lovely sister and the rest of his interesting family. Exploits was not only beautiful; it was utterly unspoiled. We were told no stranger had been there to stay except the author Norman Duncan[15] and, for a short time, an artist from Philadelphia. It was lovely painting ground, and I longed to explore the paths and bring away some tangible record of its beauty.

The *Portia* came at sunset with a long whistle and loud clanking of chains as the anchor fell and the ship's boat put off for shore. It was a lovely evening, and I looked back regretfully at the friendly little houses clinging to the rocks, the clear blue bay and the path along the cliff, but ahead lay the wide ocean, deep blue in the early evening light, and the great iceberg in the distance like a hand beckoning to the north. Captain Abram Kean[16] stood at the head of the companionway and welcomed us on board with outstretched hand as we stepped on deck, then introduced us to the purser, who stood beside him. It was like being greeted by friends.

I shall not forget Captain Kean, nor will anyone else who voyaged along the Newfoundland coast during those years. He was a remarkable man, Newfoundland born and bred, a seafarer from childhood. He once told me that at the age of nineteen he had commanded his first ship and begotten his first child. A navigator on the poorly charted North Atlantic coast requires exceptional qualities – not only expert seamanship, good judgement and courage but familiarity with the people, quick decision and resourcefulness. These Captain Kean had and also a grand sense of humor that experience soon taught me is essential to happiness in this northern country. I can see him now as he stood there directing the seamen, rather short and stocky, a short beard, ruddy, seaworn skin and very keen, grey-blue eyes often seen in men searching far distances. A man loved and respected by his crew and associates.

We went below to a small stateroom already occupied by two other passengers where a problem of hand luggage became acute, resulting in everyone walking over a suitcase-covered floor, and the process of dressing necessitated relays, those not actively engaged remaining in their berths. There were no deck chairs. I had a steamer rug and pillow which I used during the two-day voyage to St. Anthony, sitting on any structural projection or the bare deck in a sheltered corner.

There was no monotony in a run along the Newfoundland coast. Two lines of steamers sailed from St. John's: one, operated by the Reid Newfoundland Line, subsidized by the government, left weekly for Battle Harbour, Labrador, with a stop at Twillingate; while the *Portia,* privately owned by the Bowring Co., sailing bi-weekly, took the place of a train, entering all the little coves and inlets that sheltered small settlements behind the high rocky coast. With the exception of fishing schooners, these boats were the only means of transportation during the season of open water. The time between stops varied from half an hour to half a day.

There was excitement in rounding a promontory to find a little calm bay invisible from the ocean, with tiny houses and fish stages on its rocky edge reflected in the water, or to pass through a "tickle," narrow and deep, to find a settlement on the sheltered side. At one place we seemed about to strike the cliff head on, but a swift turn took us suddenly to a "hole in the wall" (to quote Captain Kean) scarcely large enough for the steamer to turn. Very few settlements had a dock. At all others the *Portia* whistled on entering the harbor, dropped anchor when nearing the shore, and the ship's boat left with the mail. From all directions little boats pulled out, bringing passengers, freight and often visitors merely to gossip with the crew and hear the latest news, for the bi-weekly boat was a social event and the only contact with the outside world. The *Portia* carried miscellaneous freight, including fishing equipment, lumber, furniture, coops of chicken, and sometimes a pig, goat or cow was lowered over the side and taken ashore in a trap-boat, adding considerable excitement for passengers.

After leaving Exploits a passenger introduced himself as Mr. Pitman,[17] a clergyman stationed at Tilt Cove, a mining village farther down the coast. He introduced the cable operator at Heart's Content, near St. John's, who showed us tapes with different transatlantic transmission forms, which were very interesting. All the next day we sailed in and out of lovely bays and fiords somewhat resembling the Norway coast: lovely scenery, but the air was thick with smoke from a forest fire in the mountains, and at times the sun was almost obscured. At Tilt Cove, where Mr. Pitman was stationed, we went on shore. It is a dreadful place, nothing but copper mines. The village is a cluster of houses surrounding a pond filled with drainage from the mines and apparently sewage from the houses. The water was a peculiar sulphurous color, and I never smelled such odors. Not a tree, shrub, grass or any kind of vegetation was visible, and combined with the smoke and stifling heat the place seemed an inferno.

We were approaching the "French Shore," its name due to French fishermen having settled there many years ago,[18] and stopped at the village of Baie Verte, where the forest fire was very near. The coal-mining village was two miles away, and the people came for their freight in coal-scuttle cars drawn by a toy engine. An American professor from the University of Michigan, who was staying at the mining camp for experience, came on board. He said the fire was not very far off, and the mining people were panic-stricken, for if it swept to the shore there were only enough boats to take a few away. He was going back with them to take his chance.

There was also a young curate from St. John's who was going north as a travelling missionary. His brother, an instructor at St. John's College,[19] was going with him. They planned to stop off at St. Anthony, then walk about a hundred miles across country and along the shore, stopping at little hamlets by the way, giving what education they could, also baptizing, marrying and burying – for in that sparsely populated country such offices, except burial, have to await an occasional visit from clergy or justice of the peace. They were fine, earnest men who knew Dr. Grenfell.

We passed the smoky region during the night, and the next morning it was so cold we unpacked our winter clothes, feeling for the first time that we were really northward bound. It was Sunday, and there was service in the lounge conducted by two clergymen from Toronto. It is said that the presence of the clergy on board is a guarantee of safety. If so, we should have no fear, for five of the passengers were clergymen – all Church of England. After service, one of them, Dr. Chown,[20] introduced himself and we talked for hours. He was an interesting man. From certain references I inferred he was a bishop whose parish extended through the lower Canadian provinces. Another attractive missionary was Mr. Richards[21] of Flower's Cove, on the other side of Newfoundland. Although this part of the journey was not easy, we could have gone on happily for days longer with our interesting fellow passengers.

2

Industries in the Far North

6 July – 25 September 1906

July 6th It was Sunday evening when the *Portia* rounded Fishing Point Head and with prolonged whistles entered the beautiful harbor of St. Anthony. From the shore in all directions little boats started out toward our anchorage beyond the low wharf, where the water was too shallow for the steamer's accommodation. We went ashore in a large, deep fishing boat rowed with very long oars and used for luggage, freight, mail and passengers. In one of the small boats I noticed a young man whose derby hat was conspicuous among the caps and sou'westers of the other men – evidently someone from the Mission dressed up for company. I realized it must be Dr. Soule,[22] the volunteer from Boston, and we shouted mutual introductions across the water as he went on to the ship for freight while we continued to the landing.

Bishop Chown, who went ashore to see the place, helped us up the rough log ladder, and we walked along the narrow landing, past an enclosure where several dogs were confined, and on to the hospital, where a woman received us rather churlishly. She did not seem to know anything about me. No one else was about. Some doubt had been expressed by Miss White in Boston as to my cordial reception on account of the would-be weavers of last year, so when I mentioned "weaving" and the woman's face hardened, I knew what was the matter. I suppose she thought to herself, "Here is another." Dr. Grenfell was "afloat" (his own expression) on board his little steamer the *Strathcona*,[23] bound here, there, anywhere during the summer, a disappointment for Miss G., who had hoped for transportation to Battle Harbour.[24] It meant her return to St. John's when the *Portia* stopped here in the night for mail. When the two nurses found us after their return from church, we were escorted to the orphanage, where I was to stay. Everyone expected to sit up most of the night to write letters

The welcoming boat at St Anthony, July 1906. *Left to right:* Reuben Simms, Dr William Soule, Mark Penny (JLP)

for the return mail, so Miss G. took a nap, and when the *Portia* whistled at 3:30 a.m. I went down to the wharf and saw her off.

It was late next morning when I joined the others and began to understand the complicated situation. Miss Storr,[25] a young English woman with some nursing experience, had volunteered to open the new orphanage and remain indefinitely in charge of it. Her friend, Miss Bayley,[26] a social worker and head nurse in a London hospital, came with her, partly as a vacation but primarily to help install seven small orphans whom Dr. Grenfell had found in isolated settlements on the Labrador coast. They arrived four weeks ago for this pioneer work.

The small St. Anthony hospital was supposed to be closed when the one and only nurse was transferred to the even smaller hospital at Indian Harbour for the Labrador fishing season, but when Dr. William L. Soule, a volunteer from Boston, arrived in June for out-patient work during the summer he found two very sick men, one of them a case of typhoid, the other blood-poisoning, and at the end of the first week two more cases needed both nursing and medical care. Since they lived in the tiny settlement across the bay, which meant rowing over there, it became obvious that the only thing to do was reopen the hospital to care for them all. Miss Bayley had expected to stay only two weeks, but schedules in Newfoundland often change without notice. She has remained to meet the emergency, and in spite of great difficulties she and Dr. Soule have carried on as best they could.

Anyone even slightly familiar with the mental and physical strain involved in nursing under the best of conditions can perhaps imagine the handicap of caring for patients without adequate assistance under primitive conditions. There is no plumbing of any kind or so-called "modern conveniences." All the water comes from little rivulets on the hill that often run dry and a spring at the foot of a slope beyond the hospital. It is brought by hand in buckets and tubs and heated on the wood-burning kitchen stove. In a shallow excavation under a part of the hospital, a small soft-coal furnace provides moderate heat during the winter. One of the village girls has been coming for part-time service as ward-maid. The nurse and doctor bear all further responsibility. The orphanage, not fully completed but livable, is a frame house without cellar, the walls bare boards, the outside clapboarded. The seven orphans, varying in age from five to fifteen years, assist in the housework. Miss Storr and Miss Bayley have done the cooking.

After breakfast I asked Dr. Soule what I could do to help. He hesitated a moment, then asked, "Can you make johnny-cakes and fish chowder?" He beamed when I nodded and explained that the English ladies did the best they could, and the food was adequate for everyone. The menu, however, consisted mostly of porridge, jam, tea, bread, potatoes and fish, all good in their way, but being a New Englander he yearned for old favorite standbys. There was johnny-cake meal [cornmeal] among supplies sent from Boston, but the Newfoundland as well as English native considers it dog food and not fit for civilized man. There was abundance of fish, but served plain boiled or fried, and he longed for what (laughingly) Miss Storr refers to as "American messes," meaning anything that could be suspected of camouflage.

That night I made fish chowder and watched for reactions. The doctor took a mouthful, smiled, and made his way through four helpings. The English girls were suspicious and played safe with bread and jam, but I hope for their conversion to Yankee cooking since I seem to be installed as cook for the summer. Johnny-cakes went better, the ingredients being more obvious, and once tasted corn-meal seemed to be a fit food for man after all. The store-room is well stocked with tinned food and staples, but the only "fresh" thing is fish, always signifying cod. Any other fish is spoken of by name. There are potatoes, but no onions. I was told whoever ordered supplies from Boston did not like onions, but that may have been merely a cavil.

July 12th Is it only a week ago today that I arrived? It seems a month since the nice bishop helped me up the wharf from the fishing boat and welcomed me to my new field of work. Since then it has been a matter not only of putting my shoulder to the wheel but my arms, back and whole body as well. I came here expecting to work hard with looms and teach others to do like-wise. I have not only done that but have cooked for the family, cared for the orphans, assisted in their baths, mended and altered their clothes, acted as waitress, chambermaid, scullery maid. It is a case of rising to occasions.

We are very merry at meals, sometimes becoming so uproarious that Dr. Soule declares he cannot say grace with the spirit suited to St. Anthony's religious standards. It is amusing to watch him taste what is set before him. He smiles, then his expression resembles that of a cat when one strokes its fur. I am continuing to experiment with available food supplies, and he says he is a scavenger when he sees before him for lunch the remains of yesterday's dinner in another form.

The day after my arrival, Dr. Soule took me to the small building near the hospital which was to house the Industrial Department. A carpenter's shop is downstairs and upstairs a rough room like a loft – a clubroom so-called – the full size of the building about twenty-five feet square. It will make an excellent place for the looms, a possible drawback being its present use as sleeping quarters for four carpenters from a nearby village who are working for the Mission. There are two iron beds and two bunks built against the wall. Three old quilts hung on a line partition them from the rest of the room. There are rough chests which cannot possibly contain clothes, for there is every sort of garment lying about. In the center of the room is the chimney and a small stove stuffed with ashes, tobacco and refuse of various kinds. When I first saw it even the hearth was full, and dirt was everywhere. I tried to find someone to clean, but it seemed to be nobody's business, and I foresaw it would have to be done with my own hands, which are becoming accustomed to handling anything, so it does not matter very much. Finally, I found a girl to assist, but it will be some time before the room is in order.

The four looms are there, just as they were left by the would-be weavers a year ago. It took time to locate and assemble the parts to make them workable. It will take time also to demonstrate their use, quite understandably in a region where hand-weaving is unknown and industry thought of exclusively in terms of fish. The first thing to do after setting up the looms was make warps,[27] and a few girls were persuaded to assist, but the loom-room was vacated whenever a fair-sized catch called all members of a family to the fish stage. It was discouraging, although I recognize the importance of fish.

One of the orphanage children, twelve-year-old Emmie [Roberts] from Labrador, is bright and so capable I have taught her to make warps and wind bobbins. There are two looms ready, one for pattern weaving, one for homespun. Emmie is now making a heavy warp for rugs. She seems interested and is willing to work for a while without interruption, which I was told might be doubtful. I am also teaching Miss Storr to enable her to carry on after I return home. When weaving was actually started on the two looms and a tangible result could be seen, interest increased and visitors mounted the steep stairs to inspect the loom-room. Very few asked questions. They looked on silently while I operated the looms, then went away as silently as they came. But by the underground, a word now and then, I recognize a growth of interest and take heart.

When cleaning a closet at the hospital, I found a set of woodcarving tools that had been terribly abused, large pieces broken from gouges and chisels. They looked hopeless, but after a determined struggle with grindstone and file they are usable, and a carving class has been started, primarily for one of the carpenters (Mr. Gushue), who wants very much to learn to carve and has never before had an opportunity. He is a nice man and will be here all winter. I hope later he will be able to teach the boys.

There has been no time for sketching except two evenings after seven o'clock, and then the mosquitoes have made it impossible. I never saw so many (since a visit at Cape May, New Jersey, long ago). They are a very large, ferocious type and sing as well as bite lustily. I am not usually poisoned by them but am conscious of lumps on my body as well as arms and legs, for clothes are no protection. The poor children are a sorry sight, with many lumps and bleeding sores. The pests bite all day, all night and in all degrees of temperature, dryness or dampness. The only relief is when the wind blows. I found some netting and tacked it over my window, hoping to sleep in peace.

My bed-room is small but comfortable. From its one window I can see a bit of ocean beyond the harbor. The high hill, with a low growth of spruce and birch below its rocky summit, is beautiful in changing lights. There is an iron bed with a new hair mattress, good linen and lovely grey blankets, but the pillow is like one I once slept on at Cape Cod and suggests goose quills. I am glad I brought my own pillow. In front of the bed, a strip of red carpet adds a note of color. An enamelled pitcher and basin stand on a box. Another box serves as a seat if turned one way and as a stand if turned the other. A chair made from a barrel and my trunk used as a table complete the furnishings.

There are two shelves. One of them serves as a dressing table with my hand mirror on the wall above it. Nails driven into the bare boards take the place of a closet. The walls are rough, and carpenters' pencil marks left on them troubled me when a bilious headache, doubtless the result of eleven days' shaking, rolling and jolting, kept me one day in bed. I tried to match the figures and wondered what they signified.

July 16th This is Sunday, the so-called day of rest. So much stress is laid on keeping the Sabbath in this Methodist community that Miss Storr worked every spare moment yesterday, getting the children's clothes in order and laid out, while Miss Bayley made puddings and we planned cold food to avoid cooking. We were to lie abed late and not breakfast until nine, for we were very tired, but at 7:30 a.m. Dr. Soule called Miss Bayley. A case for operation had just been brought in from St. Anthony Bight, and he needed her help.

It was 10:30 before they returned for breakfast, the interval being filled with household duties and enlivened by a fight among the boys. Than came preparation of the children's dinner, followed by that of the staff at 3:30, when the doctor and nurse returned from the hospital. When that meal was cleared away, the children's supper followed at an early hour to allow time before evening service at 6:30. There are times when I wish nature did not demand three meals a day, but the irregularities of daily living add spice and amusement. I defy anyone to become bored.

The service was at the little Methodist church on the hill. The church stands on a hill overlooking the harbor, with a lovely view of the sea. It is pretty and well built – quite a surprise, for I did not expect to see colored glass in the pointed windows or a red woolen carpet on the platform.[28] Since there is no resident minister and Dr. Grenfell is still away, it was conducted by one of the local fishermen, who read a sermon from a book. He was evidently no novice, and his extemporaneous prayers were better than many I have heard nearer home, but I certainly should never have thought of such metaphors and flowing sentences. One phrase (it seems to be his favorite, for he used it in each prayer) was that "all we sinners may this night be drawn from our sins, and when the white horse comes to claim us we may go sweeping through the open gates, washed in the blood of the Lamb." He was very earnest and shouted at the top of his voice. Miss Storr and I exchanged glances as she remarked, *sotto voce*, "He thinks the Lord is deaf." In the absence of anyone to play the parlor organ, one of the men in the congregation started the hymns, singing with all his might in a piercing, nasal voice. After two or three solo notes, the congregation joined in to the end of the line. Then, without a pause, he attached the first note of the following line to the last note of the foregoing to give a clue to the tune.[29] The effect was confusing until one became accustomed to it.

At the end of the service, the fisherman announced that there would be an "after meeting" which all would be welcome to join.[30] It was held for the benefit of any who felt the need of being labored with, consoled, or prayed for by the deacons and elders. The fisherman's English was strange to my ears, and his reference to "hadamantine 'earts and 'oly harms" seemed quite unintelligible until I realized he was transposing h's.

The features of both men and women are rather angular in type. The men's skin is dark with sea tan, and most of the women look prematurely old.[31] Many of the faces seem to indicate extreme reserve, sometimes even prejudice, which is not surprising since most of the people know nothing of life outside the environment of their bay. They interest me very much, and I hope to gain their confidence and friendship.

We had tea after the service, then went to the hospital, where there is a small parlor organ, to sing for the patients. The hospital is attractive and characteristic. Stepping inside the entrance hall one is confronted with a *mélange* of objects: stuffed fox heads, fur coats, snowshoes, caribou antlers, rugs and other interesting articles, while just outside are a number of wheelchairs for the patients. At one side of the hall is the small reception room, the dispensary and general office, on the other Dr. Grenfell's study, an attractive room with cushioned seats built around two corners, a large table in the center under a hanging kerosene lamp, and book-shelves above lockers on three sides of the room. There are pictures on the wall, a Morris chair and the small parlor organ, also caribou heads and antlers which seem to be used generally as decoration.

July 18th These are strenuous days, filled to overflowing. This is my program at present: I rise and prepare breakfast, care for my room and help clear away dishes, look over the larder and plan the dinner menu, then go to the shop and work on looms until 11:30, return to the orphanage and get dinner, back to the shop for work until 5:15, then to the house to cook supper. The boys' woodcarving class follows, and we work until dark, 9:00 p.m., on bright nights until 9:30. Then I am about ready for bed.

Today I have at intervals prepared special food from the milk and egg mixture I brought with me for a very sick boy at the hospital who is causing Dr. Soule great anxiety. When I offered help the doctor replied that if I could concoct something nourishing and tempting it might work wonders. It proved a fortunate experiment, for the boy has eaten all of it. He was brought in on Sunday with an advanced case of blood-poisoning and has already had two operations. Miss Bayley was with him last night, and I will be tonight. It is so sad to be able to do nothing more in spite of every effort.

Miss Bayley is wonderful, always bright and cheery, ready to help in every way, and so efficient in her profession. Last week she was called for a lying-in case while the doctor was away for the day. It meant tramping eight miles through bogs and over rocks, then back again the next day with the doctor, when they did not start until after supper and returned at midnight. Part of the way, there was no path, and they found their direction only by following a strip of sunset glow through an opening in the hills. They came home wet, muddy and nearly devoured by mosquitoes.

Miss Bayley expects to leave for England by the next *Portia*. Then I shall assist in nursing when needed. The hospital is still supposed to be closed, but if cases occur in the neighborhood they must be cared for, and we must do our best. Dr. Soule thinks I may make a fair assistant, and my experience as a patient at St. Mary's Hospital[32] in Chicago and with all kinds of patients while in charge of the sanitarium at Marblehead should count for something.

The wood-carving class is certainly worthwhile. The boys are so receptive it is a pleasure to teach them, and I hope it will mean much to them this winter, when there is no other work. They are making simple things for practice, with the incentive of something more ambitious later on. Even working on the piazza with the air fairly quivering with mosquitoes does not seem to lessen their enthusiasm. The basketry and dyeing materials have not arrived, and I am afraid they will come too late to start. There are many who could be taught to make baskets for winter work in their homes, and I want to teach Miss Storr how to dye material for their use.

Yesterday I went with Miss Storr to see a neighbor about taking my laundry. The house we visited was tiny, and as we mounted the shaky porch there was an awful odor of drying fish. The door was so low I had to bow my head to enter, and one door hinge was off; but the interior was clean and very orderly, considering the many children who crowded around us. They all looked underfed, and the skin of the six-month-old baby was like skimmed milk. The father, a fisherman, said the catch is very poor this summer, the worst in twenty-five years. That means privation and suffering for the coming winter, for fishing is practically the only industry. Dr. Grenfell has established a saw-mill at Canada Bay,[33] where men with families can earn something during the winter. If it were not for that, I don't know what they would do.

The dogs have just begun their unearthly howl, which means it is late and the time arrives to conserve my strength for the problems of tomorrow. These wolf-like creatures are often fierce and so unfriendly with each other that Miss Storr tells me three of their own pack have been killed in fights since her arrival. Each night at about quarter before twelve, they begin their unearthly wail. Suddenly, one dog starts with a long-drawn note, and the others follow in a piercing chorus for about five minutes, when they stop as suddenly as they began. I never heard such a sound before and fancy it must be like a wolf's cry.

July 20th This is more like camping than anything else, except that in camping as I have known it one has no responsibilities and a guide or someone to cook, not usually for so many. There are twelve inmates of this house, children and adults, and the cooking is done under difficulties. In the first place, there is the fire! The fuel is wood with a little soft coal, and Martha, our handmaid who is supposed to stoke it, frequently lets it go out at the wrong time or puts so much soft coal on a nearly exhausted fire that it does not kindle. There are not enough dishes or cooking utensils, and we have to improvise, waiting until one thing is cooked before starting another. There is plenty of butter but only the egg powder I brought with me, no fruit, although later there are blueberries, a large berry called "bakeapple"[34] and a few other small berries, all of them growing close to the ground. Martha makes risen bread, and the children eat large quantities of it with molasses, but it is bread the like of which I have never seen before – dark, coarse, full of flour lumps, with not enough salt, and heavy. When we weighed a loaf, the scales registered three and one-half pounds. The limitations of our kitchen equipment were evident today when I found Martha mixing bread in a pan that I had used for washing a filthy wool warp before putting it on the loom.

There is, of course, no ice, and as the temperature is often like our moderate summer weather it is difficult to keep fish that is usually delivered in large quantities, six or seven cod at a time. The English girls are prejudiced against pork and have used table butter for frying. It seemed a waste, so I tried the experiment of substituting salt pork for frying fish and waited for reactions. There were none. They did not recognize the difference and, when told of the mean trick, only felt relief in the future saving of so much butter.

The weaving industry is progressing. One loom, with the beginning of a blue and white bed-spread, is in operation. Another is being drawn in for rugs, and two others are in preparation. There are details that take much time and planning. Miss Storr assists, but the everlasting cooking interferes with working together as we would like to do. The small boys have been helping. Yesterday was rainy, and I told them they could come to assist. They strung heddles[35] on the frames, counting them carefully with much pride in their responsibility. Today, one of them brought little four-year-old Prissie [Roberts] and took great interest in teaching her to arrange the heddles evenly. It was fun to watch them. They were having such a good time, and when I finally left to go home, they all came too, revolving around me on the road. I felt like a planet with satellites!

If I only had time to paint! Everything is so beautiful – the hills, shore and water, the rocks and winding path, the masses of white flowers, odd little huts, and boats with colored sails. The people themselves are such fine studies. Well, perhaps I may come again some time. There certainly is no time for painting now.

Tonight after the carving class Sister Bayley and I took a walk as far as the church, a beautiful walk, with a silvery young moon hanging over the hill-top in the twilight sky. When we returned at ten, the sunset light was still reflected in the bay, and the outlines of little houses with tiny lighted windows showed dimly along the way, a comforting reminder of humanity in the wide, silent space. Lately, Dr. Soule and I have been telling all the amusing stories we could think of to keep up our spirits, for sometimes the atmosphere is distinctly tinged with blue, one of our worries being the *Strathcona*'s non-arrival, for she is long overdue, and we need Dr. Grenfell's advice.

The dogs are again howling, my signal that it is time to stop for tonight.

July 22nd, 9:30 p.m. The *Strathcona* has just come in! Everyone was terribly anxious, for her boiler is weak and the villagers were sure she had blown up or gone on the rocks. They did not give her up until today. Imagine our joy when, just as Miss Bayley and I were discussing the probable cause of disaster, we heard Miss Storr shout that a steamer had whistled, and it was not the *Portia.* It was then that I heard for the first time the *Strathcona*'s whistle. It is peculiar, beginning with the high, squeaky, adolescent treble of a small launch, then changing suddenly to the deep adult bass of an ocean liner. It is quite unmistakable.

As the little steamer entered the harbor, people fired guns on shore and rushed to the wharf. We did not go down but watched from the piazza as she came to the landing. A few minutes later, someone from the hospital came to say that a lady had also arrived, a sister of Dr. Soule, and would we make up a bed for her? There was no vacant room, so we turned out the oldest orphan, put her with the small children and prepared her room. We cannot think who she is, for we did not know that Dr. Soule had a sister, and how should she happen to land here? While we waited the lamp went out for lack of oil. We searched frantically in the shed for the oil-can and lighted a wood fire in the dining-sitting-room stove, for the air was as chilly as November.

Later, 11:30 p.m. Dr. Grenfell has just been here for an hour and gone. He entered suddenly, carrying a large oil lantern and wearing a sou'wester, a long oil-skin cape over his clothing and high sea boots. He is looking very well, rather heavier than when I saw him a year ago, and with a young beard which changes him somewhat. I was impressed more than ever by his strength and manliness. The lady on the *Strathcona* is the sister of Dr. Mumford[36] of Battle Harbour, and they are both on their way home to England. It was a poor, sea-sick little woman we finally tucked away in the upstairs room.[37] Then Dr. Grenfell settled down to talk.

He wanted to know if I had had a chance to make a start with the looms, and it was amusing to see his face when I told him that a pattern bed-spread was being woven on one, and another was nearly ready for rugs. He looked fairly incredulous and, when he heard we had started a carving class, exclaimed, "No! Have you, really?" He is to be here until Tuesday, giving us time to talk over the work and make plans. The rest of the evening he spent catching and mounting some of the moths that were attracted by the lamp, all of us assisting. He says there are several rare specimens in this region and that he has made collections of birds and flowers, but not yet of moths. I foresee another industry for my idle moments. This would fit in well after my carving class in the evening!

July 23rd The *Portia* came in last night but went further down the coast, so mail is again delayed. Dr. Grenfell came to see the looms today and was delighted! He was like a boy, and we had fun planning a homespun suit for him. The beginning of the pattern bed-spread brought an excited cry, and he is coming again this afternoon to plan all sorts of things. I have not yet shown him the carving but am sure he will be pleased with that also. There really seems to be a hope of accomplishing in some degree the objective of this effort, but I shall feel like taking a long breath when I start on my homeward way.

July 29th It is quiet today after a rather tumultuous week, and now that the children have gone to church the house is so still it seems breathless. The walls are only one thickness of boards, no plastering, and the floors two thicknesses of boards and no carpets. The doors are so badly warped that they have to be banged to latch, and the orphans all wear heavy shoes and scuff their feet. They also are very active children and often have scraps. Taking all this into consideration, it is understandable what this blissful quiet means.

Monday was a broken and hurried day. The wife of the missionary from Tilt Cove, who came ashore from the *Portia* in a comatose state from a dose of seasickness preventive, and the bedraggled travellers from Battle Harbour had all recovered sufficiently to drag themselves around with renewed interest in life, and their interest seemed centered in the loom-room. They came and sat, admired, asked questions, and took my photograph. They went away at dinner-time, but when I returned I found the missionary's wife begging to be put to work so that she could stay. She helped me wind the warp, meanwhile drinking in my words of wisdom. It was embarrassing to be admired so much, but the poor woman had lived at Tilt Cove during the past four years, breathing poisonous air, the miners her only neighbors, and the whole of St. Anthony seemed to her a paradise. Her little boy had never before seen green grass and called it "flowers." I am glad she is on her way to another parish with more normal living conditions. They say Twillingate is a very pretty place, but I fear it is too late to save her, for she has all the symptoms of tuberculosis, which is prevalent at Tilt Cove. I do not wonder, for the place is terrible. My heart fails me when I think of being *that* kind of a missionary.

The *Portia* came in on her return journey at three o'clock, and there was a great rush until everyone departed at six. My carving class did not meet that night, and I tried to forget the mosquitoes and paint, sitting on the piazza steps. Presently Dr. Grenfell strolled up from the hospital, and we discussed plans until the pests drove us indoors, and we went over to the hospital to look at some flower-pots the doctor had made from the clay that underlies the top-soil in this region. He had dried them on the kitchen stove and baked them inside it. They came out a beautiful terracotta red. He said he made them because he had nothing to hold his plants. Just as simple as that!

He wants me to come again next summer to start pottery and asked rather hesitantly whether the journey had been so hard I would feel discouraged about taking it again. I acknowledged it was rough but so interesting it did not matter, and with this encouragement he added that if I would come by and by to stay, he would build a house for me. Of course he must know I could not do that, but I should really like to sometime spend a winter. He loves to tease, and we had a merry time discussing the impossible. He left on Tuesday to visit his saw-mill, leaving the nurse from Battle Harbour (another Miss Bailey but spelled differently)[38] to stay with us for a vacation.

The little boy is still very ill. Miss Bayley is with him a part of every night, Dr. Soule exchanging watches with her. Dr. Soule feels his responsibility keenly and sometimes becomes so blue and discouraged we resort to foolishness in our effort to cheer him up; but still there are the facts.

On Friday, Dr. Grenfell returned from the mill and came to the kitchen to look for me just as I was struggling with supper and had begun to mix a corn-cake. He had a new plan! What did I think of taking one of the looms to Battle Harbour on the *Strathcona* to start weaving among the local people? Could the nurse in charge be given sufficient training to make it worth while? Of course, the ultimate objective of the weaving industry is to make St. Anthony the nucleus and send out looms with weavers to establish a general home industry in Labrador as well as here. I agreed it would be an excellent plan to take the loom now and I would stop on the way home and try to train the nurse sufficiently for simple weaving.

In the evening the Doctor came to visit the carving class. He thought the boys were doing very well and that it would be a fine thing for them during the winter, especially if they could make *large* things: chests, etc. I made the suggestion that it might be better to give them small things first while they learned the use of tools. Larger things they might spoil. To this he agreed, adding several helpful suggestions. It seemed characteristic of him that his vision, even in this minor instance, leaped over and beyond details to an objective.

It evidently troubles the Doctor that he is obliged during several months of the year to go about "with hat in hand" for Mission funds. Apparently the English Mission office wants him to give a year to evangelical work, but he wants to be *doing* instead of preaching, and seeing him, one can understand his point of view. I cannot imagine him an orthodox evangelist. His influence and appeal are through action, sincerity, and example rather than exhortation.[39]

July 31st At breakfast yesterday Dr. Grenfell suddenly decided to start at eleven o'clock for Battle Harbour and his northern trip, for there were patients to be taken home and business that would not allow him to wait. Of course, the loom was not ready, for we thought it was not needed till Monday, and my speed in taking it apart broke all my records. If it had not been for Mr. Bryant,[40] a young man from Boston on the trip with Dr. Grenfell, I don't know how we could have managed, for it is a heavy loom and the joints needed pounding and lifting.

The Doctor had planned to show me over the *Strathcona*, but I arrived before him and Mr. Bryant did the honors. The little hospital amidships is very interesting, with its case of surgical instruments against the wall, the medicine closet, and three swinging beds which are supposed to remain stationary when the steamer rolls. It is much smaller than I thought. The patients were brought from the hospital in wheel chairs and put aboard. One poor fellow with tuberculous meningitis possibly will not live to reach Battle Harbour but was taken on the

Dr Grenfell takes *Strathcona* from the wharf at St Anthony, July 1906 (JLP)

chance because he could not have proper care here. Other patients are to undergo operations on arrival.

Miscellaneous freight cluttered the deck. A large caribou head with magnificent antlers, destined for Battle Harbour hospital, topped the cabin roof, and on the wheel-house was a tin can filled with purple flag blossoms and roots that the nurse hoped would survive the journey. Standing against the rigging were parts of the loom and a wheel for winding bobbins. It reminded one of cartoons of the *Mayflower*, with spinning wheels and Chippendale furniture hanging from spars. Among other things on deck were lumber for building and various boxes and barrels. Half-way up the cordage hung a pair of fishermen's boots. In the bow, Dr. Grenfell's trunk stood beside four crates of white baby seals stuffed in sundry attitudes, the product of Dr. Grenfell's own taxidermy. He has given me one to take home. I hope it will go in my trunk. If I should have to carry it in an open crate through Boston and Providence, a body guard would be needed to restrain the small boys.

The party went on board and we stood on the wharf, ready to wave good-by, but the Doctor did not seem to be in a hurry to start after all. After talking across the rail for some time he stepped ashore and joined us, talking, laughing, teasing, ordering people about and making his little dog "Jack" perform his one trick of standing on his hind legs. When he finally went aboard there was not enough steam to start, and for twenty minutes the boat lay against the pier, waiting until the boilers filled, and they moved slowly away. They made an amusing group. The nurse, her face swollen and scarred from mosquito bites, sat on the lumber in line with the hanging boots. Mr. Bryant, who is really very good-looking and in real life, I fancy, quite *un jeune homme élégant*, stood on the pilot house in a khaki suit much the worse for wear, taking snap-shots of us. The Doctor, in his disreputable tweed knickerbockers and Norfolk jacket, high tramping boots, shapeless old straw hat and a half-grown beard, waved until they were far out in the harbor. The little steamer was towing a great clumsy fishing boat to be used on the Labrador. The Doctor is going north to Cape Chidley, near Hudson Bay, and will not return before I leave, for they tell me the trip takes two months with all the stops by the way, and I shall not see him again until he comes to Boston this winter.

August 1st We have tried to settle down to everyday affairs, but after the short interval of excitement life seems rather flat. This has been a gloomy day. Sister Bayley and I went to church this evening and heard the same man who shouted so lustily. His prayer was deafening and must have been heard in highest Heaven. A sense of depression seized us, and I cannot say the man's vehement praying helped matters. It is only one of the ups and downs of this experience. This day happens to be a "down."

[*Dated July 30th*][41] The "ups" have it again. This afternoon was my first half-holiday; such an unwonted event that I was nearly done up at the end. It was very warm, eighty-four degrees and very oppressive. Dr. Soule suggested at dinner that we should accompany him on his rounds, sailing across the harbor to Old Man's Neck. Miss Storr and I decided to throw duties to the winds and enjoy the perfect afternoon. We went in a brown-sailed fisherman's boat and landed at a fish stage, going through the house to the shore. While the doctor made his calls, Miss Storr and I wandered over the hill to the other side of the neck, where from a lookout on a high rock we could watch the waves far below. We longed to stay, but there were other patients to visit and we sailed back to the hospital.

Later, when the doctor was free, we started out again on a rough scramble up the high hill at the mouth of the harbor. The doctor helped me up stiff places – not so Miss Storr, who stormed ahead, choosing difficult paths instead of easier ones. We had to claw our way through underbrush and walk in wet gullies, but at the top we stretched our tired bodies on beds of soft moss at the cliff's edge. The waves were breaking on rocks four hundred feet below us. Some fishermen were trawling off shore in the clear water, and a heavenly blue sea stretched away to where two large icebergs gleamed white on the horizon. The rocky coast was untouched by light-house or human habitation. We scrambled back to the hospital, happy but rather done up after weeks of insufficient exercise.

August 2nd At last I have discovered how to find time to paint. This morning I woke early and went out to sketch at 4:30. There are no mosquitoes at that hour. It makes a strenuous day, but I want some record of this beautiful place, for who knows if I shall ever come here again?

The wool warp is at last on the loom, and Dr. Grenfell's homespun suit will soon be a reality. I wove three-quarters of a yard at odd moments today, but it goes slowly with so many interruptions. Miss Storr is a great help and has learned to prepare a loom, which will enable her to carry on after I go away. I had hoped for more interest among the village women. It was necessary to overcome the unfortunate impression left by the (would-be) weavers of a year ago and also to inspire confidence, for the people doubted the motives of a volunteer worker coming among them, suspecting some personal advantage. Someone asked me, "What are you getting out of it?" evidently fearing exploitation – perhaps a natural reaction since the only business enterprise with which these simple people are familiar is that of the greedy coastal traders.

Most of the women are reserved and uncommunicative, and we have to guess the reasons for their actions, which they seem to think we should know by intuition. For example, a woman entered our kitchen one day while I was getting dinner and, without a word, sat in a chair near the door. On her lap she held a little thin chicken. I spoke to her, but she did not answer, just sat. After about twenty minutes, she left with the chicken as silently as she had come. I found out later that hearing we had no fresh meat, she had brought her one chicken, thinking we might buy it for food. But how was one to know? They are an interesting people, and once confidence is established there could be no friends more loyal or more willing to share what they have with others.

While I write, Miss Storr is playing the parlor organ for the children, who have been singing "Jesus loves me" and are just beginning "Hold the fort."[42] There is no piano on the coast. It will seem strange to be again in a land of telephones, trolley cars – even vehicles of any kind. I cannot yet become accustomed to the lack of roads and only one footpath. I have not seen a horse since entering this part of the world, so why roads? The ground, covered with tangled low growth, is like a

Emma White with St Anthony orphans, 1908 (Yale University Library)

sponge, and water oozes up around one's shoes. I know little of the soil or geological formation of this region, but I suppose the clay and rock do not allow surface water to filter through and that this is the cause of all the little rivulets. There are still patches of snow on some of the higher hills.

Inside Mr Patey's stage head, August 1906 (JLP)

On my way home yesterday, I saw one of the fishermen coming in and went to the fish stage to watch him unload his catch. There were five quintals.[43] At $5.00 a quintal it was fair, but small compared with some catches. However, many are now larger, and the fishermen are encouraged. The boats are large and heavy with two triangular, adjustable sails. On the bottom of the boat the big cod lay in a silvery pile. Mrs. Patey, the fisherman's wife,[44] appeared in an oilskin skirt, ready for what seemed to me a very sloppy job, and when I returned later to take a photograph of the fish-house they were hard at work.

The house is built over the water, and the fish are thrown from the boat onto an extended platform [stage head], then pitched through the open door of the house, where men and women stand at a table cutting off heads and cleaning. Refuse is thrown through a small opening in the wall to the water below. The fish are then washed in a tub, split and spread in layers in a large bin, with coarse salt between them, and left for at least two weeks before they are taken to the stages. These are loosely built platforms of saplings supported on a high staging. The fish are spread out to dry, and in case of sudden rain there is a rush to the stages to turn them backs-up or pile them in heaps with tarpaulins over them. There must be anxious watching of clouds and prayers for fair weather. When finally dried or "cured," they are packed to await the traders' schooners that call along the coast to take the quintals to St. John's.

This method applies only to those who live on the shore. Many fishermen pool their resources, hire a schooner or use their own and "go shares" on expenses and profits at the end of the season. These men go far north down [to] the Labrador, which is usually the best fishing ground. They live on the schooners, using nets or fishing from small boats, and prepare and salt the fish on board, packing them "wet" in the bins, as there is no way to dry them. Sometimes the men take their families and leave them in shacks on shore while they are at sea and return south with them at the close of the season. The fishermen usually have no definite destination during the summer. They go where they think fish may be found or follow a promising rumor that may take them to islands and bays where hazards of ice and shoals await them on the uncharted coast. But they are hardy men, and the sea is the only field of industry they or their fathers have known.

I have mentioned that only cod is meant when one speaks locally of *fish*, but there is capelin, a small fish like our smelt which is used, principally dried, for dog food in winter. There is also salmon. I never thought it possible I could tire of delicious fresh salmon, but since one was brought in so huge that it had to be served at five consecutive meals to avoid waste, I am content with the lowly cod.

We have just passed through a crisis in the bread industry. The bread was so impossible Miss Storr went to Mrs. Ash,[45] wife of the Mission [construction] superintendent [Albert Ash], for advice. We found that Martha had merely dissolved the yeast in water, stirred it into the flour and then baked it without allowing the dough to rise. No one had noticed the directions printed on the box. The problem seemed solved, and Miss Storr hastened to test her new knowledge by making the bread herself. All went well. The dough rose nicely, and Miss Storr filled the pans nearly to the top. I was at the shop during the morning and, on returning at noon, found Miss Bayley picking up spoonfuls of dough from the floor, for it had risen above the pans and dripped over through the grating at the back of the stove where the pans had stood. It rose so much that when put in the oven it was exhausted and could rise no more. I never saw such dry bread, and all those loaves had to be eaten. But at least we now know the method, and experience is valuable.

August 12th A thrilling event – we have acquired live-stock! Someone has given Dr. Grenfell a fine Holstein cow. She came by the last *Portia*, and there was great excitement when she was lowered in a sling over the steamer's side to the trapboat below! Few of the people had ever seen a cow. There are also three calves. The animals already belonging to the Mission include four foxes, twelve dogs, two puppies and a cat. Yesterday five hares were added, one of them immediately producing four young ones, and today Davey [Gill], one of the orphans, came running, trying to tell me three sheep had just come. He was too excited to articulate, and all he could say was "Sheep, Sister." (The children call all of us "Sister.")

Dr. Grenfell has had a row of hutches built for the hares, with runs separated by wire netting, which also surrounds the enclosure and covers the top to discourage the dogs. The foxes are protected in the same way. The fox-farm is the beginning of a fur industry, and the hares are for winter food. The sheep are for wool as well as food. I believe the calves are also for fresh meat. One of them broke loose last night and drained the cow dry. It is well our supply of tinned milk is not exhausted.

A few days ago there was excitement in the cow-yard. We discovered Reuben,[46] the "farmer," chasing the cow furiously around the little enclosure and asked him why. He looked surprised. "Well," said he, "she was just standin' round an' not takin' any exercise, an' I thought she ought ter have some." Perhaps this explains the curdled milk of the past few days. For the first time in its history, St. Anthony is experiencing dairy problems and learning much thereby.

I found a bit of goldenrod yesterday; a different growth from ours, but the blossom looks the same. It seems odd to find it. The flowers are queer here, spring flowers not yet gone. We still have dandelions, buttercups and clover, but there are asters and goldenrod, also the brilliant fireweed and purple flag. Everything seems to come at once. I wish there were time for botanizing.

Dr. Grenfell's homespun is nearly ready. It will be off the loom tomorrow, sponged and ready for him on his return. There are eleven yards. Many of the village people have been in to watch its progress and admire, which encourages me to feel that once they realize a worth-while product is possible, especially one appreciated by the Doctor himself, they may feel the urge to make things for themselves.

We are experimenting with pottery. One of the young men, who seems much interested, brought me something he had made on his own initiative. The design was good and the clay handled remarkably well. It will be interesting to see what he can do with it during the winter, and I wish there were time to get him really started and perhaps try to build a small, primitive kiln from some bricks that are on the premises, but I fear that will have to wait for a possible second visit.

August 16th I have just come in from church at 8:15 p.m., and the service began at 6:30. It has rained in torrents all day. In fact, it has

St Anthony girls working in the loom room, 1906 (JLP)

rained ever since August began. There has been only one day without a shower. It is gloomy and rather depressing.

Care of the loom-room has been a vexatious problem from the first, for in spite of the widespread need to earn money or clothing, scrubbing has little appeal. Finally, someone agreed to come once a week for general cleaning but yesterday failed to appear. In desperation, I did it myself and at the same time hung some material Dr. Grenfell gave us to replace the ragged quilts used to screen the carpenters' beds. Now the men do not have to ask us politely to leave the room when they want to change their clothes.

I am doing all kinds of odd jobs in preparation for departure. This journey at the summer's end will be nearly as complicated as that of its beginning. It means two days on the *Portia* south to Twillingate, where I wait for the *Virginia Lake* (a steamer from St. John's used in the spring for sealing, and reputed smelly),[47] on which I will again turn north to Battle Harbour, another two days. It is a pity there is no more direct communication with Battle Harbour, which is only 60 miles from here. There is just a chance that a schooner may stop here to unload coal, lumber and supplies. If so, I can sail direct to Battle Harbour and save time. After my stay there to get the loom going, the little steamer *Home*[48] will take me through the Strait of Belle Isle along the Gulf of St. Lawrence coast to Bay of Islands, where I connect with the Newfoundland railroad to Port aux Basques, cross Cabot Strait to North Sydney by the night boat and home by rail. It makes me a bit dizzy but sounds interesting.

August 19th This is the day I expected to sail for home, and it will require all my resolution to keep me on shore when the *Portia* comes sailing in, but there is still much to be done.

Summer has come again and today is lovely, with a soft warm breeze, very comforting after continual rain, fog and chilly wind, so much like November that one day we had a fire in the loom-room. There was a terrific storm a few days ago, direct from northern Labrador – a reminder of days to come. It made me shiver. We could hear waves roaring on far-away rocks and cliffs. It is good news that a much needed light-house is soon to be built at the entrance to St. Anthony harbor, the only one on this part of the coast, which is poorly charted.

The boy for whose life Dr. Soule and Miss Bayley have been fighting died three days ago. He was apparently improving but suddenly became so much worse Dr. Soule decided to amputate his leg as a last resort. He rallied from the operation but died the next night. Poor little fellow: he had suffered so much with such patience and bravery. Miss Bayley is very sad. It was the night of the storm, and all the next morning the men were making his coffin in the carpenter shop below the loom-room. I could hear them whistling, chatting and singing as they worked and in the afternoon saw them carry him to the storehouse on the wharf to wait for the storm to subside sufficiently for them to row across the harbor in a small boat. They had covered the coffin with fiery pink cloth, trimmed it with white tape and tied a broad band of white around it. For some reason, that gaudy pink seemed to make the whole story more pathetic.

The doctor and nurse had one day of comparative rest; then yesterday a bad case of typhoid was brought ashore from a small schooner. Dr. Soule urged that the man be taken on to Battle Harbour because he needed special care which we are unable to give; but the skipper had only one man to help him and they could not manage alone, so the patient was brought ashore, and the doctor is fully occupied, also Miss Bayley. After the boy ceased to need her, she planned to leave by this *Portia*. Her work awaits her in London and she is anxious to go, but what can one do? The man cannot be left to die. I told her Miss Storr and I could take turns with the doctor in caring for him, but she would not hear of it, insisting that we had enough on our hands already. There would also be the danger of infection for the children, which would make it impossible for Miss Storr to nurse. One is certainly face to face with stern realities of life in this remote region with little of the lighter side.

August 23rd Everyone admires Dr. Grenfell's homespun. In weaving it, I timed my own work to establish a standard and found that with allowance for inevitable interruptions, three yards a day seemed a fair average of production to expect or strive for. It is also necessary to gauge the amount of wool, varying with thickness of material, required per yard of cloth.

This afternoon I saw Reuben on his way to feed the foxes and went with him to try to get a photograph. We went inside the house (it smelled to heaven!) and I tried to snap them, but they scented a stranger and were too shy to come out. There are twelve white foxes and one red. Such pretty little things, with bright eyes and sharp, pointed faces!

August 26th A heavenly day, warm as spring, and I am still here! I shall be unable to go to Battle Harbour after all. There have been cases of desperate illness at the hospital with Mrs. Ash and Jane the maid, also ill, leaving Miss Bayley and Dorcas, the ward maid, to do everything. My efforts to assist take various forms, principally keeping everyone well fed. Sister Bayley has been a trump and so has Dr. Soule. They have been through the most trying and exhausting experiences. The typhoid case died a few days ago. I was not allowed to help in nursing that case for fear of infection, but I could help to hand instruments in the operating room, hold the legs of a patient under ether, and feed the child who has a tuberculous leg. Sister Bayley, in spite of her own deferred work, has decided to remain until the patients are better or until help arrives from the other hospital. The doctor will of course stay until someone relieves him. I will stay two more weeks, then must go as my own work is waiting.

The delay will have compensations, for it will mean more training of the enthusiastic boys of the carving and pottery classes who are doing such good work. One of them seems to have real ability and is so interested I think he will be able during the winter to help the little boys who want to learn. In addition, the long-delayed box of dyes and basket material has at last arrived. Miss Storr and one of the girls will now have time to learn how to use them. This will practically complete the summer's plans.

10:00 p.m. This is a glorious moonlight night, almost too beautiful to be real. Moonlight floods the harbor shore and defines the little path leading off to the dim hills. Over it, from the north, a wonderful aurora throws its great arc, with pulsing, gleaming streamers to the zenith. The air is filled with quivering light, and the great hills are glorified by its weird radiance. There is no sound but the rushing brook and far-off ocean roar. Dr. Soule and I walked to the end of the one path and back again, slowly, with frequent stops to absorb the night's beauty. We spoke rarely. There is no need for words when nature and its great forces are so eloquent.

Not a soul was stirring in the little hamlet, and only one of the huts showed a lighted window. We did not even see or hear a dog. Since hearing a tale of the dogs, I have not wandered abroad at night as I used to do. One evening I told the carving class that I often took a walk alone before going home, and to my surprise they looked startled, saying it was not safe. I assured them I was accustomed to dogs and they were friendly with me, but Mr. Gushue, the carpenter, said these dogs are different. Many of them have a strain of wolf. Usually they are harmless in the daytime, but at night they sometimes trouble strangers, and he told a gruesome tale of a passenger from the *Portia* who went ashore at Jackson's Arm (a nearby settlement) while waiting for the steamer to unload freight. As he walked away from the coast, seven dogs attacked him, and in ten minutes he was "stripped."

"Of his clothes?" I asked.

"Oh no, Miss," said he. "His flesh was so torn that before anyone could reach him he was dead." He added that he was there and could vouch for the story. Otherwise, I might have doubted it.

Our dogs are usually kept in the compound but sometimes break out. The conversation started that night from a remark by one of the boys that three of the dogs were at large. It was a black, misty night, and as I left the shop and started on the path to the orphanage, three forms crept out of the darkness and followed silently at my heels. There was no sound, but a furtive glance over my shoulder showed them still there. It reminded me of my terror when, as a child, I made myself go upstairs in the dark without running from the "thing" which might be lurking behind me in the shadows. Summoning all my will power, I forced myself to walk evenly, for I knew running as well as falling invites attack, and with my heart beating wildly I arrived safely at the orphanage. I am told no one has ever been hurt in this village, but my confidence is shaken.

August 28th Last Wednesday St. Anthony had its annual picnic. The entertainment always consists of marching by the children and some grown-ups, then games and foot-races, followed by a "feed" out of doors. It is a general holiday and was announced from the pulpit on Sunday. At eleven o'clock people assembled at the church. The procession formed and marched around the village carrying gay banners and led by a man with the British flag. The other flags were of various sizes, made of turkey red[49] cotton with designs of black or white sewed on them or pieces of gay calico stitched together like a patchwork quilt, and very effective. The children wore their best clothes, most of them thin white or colored dresses, although the air was cold like our November days, and I wore a winter jacket. I suppose their pride kept them warm.

In the midst of preparing dinner I ran out to photograph the procession, but we did not join the picnic until after everyone had gathered at the church, which seemed to be the head-quarters. By the time we arrived, the mid-day meal was over and the women were washing dishes, but another table was being prepared on the sheltered side of the building. In a tent, some of the older women sat, guarding the provisions and prizes for the races.

The minister asked us to start the races and distribute the prizes, then invited us to tea, and we joined the women and their children on the long benches placed before the table, which was covered with bath towels varying in size, color and condition. There were plates and trays of buttered bread and crackers, but the feast was mostly cake of many kinds. We were asked if we would have tea or cocoa. I took tea but Miss Storr rashly took cocoa, and I watched her struggle with the mixture, which was poured from a big iron tea-kettle. She told me afterward that it was made with very little tinned milk and sweetened with a great deal of molasses. Some of the cakes were highly colored, the one with brilliant crimson icing and jelly filling being very popular. About a dozen kinds were passed. I took one piece that proved to be a happy choice, for it was made by my neighbour on the bench, and we became quite chatty over it. The attitude of the women was very different from their reticence and suspicion when I first came among them only a few weeks ago. They will be friends whom I leave behind.

Dr. Soule has been fishing at odd moments lately, and we have had trout for breakfast or supper: a rare treat, for there has been no fresh fish since the cod season closed. Our potatoes gave out a month ago, and we have used macaroni as a substitute, fortunately with the approval of the family, who have also accepted baked beans and other items formerly scorned. We are all growing fat since my "American messes" have proved generally acceptable.

The orphans have misbehaved lately. Jimmy [Hedderson] and Hayward [Patey] were found almost entirely undressed (why, no one knows) in the front porch on a cold day and promptly sent to bed with bread and water diet. Tom and Davey monkeyed with the water

Orphanage children, Emmie Roberts and David Gill,
July 1906 (JLP)

pipes, sending a flood over the kitchen stove. One was shut in a dark closet, the other sent to bed. Davey had the best of it, for he did not mind the disgrace and had a nice sleepy time instead of "doing the water," which however he found was a chore merely deferred.

The children are very funny. I heard a great outcry tonight and saw Tom with a little fuzzy caterpillar curled up in his hand, frightening Lizzie [Hedderson] and Prissie. I told them it would not hurt them and asked Tom to take it outside and put it down while the other children watched it uncurl itself and crawl away, but not to kill it for it did no harm. (I have been talking to the children about killing things.) They went away, but Lizzie came back to say that May had killed that "machine."[50] The others simply called it "that."

It is very late. I have just closed the house – that is, shut the outer doors (we never lock them) – and stepped outside for a last look at the moonlit hill and to hear the rushing stream before going to bed. The main drawback in returning to urban life will be having to hear the trolley cars, the whistles, the sounds of traffic, see electric lights and smell the smells. It is so beautiful and restful without them. Here is the open country, the wide sky, the stars, the great silence, a country where nature's dominance is overwhelming and man is small indeed.

Sunday, September 2nd The *Portia* has not yet arrived. She is probably delayed by the storm. We amused ourselves this morning at breakfast by imagining what our respective families would have for dinner. Miss Storr thought hers would have roast beef, mashed potato and fruit tart (in other words, pie). My family, I thought, might have roast chicken, potato, corn and lima beans, with possibly peach ice cream for dessert. It was a lovely dream.

Fishermen on the coast rarely sail on Sunday in rigid observance of the Sabbath. They usually lie at anchor in any harbor near their fishing ground. Every week we see schooners in our harbor, and the men often come ashore during the day. Dr. Soule went to church this morning and reported that the minister had gone to one of the small outparishes where he preaches. It is the custom for anyone who is willing to do so to take his place and read a sermon from a book. This morning the service was taken by the fisherman from one of the schooners, and Dr. Soule said it was the best he had heard since he came. The man was caught without collar and tie, but that did not trouble him or the congregation. Tonight, Sister Bayley and I joined the doctor at the evening service, and the same man officiated. This time he had found his collar and tie. He gave a list of the hymns to Dr. Soule, who played the organ, but of those chosen only one was familiar to the doctor, Sister Bayley and me. We stumbled through, Sister and I trying to follow the tune with no music score to guide us. There were seven verses, and we sang them all, not one soul in the congregation uttering a sound. The hymn that everyone knew begins, "Work, for the night is coming!"[51] They sang it one night when Sister Bayley and I went to church after we had worked sixteen hours out of the twenty-four and were nearly worn out. We felt a reminder was not necessary. There was, however, comfort in the last lines [of the second verse]: "Work for the night is coming, / *When man works no more.*"

We have correct time here only when the mail-boat brings it. Some of the coast people, we are told, keep tally on the day of the week by the food they have for dinner. It is a routine: Sunday, fish and brewis (pronounced [like] *bruise*), composed of boiled fish, soaked hardtack and fried salt pork, the fat poured over the whole. It sounds uncanny but is good. Monday, it is fish and potatoes; Tuesday, salt pork or salt beef boiled with beans; Wednesday, soup with dumplings; Thursday, pork (or beef) and duff; Friday, fish and brewis; Saturday, pea soup. Of course, there is always tea, and the hospitable tea-pot is almost invariably found on the kitchen stove, ready for an additional pinch of tea and boiling water on the steeped leaves when a visitor drops in to call. Tea also figures in the open, and one of the interesting accompaniments of excursions is a stop to boil the kettle and have a "mug-up"[52] when one feels inclined.

At sunset this evening, Sister, Dr. Soule and I rowed over to the post-office across the harbor to get stamps and carry letters. It is a little hut with a bare front room and post-office behind, perched on a rock at the water's edge. St. Anthony is divided into two sections on opposite sides of the harbor, the Methodist and Church of England. No one knows why the post-office is located in the smaller (Church of England) community. It may be that before the Mission's advent, the two sides were equal in population.[53] Anyway, there it is.

While we were waiting, a trap-boat came in rowed by three fishermen. A long boat with furled brown sails, and the sturdy young men pulled together like a machine, a sight characteristic of the coast. The silvery pile of their day's catch lay in the bottom of the boat. The traps are up for the season, but men will fish with hook and line as long as weather permits.

September 3rd The *Portia* came in today in the midst of pouring rain and a howling gale, and not a passenger came ashore. They have usually landed this summer through curiosity, and today, after having polished our respective departments, ready for inspection, we felt deflated. The shop was in perfect order, with all the products displayed. I even wore my last good shoes. The hospital was also spick and span, but there was no one to inspect. A child patient and her grandmother went on the *Portia* today, also the woman in a plaster cast. They must have been soaked during the long row to the steamer in the trap-boat. That leaves only a convalescent typhoid case and Jane the maid, so if no other patient comes Dr. Soule will sail with Miss Bayley and me by the next *Portia* and be able to take his exams at Harvard.

Mr. Ash had a card from Dr. Grenfell saying he would be here by the 15th, and Reuben thought that he might arrive even sooner because he had with him on board the *Strathcona* two black bears, three foxes and perhaps two reindeer. He would not want to keep that menagerie afloat very long, so I may see him again before I go after all.

September 6th The basket class is quite a success. The girls seem to enjoy it, and their work is generally good. It will be an interesting occupation for them during the winter and I think has been worthwhile. The class met this afternoon in the kitchen while I made blueberry pies. It was rather distracting and complicated, but the pies came out fairly well, a convincing proof of my theory that (under pressure) it is possible to carry on more than one occupation at the same time.

Mark Penny,[54] the boy who is making excellent pottery, showed me today with justifiable pride a piece he had just finished. He seems to love the work, and his handling of the clay is surprising. He told me something of himself and his family, of his young father [Richard Penny], lost on the "Banks" when Mark was two months old; of his mother's later marriage to her cousin [Allen Pilgrim] and of his six half-brothers. He says he taught his mother [Ann E. Pilgrim] to knit after having learned from one of the native women, several of whom have spinning wheels and use all available wool to spin yarn for stockings and mittens, which are an absolute necessity during the winter. The woman also taught him to knit patterns, and I have ordered a pair of mittens for myself with gay red and black checks.

September 11th Dr. Soule continues to visit the nearby trout streams, sometimes starting at 4:30 a.m. and returning with a string for breakfast. He and Miss Storr, also an enthusiastic angler, have discussed an excursion for all of us to a larger, more distant stream, but it has been an unfulfilled dream until today, when for the first time everyone could get away and the weather was perfect. Directly after breakfast Dr. Soule began digging earth-worms for bait from the small area of loam near the hospital and, by the time we were ready, had filled a baking-powder tin. Miss Storr, as an expert, was fully equipped with rod, net, basket and a case of beautiful flies. The rest of us had improvised gear: sapling rods, fish-line with trout hooks attached.

We started off in gay spirits to the trout brook two miles away. It was rough country. The interlacing branches of small alder bushes and tuckamore (a low growth of spruce springing from the boggy land) made passage difficult. We knew the general direction, but no trail was visible. We often stepped on low branches where the growth was too crowded to push through and slipped off, sinking even to our knees into the bog beneath. Our clothes were torn, our stockings in shreds, our hands and faces scratched by sharp branches. Moreover, there were black flies. Ragged, weary and dishevelled, we finally reached the brook and prepared our lines in a lovely spot where cascades fell into a deep pool.

Miss Storr selected a fly and adjusted her rod. The rest of us unwound our fish-lines and prepared to bait our hooks. Someone called, "Where is the bait?" Dead silence! Everyone thought someone else had the tin of worms that still sat on the hospital porch. We tried to dig more with a pocket knife, but the leaf mold was deep and the earth beneath it not the sort to harbor worms. We took turns using Miss Storr's equipment and fought flies while catching three small trout. And we called it fun! It is remarkable how much voluntary discomfort one will undergo in pursuit of sport. What would be our reaction were similar effort and conditions forced upon us against our will?

September 12th This is truly "the day after." We staggered in last night after our rough picnic, muddy and wet from walking at times through shallow brooks. Miss Storr's face was bloody from fly bites, Sister's beginning to swell. This morning she is almost unrecognizable and can scarcely see. The doctor seems to be immune.

Today we tried as an experiment to fire a piece of pottery in the blacksmith's forge, but it went to pieces. We shall have to wait until a real kiln can be built, which may not be until next year, when I may come again. Meanwhile, we have finished articles ready to fire, and the boys have had pleasure and experience in making them. That is something.

A new patient who came to the hospital two days ago for tooth extraction has developed an abscess and is in a serious condition with temperature 102. His face is so swollen he cannot see or open his mouth. Dr. Soule, in addition to his anxiety, is worried on account of his delayed departure, and Sister Bayley is again in doubt about leaving. We hope the *Portia* may be late and not arrive until this crisis is past or Dr. Grenfell returns to take charge. Such doubt is very upsetting. As for my own plans, I expect to sail on the next *Portia.*

Lewisporte Hotel, Lewisporte, N.F.L.D. September 23rd, 1906 At last on my way home! after having waited, fully clothed, more than two days, momentarily expecting the *Portia*'s arrival. There was no word, and no one knew what had happened. Dr. Soule wanted to go too, and if Dr. Grenfell arrived before the *Portia* he would be free to do so. Sister Bayley was ready, and we were to travel together as far as Exploits, two days' journey. We heard vague rumors of wrecks but knew nothing authentic until the *Portia* steamed in. The *Strathcona* had not come, and we departed with Dr. Soule.

We had a great send-off. Everyone, including the orphans, came to the wharf. Mrs. Ash wept, and my boys, Miss Storr and the doctor all went out with us to the steamer. We found the *Portia* crowded with wrecked fishermen going home after having lost their schooners and their summer catch besides clothing and fishing gear. There were eleven wrecks on Bell Isle. Every one of the fleet went down. One old man told us of reaching shore on a spar after drifting on the capsized vessel. He was wearing a cap belonging to Dr. Grenfell, who happened to come along at the time and was able to provide food and clothing for the men.

Two days later we reached Exploits, where I bade a regretful good-by to Miss Bayley, who continued on the *Portia* to St. John's while I waited to finish my journey by rail. One of the village women offered me accommodations for the night, and I shared the room with another passenger, a young girl from Pilley's Island. It seemed strange to have an utter stranger for a bed-fellow!

Last night there was a wonderful aurora. The entire heavens quivered and danced with red, green and yellow light. Long streamers, falling like a vibrant curtain from a great crown overhead, seemed to reach the earth. Stars glimmered faintly as through a veil, and masses of color concentrated and dissolved, to reappear in another part of the sky. The natives speak of such a display as "lively" and claim they can hear a hissing sound. It seemed a wasted opportunity to leave it for bed in a stuffy room, but one must sleep in spite of electrical phenomena.

The *Clyde* came this afternoon. The steamer was crowded. There were no seats, and everyone was wretched, as it was too wet to stay on deck. We arrived an hour ago and find the train leaves at three tomorrow morning. It is now 11 p.m. We expect to spend the night downstairs in a little sitting-room which is quite comfortable and has a fire.

Petrie Hotel, Point Pleasant, Bay of Islands, Newfoundland. September 25th, 1906 At 3:30 a.m. we left Lewisporte, all crowded into one little car for the ten-mile run to Notre Dame Junction. It was dark, but the storm was over. Stars were beginning to appear through rifts in the clouds, and when we reached the station, dawn was breaking. On arriving at Bay of Islands, I found that by mistake my luggage had been put off at another station, and there was no way to get it until morning. It was still raining: an open wagon waited for passengers, and the only bright spot while driving over muddy roads to the hotel was the thrill of again being drawn by a horse. The hotel is comfortable but chilly. I shall stay only long enough to connect with my wandering luggage. My thoughts and interests are now centered on home!

September 30th, Providence For the benefit of those who may have felt interest in the rumored menagerie with Dr. Grenfell on the expected *Strathcona*, I report the arrival of foxes and hares, but the bears were their skins only, and no reindeer were seen or even expected. Such is rumor!

3

Interim

During the winter following my return from Newfoundland, Dr. Grenfell came to the States for lecture engagements, and Providence was on his itinerary. He stayed at our home during his visit and renewed his plea for my help, meaning a full year at St. Anthony to establish on a permanent basis the industries hardly more than begun. Being already engaged in developing occupational therapy at Butler Hospital, with the encouraging co-operation of its superintendent, Dr. G. Alder Blumer, I told Dr. Grenfell it was not a question of my interest or inclination, but the matter rested entirely upon Dr. Blumer's willingness to allow me a year's leave of absence, for after all, occupational therapy (afterward defined as "A valuable adjunct to the medical profession") was my permanent interest and responsibility.

A meeting with Dr. Blumer was arranged, and the morning after Dr. Grenfell's lecture I drove him to Butler Hospital and left him in the Superintendent's office while directing my patients with a divided mind, for I did want to go. After a while, Dr. Grenfell came with Dr. Blumer to see the class. He was smiling and with his characteristic chuckle announced, "You can go!" And so it was arranged that I should return the following September for a full year at St. Anthony. Dr. Blumer's interest in the Mission was deeply sincere, but not everyone would so fully have understood Dr. Grenfell's vision and point of view, and I have always been more than grateful for his encouraging and enthusiastic co-operation, without which my part in the Mission's early development would have been impossible.

The possibility of developing pottery from native clay had taken a strong hold on the Doctor's imagination. We had also discussed making brick, much needed for chimneys to replace iron stove-pipes through roofs, a distinct fire hazard. Before launching such a venture, however, some experienced advice was needed, and on leaving St. Anthony in the fall I brought samples of the clay to be tested by experts. It was found to be suitable for both pottery and bricks, but funds were needed. The Industrial Department has no regular appropriation, and its development (until there is income from sales of products) is dependent on special donations, Dr. Grenfell's appeals, his contribution of the proceeds from some of his lectures, and from my own few lectures and appeals in pamphlet form for specified purposes or equipment, which included a potter's wheel and material for a brick pottery kiln. Both were generously donated by Mr. and Mrs. William Hodgman[55] of East Greenwich, Rhode Island, as an expression of their interest in the entire Industrial Department.

Mr. Grueby,[56] of the Grueby Faïence Company, supplied plans and instructions for building the kiln and some of the material. Also, with brick-making in view, I looked up a one-man brick-yard on Cape Cod and found it in a region where clay was abundant. It was truly a one-man affair and very interesting. The home-made pug-mill (a large wooden cylinder containing a revolving shaft with wooden blades) was used to grind the moistened clay, mixed with fine gravel, to the proper consistency. It was operated by a horse that walked round and round at the end of a long bar attached to the top of the revolving blades, like the handle of a coffee grinder or an old-fashioned ice-cream freezer.

The softened clay was pressed into wooden molds to form the bricks, which were then turned out and dried in the sun on a level space before being piled to form an oven and baked by a wood fire built inside it. It was very simple and seemed quite feasible for our local use. Moreover, it required no expensive imported equipment or material, therefore involving little financial risk. There was plenty of clay at St. Anthony. A little stream that trickled down the hill would supply water. There was fine gravel nearby and open, level space for drying the bricks. "Harry," the Mission horse, would (or could) supply motive power, and Uncle Joe Pelley,[57] who was much interested when I talked with him about it before leaving St. Anthony, had agreed to take the job. It did seem a worth while venture, especially when imported bricks cost five cents (someone said even twenty-five cents) apiece. I was keen for it and optimistically made designs and working diagrams drawn to scale for the pug-mill and other equipment.

In our zeal for making use of natural resources, we are also considering copper (there are copper mines in Newfoundland) and Labradorite, known locally as "blue stone," a variety of feldspar which when polished reveals a flash of brilliant blue. This is found in some localities on the Labrador coast and can be used for paper weights and for jewelry combined with copper. Appeals have therefore included requests for metal tools and equipment necessary for making bowls, trays and other small articles from sheet copper and a machine for polishing Labradorite. The latter valuable article of equipment has been made possible by Professor Charles W. Brown,[58] geologist at Brown University, whose generous advice and information resulted in the acquisition of the desired polishing machine, which with the potter's wheel will be shipped to St. Anthony. The brick and other material for the kiln will be sent down during the summer, in readiness for construction whenever working conditions and temperature will permit. The clay is to be dug before frost hardens the ground, and pottery can be made during the fall and winter to be fired when the kiln is ready. At least this is the plan, a stimulating objective to work for.

4

Via St John's to St Anthony

31 October – 13 November 1907

On board Reid Newfoundland Railway. October 31st, 1907 Here I am again on my way north! This second journey to St. Anthony includes no "fancy trips." It is by the most direct route. A year ago, when passing through the provinces in early July, I overtook the spring-time. On Cape Breton Island, lilacs and apple trees were in bloom, and frogs were peeping in the marshes. Now, in late October, I am overtaking the winter. I have seen from the train bare trees and brown fields. On higher land, the ground is white with a light fall of snow, and little ponds are frozen.

The really interesting part of the journey begins on leaving the train at North Sydney Junction, on Cape Breton Island, when passengers change to a small crowded car ready to take them to the wharf and the little steamer *Bruce* waiting to cross the Cabot Strait to Port aux Basques, Newfoundland. A strong current flows through Cabot Strait, and crossings sometimes suggest the English Channel, but this time it was mercifully calm. When we landed, the well remembered train was waiting with its diminutive cars, which I am sure must be a part of the original equipment of the Reid Newfoundland Railway. At the custom-house, the name *Grenfell* inspired confidence, but one of the inspectors spied my small packages and pounced on a box of plated silver for use at St. Anthony, brought to the train by Miss White just before I left Boston. I explained that although it had not been used, it soon would be in a good cause, even by Dr. Grenfell himself. The man smiled, saying it was all right, and when on opening my trunk I told him the only new and valuable articles in it were a pair of high rubber boots, he grinned and marked [passed] everything.

On boarding this untidy, ill-smelling train, passengers having once taken the trip realize the curtain is about to rise on a play. It may be a tragedy but is usually a comedy and sometimes a howling farce. The ex-

En route from Port aux Basques, a switch is made to allow two trains to pass, October 1907 (JLP)

citing thing about it is that one never knows what the acts are to be or how long or who are to be the actors. One might even become a member of the cast. It is well to start in this philosophical frame of mind.

Each section of the roadbed is supposed to be inspected by a man in charge, but apparently care rests lightly on his shoulders. The whole railroad is really a joke, for it runs up and down grades, makes long waits, and is liable to break down at any time. According to rumor, it has even been blown from the track, and if a gale develops the train waits before reaching the perilous spot until things are calmer. At least that is what fellow passengers told me, and anything seems possible after our experience this morning, when an unexpected freight train appeared ahead, waiting for us beside a short siding on the single, narrow-gauge track.

There was a switch, but something was wrong with it. The trains faced each other while our conductor and the engineer of the freight conferred. Then brakemen brought crow-bars and transferred a rail from the main track to the siding for a temporary switch, but when our train had run over it to the siding and the freight started to pass, it was discovered that the space between the tracks was too narrow to avoid collision. The freight retreated, and again the conductor and engineer conferred. The trainmen again took their crow-bars and this time shoved the main track, rails and sleepers, a foot away from the siding, the full length of the train. The rail was transferred to the main track, and the freight scraped cautiously by while the passengers, who had left the train during the show and lined up along the track, looked on and offered advice. When the freight was safely by, the rail was readjusted to the siding. The train backed onto the main track, the rail was again changed, and we proceeded on our way. I think some of our railroad magnates should know of this and congratulate themselves on "progress."

The night was long. In the morning I watched from my window the daylight brighten on a land covered with glistening hoar-frost. We were running through one of the wide tracts swept by forest fires, where ghastly stumps and whitening trunks of great trees suggested tombstones of the departed forest. It is odd that so little new growth has started. Often the ground is as bare of vegetation as if the fire had recently swept by, and boulders of all sizes among the ruins add to the impression of desolation. Within eighty miles of St. John's, the character of the country changes to rolling hills without trees but covered with vegetation, and farther on, as one nears the coast and approaches lovely Bonavista Bay, a gleam of water appears in the distance – a lovely part of the country that I am seeing for the first time. We arrive at St. John's at noon.

St. John's, Newfoundland. Sunday, November 3rd A perfect day: still, clear and warm. A lovely blue haze lies over the high hills guarding the harbor, and the bay, like the sea beyond, is a softly brilliant blue. The hotel and even the streets were silent until past seven, and one could easily imagine oneself in the wilds instead of a city of fair proportions.

The English cathedral[59] is near by, and on entering it this morning and being shown to the very front pew I almost felt myself a part of the service. With its grey stone walls and absence of stained glass, the large church is very attractive and restful. The choir boys wore purple cassocks and the officiating clergy their doctor's [academic] hoods. The address was short and excellent, and I am glad I went. The day was so tempting I explored and found an attractive residential quarter on a hilltop overlooking the beautiful harbor and great headlands. The streets ran any sort of way, and there were "short cuts" and rocky stairs leading to streets above and below.

In the harbor, a government ship lay at anchor,[60] and after dinner, when again starting out for a walk, I met a marine funeral with a band on its way to the Catholic cathedral.[61] It was so impressive that I experienced a shock when, later, I met the same band (now playing a rollicking tune instead of dirges) as it led the company of marines, marching at a quick-step, trundling the now empty gun-carriage, a reminder of how soon the waves of today obliterate the footsteps of yesterday.

November 4th After breakfast, I went to the St. John's office of the Grenfell Mission to report my arrival to Mr. Peters (the agent)[62] and had been in the office only a few minutes when to my surprise Dr. Grenfell walked in. Neither he nor Mr. Peters knew I had arrived. They say the *Portia* is overdue from St. Anthony and will not be in until the last of the week. That means a longer wait in this place. The hardest part of the trip is not over, and for once in my life the day has too many hours instead of too few. I have repacked my trunk, written letters and read even the advertisements in my one magazine.

November 5th This morning an invitation came from the Governor's wife, Lady MacGregor, for four o'clock tea at Government House. It was through Dr. Grenfell, of course. I did not know if it were a "function" or simply a social cup of tea and decided it must be the latter, as the invitation would otherwise have been more formal. It was well, for my wardrobe consists only of street and work clothes with one real gown.

On making inquiries, I was directed to a great castle-like stone house surrounded by a dry moat that stands in a park on a hill-top. I walked up the drive past the lodge and, not seeing any imposing entrance, stopped at the first one in sight. A functionary answered my ring and on receiving my card told me the house was not open to visitors that afternoon. When told I was expected, his manner changed and I was ushered through a hall to the drawing-room, where Lady MacGregor and the Governor[63] received me. A fire burned on the hearth. The tea-table stood near it, and I spent a most interesting hour with my gracious host and hostess, who asked many questions concerning plans for the winter at St. Anthony and the various branches of handicraft to be undertaken. The Governor then showed me his unusual collection of curios, beautiful and interesting things that I longed to examine one by one, but my visit had already been long and I reluctantly made my farewell. It was a delightful hour to remember.

November 6th The *Portia* came in early today, and I joyfully recognized her whistle as she entered the harbor. This morning Dr. Grenfell came to see me and made a long call. On my way to the drawing-room to meet him, a tall man preceded me and exchanged cordial greetings with the Doctor, who introduced his cousin, Colonel Sawyer,[64] just arrived on the *Portia* en route to Montreal after a short visit at St. Anthony. He sits next to me at the table, and I have asked him all sorts of questions about the staff members already at St. Anthony for the winter. He says there is an Irishman, an unkempt creature with very bad manners who has a deadly fear I will try to boss him and upset the cherished arrangements for his comfort. If that is true, it is unfortunate that he has appropriated the one room in the St. Anthony Guest House that Dr. Grenfell had designated for my use long before the Irishman emerged from the unknown. He has assigned to me a tiny box of a room which I shall have to pass through another room to reach.

Dr. Grenfell confirmed that my room had been taken by the Irishman but was sure he would not object to letting me have it, especially as it was previously arranged for, and he drew a diagram of the house, designating which room was to be occupied by whom. Colonel Sawyer and Mr. Morgan, the Doctor's secretary, are by no means so sanguine in regard to the Irishman's willingness to relinquish the room and prophesy trouble. They say he has to be handled with great tact, and I surmise the possibility of a test case, so to speak, fills them with glee. I know they would like to be there to see what happens. They also say that the other men of the household are very rough. To what am I going? Thoughts of the possible difficulties fairly make my hair curl.

Colonel Sawyer's ingenuous solution of the difficulty is for me to enact the role of poor, weak woman in need of help, advice, direction, etc., and dependent upon the Irishman as a friend in need, a most inconsistent suggestion since, in fact, I do not need the man at all, unless he will lay the bricks of the new pottery kiln under my direction. Even if I tried to enact the meek and mild character, it might not be convincing. I must say I am rather upset, but the men, perhaps even the Doctor, are I am sure enjoying the situation. They won't be there during the first *mêlée.*

November 7th The *Portia* sails tomorrow at ten o'clock. This has been a busy morning, including financial transactions at the bank and last purchases before disappearing beyond icy barriers for many months, far from a source of supplies. Colonel Sawyer and Mr. Morgan still tell me tales of the St. Anthony colony. It now appears that *all* the men are rough and unmannerly, but when I fairly gasped at the idea of keeping house for a party of boors, they said that there was really much good in them and, if taken in the right way, they would probably do anything for me. I can see that these baiters are, to be slangy, trying to get my goat, and I shall probably know no more until I come face to face with the creatures. What may complicate matters is that the Irishman and Dr. Little, the Boston doctor, are on a hunting trip and probably will not be there when I arrive, so I have visions of sleeping in any sort of place until the room question can be settled.

November 9th. S.S. Portia We left St. John's yesterday morning. Mr. Peters and Colonel Sawyer came to the steamer, Mr. Peters attending to my luggage, even coming for me with a cab, a courtesy which, after fending for myself all the earlier part of the journey, gave me a sensation of luxurious comfort and relaxation. Is it possible I may be, after all, a clinging vine at heart?

November 10th I have been on deck all morning. The stewardess found a camp chair with a back (there are no steamer chairs), and I tried to find a sheltered spot, which was difficult as all of it was literally hurricane deck. My seat at the table is beside the Captain, with a man opposite who without wasting words simply pointed at any desired food, a surprising gesture but one that brought results. He had long, oily hair and wore a shiny celluloid collar. It was a relief when I saw him drop into a boat and go ashore.

In the absence of a clergyman there was no service on board this Sunday morning, but a group of passengers sang Moody and Sankey hymns[65] during the afternoon and evening. The Newfoundland coast people love to sing. The voices are frequently off key and have a nasal quality, but there is no doubt of their enjoyment.

We have been at anchor each night of the voyage thus far, for the runs between ports of call are so short (only two or three hours) the stops would often be during the night, and as there are no light-houses it is not always safe to enter the narrow harbors in darkness.

This trip would be restful if my mind were easy concerning that wretched Irishman at St. Anthony.

November 11th We reached Exploits just at daybreak. As we anchored in the harbour, I could see the path along the rocks that I remembered so well, but there was not time to go ashore to see the Manuels, who were so good to me last year, and we were soon on our way. A few passengers came on board at Fortune Harbour with their simple luggage, an old-time carpet-bag or only a box or bundle. One old man had a small trunk with a heavy boot tied to each handle.

At one of the few villages where there was a wharf, a small cow was taken on, but it was a very uncertain wharf, and the poor animal had to walk up the narrow slanting gangway. She was terribly frightened, and it took the combined pushing and pulling of seven men to get her on board. As the man who brought her was leaving the ship, mopping his face, Captain Kean asked, "Where's the hay to feed her?" He replied he hadn't brought any, that he had all he could do to bring the cow, but just as the steamer was drawing away he came tearing down the wharf with an armful of hay, which he threw at the steamer, most of it falling into the water. Sometimes the ways of these people seem more like the ways of children.

I asked the Captain today to tell me just what was thought of the Irishman by the people in general (it is plain the man is on my mind). He said he was considered an adventurer. He will answer no questions and give no account of himself, and he is understood to be a wealthy man who has "gone to the bad" and is now, as the Captain says, "taking a rest after spending his portion like a prodigal."[66] The Captain's description of his first appearance was very funny. He had on "trousers buttoned tightly to his knees" (probably riding britches) and wore a monocle. The idea of a monocle at St. Anthony convulsed me. On his last appearance he wore a red, white and blue knitted cap and overalls much too large for him, turned up to the knee. I had a faint hope the Colonel was drawing the long bow to tease me, but now I doubt it. There plainly are breakers ahead!

November 12th We are in our first snow-storm. Last night we anchored in Little Bay, and on looking from my port-hole this morning I saw the coast, white and ghostly. The Captain laughed at breakfast, remarking that it was merely a foretaste of winter, and I would probably find a foot of snow at St. Anthony.

Mr. Pitman, the clergyman of Tilt Cove, a fellow passenger on the *Portia* last year, came on board at Nipper's Harbour, and when he recognized me I told him we were about to start a pottery at St. Anthony. He was interested at once and said a man at Nipper's Harbour had tried to make pottery from native clay but found the cost prohibitive for anything beyond the experimental stage. He had, however, made some bricks, flower pots and pipes and fired them in the kitchen stove. Mr. Pitman offered to go to the man's house for samples and has just brought them aboard. I find the clay in its natural state, as well as the partly fired pieces, has the same color and texture as that found at St. Anthony. This offers a suggestion for extension of the craft to other localities where clay may be found. It is possible men and boys, during intervals of the fishing season, could start the industry in a small way after instruction at St. Anthony. Even if the effort were confined to making bricks, which are so much needed on the coast, it would be in line with the development of native resources that has from the first been one purpose of the Mission. We hope the St. Anthony industries may provide a nucleus of instruction for the benefit of other centers in this country of scanty and widely scattered population.

November 13th The purser says we shall reach St. Anthony some time tomorrow. It has not been a comfortable voyage, either physically or mentally, its brightest side being the kindly attention of Mrs. Brown, the stewardess, who has done everything possible for my comfort.

We had just stopped at a little settlement and were going to dinner when the constable came in excitedly, saying that the son of Mrs. R., whom everyone seemed to know, had just died. Captain Kean was behind me on his way to the table. "What?" said he, as if he could not believe it, "and Mr. R. away at Pilley's Island, poor woman, poor woman, and such a likely boy!" He could think of nothing else and after a few moments turned to the purser. "Is the poor boy lying dead?" he asked. The purser nodded and the Captain called the steward. "Tell Bill to put the flag at half mast," he said. It was like the Captain to order it. When I asked for details, he told me the boy was a fine young fellow and died of pneumonia. I asked if there was a doctor. He said, "No, but the minister was quite a doctor and thought he was pulling him through." Mrs. R's only other child is a cripple. Nothing could illustrate more clearly the need of good medical service on this coast.

5

We Prepare for Winter

14 – 25 November 1907

November 14th We arrived at St. Anthony after a comfortable breakfast on board, instead of landing, shivering and forlorn, in darkness as we had expected. With a thrill of excitement I recognized familiar landmarks, but a curious misgiving hovered somewhere in the background, and before the day was over I realized it was prophetic. As we came to anchor, little boats put out from shore, and I saw the big trap boat start from the Mission wharf. Soon it was near enough to recognize Mark Penny, Albert Ash and Reuben and in one of the smaller boats Miss Storr and Dr. Stewart,[67] whom I was eager to meet as I had important communications for him from Dr. Grenfell.

After landing, I went with Miss Storr to the orphanage, which is improved and more comfortable. The children greeted me with enthusiasm, and I noticed several Eskimo additions to the family. The hospital is practically unchanged: the only thing lacking is a certain picture of a lady in pink that offended our eyes all last summer. Only an unfaded place on the wall-paper remained. I was told she had been cremated. The only visible addition is a phonograph with records.

During the past year, Dr. Grenfell, in view of the need for accommodation of an expanded staff, bought a small house near the hospital and enlarged it by adding several rooms and a glass-enclosed porch. This he calls the Guest House.[68] It will be the home of all staff members not resident at the hospital or orphanage and will as well be the Doctor's home while he is at St. Anthony. Miss Storr took me there after leaving the hospital, and the first problems of my new field of work were immediately self-evident. Some of the rooms are finished and already occupied by the three men: Mr. Lindsay, Dr. Little and Mr. Lee and by Miss Ruth Keese,[69] the young schoolteacher from Massachusetts. But the kitchen, although furnished with a cook-stove, a small table and a few chairs, was cluttered with lumber. A carpenter's bench

The Guest House living room, St Anthony, November 1907 (JLP)

partially filled the room and shavings littered the floor. Two village girls engaged as maids are sleeping downstairs until the third storey is ready. Supplies are not unpacked. There has been no supervision, and the hospital has furnished meals for everyone.

Dr. Grenfell's study is a lovely room, just what it should be, with his books, writing table, comfortable chairs and rugs. Another small room, planned by the Doctor for a sitting room, has been appropriated by the men as a "gun-room" (otherwise, smoking room) and is filled with a variety of sporting equipment. There remains only the dining-room, which is very ugly. Its decoration, in spite of plans made in Boston, has somehow gone wrong, and I find a yellow wall instead of light green, bilious yellow paint instead of white, and a wooden mantel behind the stove is painted and grained to "resemble" yellow marble. A large shiny golden-oak sideboard with a mirror at the back nearly fills one end of the room. A table of unfinished wood and eight oak dining-room chairs complete the furnishings, and on the walls are several large framed engravings with wide white mats which, in the small low room, seem ready to topple on one's head. There is a small wood-burning stove, but as the stove-pipe has been taken for use in the Doctor's study and not replaced, there can be no fire. The bedroom assigned to me is too small to use as a sitting room, and Miss Keese has to pass through it to reach her own, which is smaller. It is evident the dominant males do not consider the comfort of ladies a part of their scheme of things, which is unfortunate, for some adjustment will I fear be necessary.

The enclosed piazza will be very nice when cleared of lumber and other construction material, besides trunks and the still unopened packing boxes. Only a few of the boxes containing industrial materials have been located. This is disturbing, for the time is growing short before ice closes navigation for the winter, and I have visions of essential equipment and supplies going astray, with consequent failure of the work, which the people anticipate so eagerly.

November 15th I spent an uncomfortable night in the unheated Guest House. The mercury fell nearly to zero, and this morning the window-panes were opaque with heavy frost. The sheets were like an ice bath. After breakfast at the hospital, I took two of the maids and three of the larger boys and attacked the house. They were fine helpers, and already the results surprise me. When the carpenters finish the shelves and supplies are unpacked, it will be quite orderly. Moreover, it will be clean, for to my amazement the girls love to scrub!

This morning I talked with Dr. Stewart and told him some adjustment will have to be made about rooms. He acknowledged the lack of provision for our comfort, and when I found he had received by this mail an assignment to spend the winter with the Eskimos in the far north and Dr. Little will therefore live at the hospital, I suggested his room might be given to me, and I would share it with Miss Keese, for it is the largest room in the house and quite ample for two beds. Thus, the delicate bedroom problem is amicably solved and we shall move at once. The dining-room adjustment will come in time.

At about ten o'clock Dr. Stewart came in to say he and the hospital nurse, Miss Kennedy,[70] were going up the coast in the motor launch *Daryl* to take a patient home and would return at about four. After the rush of work was over, I went to the hospital in the late afternoon to get warm, and Miss Keese joined me on her return from school. The nurse and doctor had not come, and after waiting and waiting we finally had supper, a rather meagre meal as Miss Kennedy had the keys of the store-room and the maids could offer only tea, toast and jam. There are nine patients in the hospital, some of them newly arrived, and Miss Keese and I are spending the night, for someone must be at hand. I wanted Miss Keese to go home, but she refused and is now trying to take a nap on the hard little seat in the corner. It is frightfully cold, and I am fast emptying the wood-box. It is 1:30 a.m. and the lamp is going out.

November 16th I have had an awful day, for the doctor and nurse have not returned. After breakfast at the orphanage, I went to the Guest House to direct the maids, who are still cleaning, then returned to the hospital, where some surgical patients needed fresh dressings. One is a case of appendicitis, one a bad gun-shot wound, one an amputated portion of a hand and one an abdominal abscess. I waited until ten o'clock, then put on a nurse's gown, summoned all my courage and went to work, assisted by the ward maids. I did all the dressings and left the patients more comfortable. The gun-shot case was the worst, and my hand shook so much I could scarcely hold the antiseptic bottle while I poured.

As an added complication, Kate, the newly arrived cook, ran a nail into her foot and said nothing about it until she was unable to step on it. Upon examination, I found it in bad condition and knew it should be lanced but dreaded to do it, so compromised with a boric acid dressing and waited. When the doctor had not returned at two-thirty, I collected antiseptics, selected a lancet from the surgical cabinet and prepared the girl. Miss Keese held her foot while I used the lancet, a nerve-racking job for a novice, but when it was over and the girl more comfortable I felt more lighthearted myself and shall not mind doing the dressings again tomorrow if necessary, but my fear is that a new patient may be brought in who requires treatment beyond my power to give. I had recognized the possibility of being called upon to assist the nurse in an emergency, but taking charge of a hospital with nine patients and performing even a minor operation is more than I bargained for.

We are terribly worried about Dr. Stewart and Miss Kennedy. The wind was very heavy yesterday and may have prevented them from returning. The launch may have broken down. In that case, we hope it happened where they could go ashore and spend the night in some native's house. Last night the cold was severe, and today ice is forming in the harbor. If it thickens, then the launch may have difficulty in breaking through. Visions of being without medical assistance until Dr. Little arrives on the next *Portia* quite overwhelm me, for there is no possible means of communication, even with Battle Harbour.

7:30 p.m. They have come, gay and untroubled, and think it strange we worried. They had a lovely time. The launch broke down, as we surmised. They attempted to start but gave it up and spent the night on shore. Fortunately, the patients have not suffered, and with joy I relinquish an unwelcome responsibility. At least the winter's beginning has not been prosaic, and I have gained unlooked-for experience.

November 19th The struggle for order and cleanliness at the Guest House continues. The maids are fine workers, and all day I have heard the joyful sound of scrubbing brushes. At one time they cleaned the kitchen stove so vigorously I feared the pipe which runs through a drum in my room above would come down. Miss Keese and I will have our first meal in the house tonight, but it will have to be in Dr. Grenfell's study, for the dining-room stove-pipe is still lacking. Perhaps the use of the study as a dining-room may bring results.

Another member of the household has just appeared, young Mr. Lee,[71] a prospective Harvard student. He seems jolly and nice, not at all as Colonel Sawyer depicted him: just a round-faced, husky boy with a nice voice. It is possible the others have also been maligned and may prove a pleasant surprise after all.

November 20th Supper in the Doctor's study did bring results! The table being too small to accommodate more than Miss Keese and myself, we all had breakfast in our rooms, and before lunch-time the dining-room stove had a pipe and a fire was started. We can now put the room in order and use it regularly.

This afternoon we moved into Dr. Little's room and will soon be settled for the winter. In changing his belongings from one room to the other, I found some articles that needed mending, as well as some of Dr. Grenfell's, and have taken them in hand. It does seem queer to be mending the clothes of a man I have never seen, and without intention I find myself gauging his personality by what he wears. Colonel Sawyer said he was more possible than the others, but after meeting Mr. Lee my spirits are rising.

A woman came today to sell some tern (a kind of seagull), and I bought them for dinner. If one can imagine eating mackerel with a meaty fibre, it is possible to enjoy them, for they are tender and tasty. The two maids, Mary and Lizzie, are responding very well to household economics. I am trying an experiment by training both to do the work of waitress and chambermaid on alternate weeks, with the result that both become expert and neither has cause to complain of an unequal division of duties.

Table linen is limited, and after several meals at a bare pine table with large steel screw-heads in evidence, I made a tablecloth from two widths of unbleached sheeting and with a little imagination feel quite tidy.

November 21st In unpacking supplies sent especially for the Guest House, I find all our milk powder has vanished. No one knows where. Mr. Lee says some of it went to Dr. Grenfell's "tilt" (a shelter near the trail), and Reuben says some of it is on the *Strathcona*. Anyway, there is none here now, and we are facing the winter with only condensed milk to depend on, for the cows have not been properly cared for, and real milk is limited. There are fifteen people at the orphanage, about the same number at the hospital and five here, so our share will be small in any case. A really serious blow is the loss of our coffee pot that someone appropriated and several cooking utensils, including a pastry board (they left the rolling pin), and we have to use the cupboard door, which has not yet been attached. There are barrels of salt pork, salt beef, and corn-meal, but all the other supplies for the winter, besides material for the industrial work, are coming from St. John's on a special schooner which has been on its way for weeks.

November 24th We went to church this morning and heard Mr. B.[72] preach. I wonder if any of the people really knew what he was talking about. It was a temperance sermon. He used words I am sure the congregation never heard before, and his argument would have been unlikely to induce topers (had there been any in this place where liquor is unobtainable) to turn from their wicked ways. On the way home, dear old Mr. Pelley (familiarly known as "Uncle Joe") joined me and told me that everyone is glad to have me again with them, a heartening assurance, and a bit of cheer to offset discouragement.

Mr. Lee is helpful. He has put up some of the shelves and is thankful something is being done about the ugly dining-room, where we are planning a metamorphosis. We endured the overpowering pictures as long as possible; then I laughingly told Dr. Stewart they were to be banished to the hall, gun-room and attic, and he assented without a murmur. There is also white paint in prospect and curtains to be hung. We shall soon be quite lovely.

November 25th After several trials in household adjustment, everything is running fairly smoothly and everyone is much more cheerful. In a few days all my time can be concentrated on the industrial work, my real reason for being here, but it has been a case of first things first. The housework is so systematized that the girls will have some time each day for weaving. The others are anxious to begin, and I foresee large classes.

We are very merry at the Guest House. To my relief, both Miss Keese and Mr. Lee have a strong sense of humor, and we find amusement in the many daily incidents. I suggested that we edit a weekly chronicle, and we had a gay time this evening reading our first edition, laughing until we wept. It is not an intellectual effort – far from it – and is not intended as such. Indeed, much of it is in the form of grinds [personal jokes], personalities only appreciated by the parties concerned. I hear Dr. Little has already written a poem on the fish industry, so we hope for his support. In making the suggestion I thought it might vary the monotony of winter life, but the possibility of monotony already seems remote. The next excitement will be the *Portia*'s arrival with the remaining members of the Guest House family (except Dr. Grenfell, who is not expected until Christmas). I must say, I dread it and hope they will not be rude. If they are, it will be so trying to have them around.

Mr. Lindsay, the Irishman, is to have charge of a herd of reindeer after their arrival next month, and Miss Storr says he is trying to inure himself to the discomforts of sleeping in the open. To do this he plans to use his sleeping bag on a sort of portable cork life preserver mattress placed on his bed. It makes his room look rather queer, so I have made a cover of chintz to match the curtains that will cover any irregularities. We are anxiously awaiting his reaction. They say he does not like "frills."

The introduction of reindeer for use in Newfoundland has for some time been one of Dr. Grenfell's cherished plans, its purpose being not only to supplement – possibly even to supplant – dogs as a means of transportation in winter but to supply other needs as well: meat (always difficult to obtain), skins for warm clothing and (of prime importance) milk for babies, a rare luxury on the northern coast. The project, similar to one already operating successfully in Alaska,[73] seemed especially applicable to this part of Newfoundland, where native caribou are found about fifty miles south of St. Anthony, but they do not frequent the Northern Peninsula. It is hoped the reindeer may eventually breed with the caribou and bring them north to benefit the entire region, where as an added asset there is an abundance of the kind of moss on which the caribou subsist.[74]

There was so much to recommend the venture that friends of Dr. Grenfell and the Mission, led by Mr. William H. Reed of Boston, sponsored a Reindeer Fund with such success that arrangements were made with the Lapland [Norwegian] government for the shipment of a small herd of reindeer and three Lapp [Saami] families to care for them. They are expected before Christmas, and I am told Mr. Lindsay is already feeling the weight of responsibilities that are more or less vague in detail but seem to grow in importance with increased publicity. The Harmsworth Company[75] at Grand Falls, Newfoundland, which supplies London newspapers with pulp-wood from the rapidly diminishing forests of the interior, has become interested and purchased fifty of the reindeer. They are to be delivered at St. Anthony, and we hear that two men are expected soon to await their arrival and guide them and one of the Lapp families to Grand Falls.[76] Mr. Lindsay is glad that is not a part of his responsibility.

During the past two days we have had a "mild" (otherwise thaw), and for probably the last time this winter the windows have been washed. The shelves are up in the dining-room, and we are building a corner seat. Some green material has been found for it, and the room, which is also our living-room, will soon look more homelike. Some of Dr. Grenfell's large silver athletic trophies add brightness to the simple little room, and his silver tea-service and other dishes gleam on the ugly sideboard. I am glad everything will be running smoothly by the time he arrives. The loom-room is neat and orderly: lamps with reflectors are on the walls and many women and girls anxious to begin. Dr. Stewart has finally given an order for digging the pottery clay and we have several barrels full – just in time, for the snow will soon be too deep to reach it.

Several of the women have been to call. They are very friendly and I am planning to start weekly clubs, one for the older women and one for young people to bring them together for some social objective. Miss Keese and Mr. Lee have offered to help. They say the coast people have never had anything of the kind, and it is much needed. The forms of social activity will be interesting to work out, for some of our accepted types of entertainment (dancing and card-playing) are absolutely taboo and considered work of the devil. In some ways the inhibitions and prejudices of the people are those of a hundred years ago.

I have just seen dogs drawing a *komatik* (Eskimo for sled), my introduction to winter transportation. It gave me quite a thrill and an immediate desire to ride on one. Other means of transportation will be the much heralded reindeer (when and if they arrive). We may have ample opportunity to study them, for Reuben and the local men report that they are to be kept in an enclosure surrounding our house, but that is not authoritative. It may be merely local gossip.

6

Some Britishers and a Bostonian

26 November 1907 – 1 January 1908

November 26th At 7:30 this morning, just as I was out of bed, the *Portia* whistled, and immediately everyone was electrified. I suddenly thought of the window curtains not yet hung in Mr. Lindsay's room and realized it might be difficult to make any change after he was settled. Still in wrapper and slippers, I hung the curtains and spread the chintz cover over the uneven bed; then with a watchful eye on boats approaching the shore we snatched blankets from other beds, borrowed sheets from the hospital and were ready. But Mr. Lindsay did not appear, having gone with Dr. Little to the hospital for breakfast. Mr. Lee says both men have grown beards while in the brush and Mr. Lindsay is a sight. He departed in overalls and sweater, a gay toboggan cap with a long tassel, and a monocle. The beard will add the finishing touch. I am dying to see them.

Later. This has been a day of shocks. As if the men's arrival on the *Portia* were not enough excitement, the mail has brought news that the British government steamer will be unable to go north on account of ice. Dr. Stewart is much disappointed, and no wonder, for this was an opportunity that may never come again. He will be here at the hospital after all and Dr. Little at the Guest House, which will please Dr. Grenfell.

There is also disturbing news that the steamer bringing equipment for the industrial work, food supplies and everything needed for the winter has been damaged in a gale down the coast and has had to return to St. John's for repairs. It is uncertain whether she will start out again or unload her freight and send it by the *Portia* on her last trip. In that case, everything would have to be landed on the ice, with possible damage or loss. We are liable to be frozen in by that time, for there was a skin of young ice across the harbor this morning. It was just sunrise, with a warm light on the water, a lovely background for the big black hull of the *Portia* and little boats going out from all directions, making ripples through the light coating of ice.

Later. The Irishman has just come. I caught a glimpse of him on his way to his room. He is a sight! Miss Keese, Mr. Lee and I are sitting here expectant, wondering what he thinks of his chintz curtains and bed-covering. I am really a little scared. They say he can be very rude and disagreeable, so I am bracing myself.

Later. I have seen the Irishman. He came to the dining-room and sat with us after having brushed his hair and washed his face. He told us about his trip, and we conversed amiably for a while. He even laughed. I told him if he did not like the chintz in his room he could remove it, but he said he did not mind at all. Presently he went out and returned smoking a cigar. I told him it seemed very nice to smell cigar smoke again, and his smile was a beam. Since then he has been off to trim his beard and now looks like a civilized person. I foresee friendly relations and no further trouble. There is now only Dr. Little to conjure with.

Later. Dr. Little[77] has just come in and is very nice. His half-grown beard and disreputable clothes make him look like a scarecrow, but one can see at a glance the kindly, attractive gentleman beneath them. I breathe a sigh of relief. His beard is awful, what there is of it, mostly a hair hither and yon. He evidently was never intended to be a hairy man.

November 27th We have had our first dinner party and the Guest House is officially "warmed." The party taxed the seating capacity of the dining-room, and Lizzie when serving sometimes became wedged between my chair and the stove or Mr. Lindsay's and the sideboard. Dinner was at night by popular choice, and with coffee afterward in the Doctor's study we felt really civilized. We now have a real linen table cloth to replace the unbleached cotton. The sideboard has lost its mirror, to everyone's relief, and the bilious yellow paint and "marbled" mantel will soon be white. We are progressing.

November 28th This morning Dr. Little returned from the hospital with a large tin bath-tub on his head; he had borrowed it from Miss Storr, who now possesses a real tub with hot and cold running water and does not need such a primitive article. Dr. Little's interest in it is intensified after three weeks of camp life with few opportunities for baths of any kind. We are also interested and hope he will not consider it his exclusive property.

Dr. Stewart has just attended a wedding on the other side of the harbor. It was a grand affair. The ceremony was at the church with a wedding feast at the bride's home, where it was prepared, eaten, and cleared away all in one room. The table seated only fourteen, so that number ate; then one of the men guests took off his coat and washed the dishes. Fourteen more feasted, then once again. The feast consisted of part of a sheep, turnips, cabbages and potatoes. Dr. Stewart, as guest of honor, was expected to ask the blessing at each sitting. We asked him if that included eating with each group, but he said they acknowledged a limit in spite of expectations.

I gasp when I think of all that company in the one living-room of a typical fisherman's house! The ceilings are usually so low a tall man can scarcely stand upright. The windows are tightly closed, often not built to open. There is always a cook-stove with a roaring fire, and hooked mats (the only floor-covering) are frequently used by the men as receptacles for expectoration. It is an abominable practice in this part of the country, and Dr. Grenfell is waging a vigorous campaign against it, for nothing could contribute more to the spread of tuberculosis, so prevalent along the coast. He has had small posters with the warning "Don't spit. It spreads disease" distributed among the people. They are referred to among the staff as "calling cards" (with no frivolous intent of minimizing their purpose or importance). Dr. Little, in his zeal for impressing the warning, has designed a rug with the device "Don't Spit" in red letters on a white background, bordered with a pattern in red and green. Several village people have used the design on door mats, and one is confronted with the reminder on stepping over the threshold.

November 29th Our supplies are low. This morning the cook told me we were out of butter and potatoes, having waited till the last minute before mentioning it. The hospital supplied butter, and Mr. Pelley loaned us potatoes. For some reason a supply of potatoes is taken for granted, and it seems strange to be without them.

As the days pass, Mr. Lindsay is gradually revealing his true character of a kindly, considerate gentleman under a sometimes brusque exterior. It is a character so at variance with Colonel Sawyer's report that I realize the Colonel was either blindly unimaginative or he indulged in fiction to stimulate my reactions. This has been a glorious day, clear and still, which is something to chronicle, for the wind usually blows. We went to church this morning and were stunned by the appearance of Mr. Lindsay wearing a white stock and even a scarf-pin, also trousers without knee patches, his one remaining pair. He is nearly out of clothes, awaiting a supply from England.

We walked with him this afternoon to the peat bogs which Dr. Grenfell, in his effort to make use of all native resources, is trying to develop as a source of fuel supply. Mr. Lindsay is in charge of this enterprise, having been given the appointment on his arrival at St. Anthony, which somehow seems appropriate since peat bogs and Irishmen have traditional association. The bogs are quite extensive in this locality, and Mr. Lindsay is proud of his progress in developing them. A small pond has been drained, and the peat beside it was found to be eight feet thick. We took some pieces for drying and plan to burn them in our stove to test their heating capacity.

Our Sunday program now includes supper at the hospital, followed by music. By this time I have reached the point when even a hurdy-gurdy would be welcome, and the parlor organ's Aeolian attachment[78] and the gramophone are real treats. We have also just read our "weekly chronicle" and waxed hilarious over the contributions: limericks, doggerel, odes, items of local and personal interest, suggestions and advice. As before mentioned, we disclaim any literary, educational or intellectual effort in this periodical. It is for amusement and relaxation only and serves its purpose.

December 4th Mr. Lindsay wants to learn to weave, having in mind the stimulation of such interests among his tenants when he returns to Ireland. I have taken him on as assistant, which at present means disentangling threads of a badly made warp. He is very patient, but I fancy such practical steps leading to the art of weaving were not anticipated. He had only finished results in mind. He has entertained us all this evening with tales of his adventures, thereby revealing some of his background. As the eldest son, he is heir to the family estate but is unable to manage it or enjoy its financial benefits during the lifetime of his mother, who provides him with an allowance, a situation far from satisfactory for a man of independent spirit and individual interests. He has been engaged for fifteen years, but his fiancée's father will not allow their marriage under present conditions.

One can understand his revolt, which had led him to seek both adventure and business enterprise. He was a major [lieutenant] in the Boer War. A business venture in the United States failed through complications and lack of experience. In the intervals he returned to Ireland for a while, then started out again, often to America. His latest venture was to offer his services to Dr. Grenfell at the suggestion of a friend. Knowing little about the Doctor or where to find him, he crossed from Ireland to St. John's, where he boarded the little steamer for Labrador, and the most amusing part of his story was his inquiry, on arriving at Battle Harbour hospital, if they knew a man called Dr. Grenfell. The Doctor was "afloat" on the *Strathcona*, and several days passed before Mr. Lindsay located him and climbed aboard, announcing that he had come to help. The Doctor was naturally surprised. "Who the devil are you?" he asked. "You may be an escaped convict for all I know." But membership cards in certain well known British clubs saved the day. Mr. Lindsay not only stayed on board but was forthwith engaged to take charge of the expected reindeer herd. His superintendence of the peat bog development will be of minor importance after the arrival of the deer in the near future.

Dr. Little and Mr. Lindsay are taking lessons in managing the long dog whip used with the komatiks. It is 20 to 30 feet long, and Reuben, who is the instructor, can strike a small stone at a distance. It must sting the dogs terribly. For a novice there seems to be a great deal of lash that has a way of becoming a boomerang. Today we have all practised on the snow. It is a graceful motion in the hands of Dr. Little and Mr. Lindsay, who are fast becoming expert, but I am very clumsy, and the men, though encouraging, warned me not to lash myself.

The piazza was cleared today by Dr. Little and Mr. Lindsay and it looks like another place. We have amused ourselves by suggesting to Mr. Lindsay and Mr. Lee a plan for bringing the hospital patients to enjoy the piazza, since it is now so attractive. We suggest septic cases at one end, surgical cases at the other, and the tuberculous in the long southern side. We added that on certain days they could have space for acrobatic performance to entertain the patients. They are considering it!

This morning I was up before sunrise to sketch the harbor, practically covered with "young ice," lovely with iridescent color in the early light. There is so much that is beautiful to paint, but I had not taken into consideration window panes so thick with frost that I have to scratch a hole to peep through. Pastel is the only practical medium to use. Watercolor is, of course, impossible. One cannot paint with ice. Color and light effects are too fleeting for preparation of an oil palette, but pastels are always ready for use.

December 5th This afternoon I visited Miss Keese at the school-house and wish I could have photographed the room, so typical of the primitive, ungraded country school. Benches serve as seats. Desks are shelves that run along the wall. The children who sit between the windows have little light on their books, and all are so crowded that they fairly rub elbows. A wood stove fills the center of the room, the iron stove-pipe projecting through the roof. Miss Keese says the pipe sometimes becomes overheated, setting the roof afire, and a ladder resting outside against the building is always ready for emergencies. After an interesting spelling class, Miss Keese asked the children to write letters, and one of them was addressed to me, beginning, "Dear Friend, I am well, thank God. I hope you are well." I want to visit the school again.

December 6th It has rained all day, a freezing rain that has coated everything with over an inch of ice. We are concerned about the angora goats that someone sent Dr. Grenfell last summer (thinking their long hair could be used for weaving). The question of feeding our animals is serious, with three cows, a pony (also sent last summer), seven goats, a sheep and a bull to be cared for, and the winter food supply is still delayed.

The reindeer will feed on native moss, and in any case their ship will bring a supply of food to last until they can be taken into the interior. They may appear at any time, and Mr. Lindsay is preparing himself by reading all available reindeer literature. He finds they can subsist on spruce buds besides reindeer moss, but his first concern is how to get them ashore from the steamer. That is a problem in itself. When one considers the importance of the reindeer venture and its widespread publicity, it is really appalling: arrangements are so vague and so dependent on conditions that nothing can be definitely planned. There is growing excitement as the days pass, and Mr. Lindsay is asked many questions, to which his answer is "We don't know." Meanwhile, he sleeps in his sleeping bag on top of his bed to become accustomed to the sensation, uses cold water and keeps his window open day and night. A tent has come for his use. We have not seen it, but he says it is a regular bungalow tent with four or five cots and is so big its transportation on komatiks is a problem. The arrival of the deer will be exciting. It is too bad Dr. Grenfell will not be here in time to have a part in it.

December 7th The harbor has frozen over but is not yet hard enough to walk on. The hills are covered with snow, and the path is waist deep in drifts. I have worn my knee-high sealskin boots with only qualified enjoyment. One seems to be walking in stocking feet. They are perfect for deep snow and very warm when worn with several pairs of woolen stockings inside, but at present one is conscious of stones, stumps and rough ice under the soft snow. They are shapeless and tied below the knee over heavy knickerbockers. All contours will be lost when our winter costumes are complete, and no one will know whether we have "figgers" or not.

The house is nearly in order. We are trying to make it as attractive as possible, for it is to be Dr. Grenfell's present home, and we hope an atmosphere of comfort, order and peace will mean much to him.

This afternoon I saw a komatik with two men on it drawn by three dogs. The dogs were galloping, and one of the men was standing with hands on upright stakes at the side. It was quite like the dog-team of my imagination. Miss Keese came in, jubilant after a wild ride with a team down what is known as the Gully Bank, unusually exciting as it was covered with glare ice [smooth and glassy], and the dogs raced to keep ahead of the komatik.

December 14th Lydia, one of the hospital maids, has died and was buried today in the little cemetery near the church. The service was at the hospital, and most of the staff, the maids, and village friends followed her to the grave. It was a strange sight, the plain coffin covered with white cloth on a komatik drawn by four young men, who guided it down steep grades onto the path, which was very slippery. An icy wind was blowing across the snow. On reaching the cemetery we waded through snow to the shallow grave, which was only three feet deep. The land is so rocky it is difficult to go deeper. Dr. Stewart read the service of committal, and he, Dr. Little and the boys stood with bared heads while the white coffin was lowered and Mark threw a handful of earth on it. For some reason it seemed to me more moving and real than similar events in our civilized country. The same impression was strong during my first visit here regarding the life in general: one is face to face with actual realities of living, with little to soften the circumstances of daily life. That company of simple people, the plain coffin drawn to its resting place in such a fashion, the grey cloudy sky with a gleam of cold, pale yellow sunset light above a background of snow and ice-covered hills, made a picture I shall not soon forget.

I talked with Dr. Little tonight about the industries and submitted a plan for immediate development, to which he added valuable suggestions. The men and boys are to meet me Wednesday night in the loom-room to talk about pottery, wood-carving and drawing, the women on Thursday to plan weaving, spinning, basket- and mat-making. Classes will be formed for the winter and the work begun after Christmas. The announcement will be made in church on Sunday.

December 15th Miss Keese, Mr. Lee and I called on neighbors this afternoon and stayed a long time with Mr. and Mrs. Ash. Mr. Ash is one of Dr. Grenfell's earliest friends in St. Anthony as well as one of his chief helpers. We spoke of the phosphorescence so often seen in the water, but Mr. Ash did not know it by that name.[79] He said the natives call it "water fire."

This has been rather an exciting day. This morning we heard shouts outside and, on going to the window, saw the pony and all the male population of our end of the harbor pulling on a long rope that was hauling the yawl *Daryl* on shore for the winter. It was a long pull, and the rope extended half way across the Guest House "lawn." I rushed for my camera, slipping over icy hummocks, trying to get into position to take a picture.

It is a glorious night, almost as light as day, with a nearly full moon directly overhead. One can see details of the land at Fishing Point, and the white hills are wonderful. The heavy storm ten days ago covered even twigs and blades of grass with ice. The natives call this kind of storm a "glitter." It has been too cold to melt, and the world is one glorious sparkle. The air is clear and very still; the mercury is five degrees above zero. It is difficult to describe the beauty of winter in this place, where at this season the sun rises at nine, sets behind the hills at two, and during the five hours of daylight the arc of the sun is so low the light, except at mid-day, is like late afternoon.[80] The sky at sunset is a curious green near the horizon, and the pinks, blues and purples on snow and ice are beyond description. It is interesting to see the winter gradually closing in and the low growth of spruce slowly disappearing. Every snowfall adds to its submergence, and the hills, which were covered with a dull, olive-green growth, are now whitish grey. Soon they will be pure white, and we can walk over the tops of small trees. We are walking on a level with some fences now, though the general snow level is only a little over a foot.

December 18th There is an epidemic in the harbor and half the people are ill. They say it is grippe. Miss Keese reports only fourteen of her forty children came to school, and four of Miss Storr's orphans are down. Kate, my cook, shows signs of weakening, and I foresee acting as cook tomorrow. The epidemic also affected last night's meeting for men and boys, called to hear me explain the industrial work, Dr. Grenfell's idea in establishing it, and what we plan to do. Only fourteen came – some stood against the wall, some sat on a bench near the stove, and all were absolutely mute when I asked how many would like to join the classes. Finally, a few found their voices and agreed to begin tomorrow. I hope others will join when stimulated by visible results. It is unfortunate the potter's wheel and metal supplies were not ready for demonstration, for I think those crafts would interest them.

In re-reading Dr. Grenfell's letter, I think he means to be here for Christmas. In that case he is already on his way on the *Portia.*

December 19th The epidemic has spread over the whole harbor. Miss Keese came home at eleven, as only eight pupils were at school and three of those became ill during the morning. She took them home and they cried all the way. This house is next in line. At dinner last night, Dr. Little bet me that one of the maids would be ill before morning. The stakes were an hour of assistance on looms against darning his stockings. In two hours Mary was down, but I should have darned his stockings anyway, *comme toujours.*

Today, Dr. Little and Mr. Lindsay have succumbed. Dr. Little calls his trouble "indisposition," and signs point to a desire to be left alone. Mr. Lindsay refers to his ailment as a "bilious attack." He is rebellious that such a thing should happen to him and tries to trace the cause to some fault or omission on his part. He would like to go home for a few days and longs for a big easy chair and an open fire. The cases in the harbor are called plain "grippe."

December 20th The epidemic is spreading. The church is closed from lack of congregation, the school from lack of pupils. Our household is demoralized. The fire was started late, and the dining-room was like a refrigerator. Fortunately, this form of grippe is of short duration and not usually serious, but some of the village people are having a hard time. Often, all members of a family are ill at once. In one case a ten-year-old child was doing all the housework. All the Greens (a half-Eskimo family) are very ill. Dr. Little tells me the Eskimos as a race are susceptible to grippe, and when some years ago an epidemic swept the Labrador, fully one-third of the population died. Dr. Stewart is beginning to have calls from settlements nearby.

Mr. Lee has just shouted through the wall to ask what day of the week it is. It is really hard to remember, for we have not adopted a native custom of rotating certain foods in our daily menus during the winter months, when the mail-boat no longer brings its bi-weekly reminder. Correct time we get in summer from Captain Kean of the *Portia* – real St. John's time – but in winter I don't know what is done. Last week, there was a difference of half an hour between the hospital, the church and the orphanage, which was inconvenient.

There was a meeting tonight in the village "hall" to discuss the reception of the reindeer. The dogs of the neighborhood will have to be kept chained when not in use, for many of them are half starved, and all may consider the deer their special food. Mr. Lindsay went to the meeting prepared to make a speech, but to his surprise Uncle Joe Pelley said just what he had intended to say, and Mr. Lindsay could only agree with him. It was well the initiative came from a harbor man. The audience was responsive, and the men agreed to tie up their dogs and co-operate all they could. Any dog seen attacking a deer will be shot. It seems very hard but the only way to safeguard the deer.

Mr. Lindsay is rejoicing that his so-called "indisposition" came before the beginning of his life in the open. It is amusing to hear the men tell of their sensations during their illness and how it affected them. We told Mr. Lindsay we longed to hold his hand and smooth his brow but did not dare.

December 21st Mr. Lee has succumbed after looking for symptoms all the week. His case is mild, and he is already recovering, even asking for food and society, but insists on finishing the pills Dr. Stewart left for him. He has arranged them carefully in groups: two white, two brown.

Miss Keese and I have gone about among the people all day, visiting those whom Dr. Stewart reported most in need of help. In one home, we found the mother just sitting, looking utterly miserable. The father was holding the youngest child in his arms while the older children huddled on a bench. He told me he had just washed the dishes for the first time in four days, and the one basin still held a rag soaking in black water. The problems of wood and water are acute, and economy in the use of water seems general. When one realizes that water for the community is drawn in barrels or buckets, by hand or by dog-sled, from a hole in the ice over a spring and that a kettle over an uncertain fire represents the hot water system, one must be lenient in criticism of local house-keeping, further complicated in this case by only one basin in which to wash dishes, food and the persons of the entire family and only one cloth for wiping everything.

Some of the houses are neat and well arranged. This one is not, and the people were evidently fearful lest we include the loft in our cleansing process. I saw at once that it would be an embarrassment, so only swept the two downstairs rooms and shook the mats, remembering all too well a visit at the same house a year ago, when the father and another man sat near the stove and spat continually on the floor.

It is hard to realize that the day after tomorrow will be Christmas Day and at home people are hurrying about, making last purchases. Here, many are recovering from the epidemic, but the general atmosphere of illness is depressing. I fear it will be a dull Christmas Day.

December 24th Christmas Eve! As far as weather is concerned, nothing could be more typical, plenty of snow, icicles in fringes from the roof, real winter everywhere, but where are the hurrying throngs we city people associate with it, where the preparation of gifts up to the last moment, the tying of red ribbons? There have been times today when I have forgotten whether it was December or March. It has been what the natives call a "nasty" day – heavy snow, damp enough to stick but too heavy to drift, and a furious wind. In making neighborhood visits, I found several families needed soup as well as house-cleaning and asked Kate to prepare a large kettleful. Miss Keese and I spent the morning carrying it in pitchers through deep drifts and down icy slopes, in momentary fear of sliding or stumbling.

Kate is much better. She attributes her recovery partially to a local remedy recommended by a friend, a mixture of molasses, pepper and kerosene boiled together and taken by the spoonful.[81] Kate said it helped her very much. She must have had courage and faith.

We are having a really dreadful time with wood fires. There is only wood and soft coal anyway, and wood is burned in the study and dining-room. The stoves are little iron affairs that can be opened in front to delude the occupants of the room into believing it is an open fire. This gives a little firelight, but the fire goes out when the front is open. Since the sleet storm, the wood is so covered with ice that when pieces are brought into the room and placed under the stove to thaw and dry we soon are mopping great pools of water.

The harbor is frozen over. Its smooth sheet of ice is covered today with yesterday's fall of snow. When the *Portia* comes, passengers and freight will be landed on the ice, and dog-teams will take the place of boats in bringing everything ashore. Life is daily growing more exciting. Those at home who picture this a long, monotonous winter of drudgery and self-sacrifice (with a halo attached) should know that it is more interesting than any winter life I have experienced or even imagined. One is daily faced by the unexpected and a challenge to effort.

Christmas Day This has been a happy day, though very different from any Christmas I ever knew. Festivities for the community were postponed until the people are better, but we hung stockings for some of the children. This morning we visited the Greens, who still need attention. This entire family sleep in one small room with two bunks, one large, one small, each with some sort of thin mattress. Mrs. Green lay huddled in the large one with a thin blanket, an old table cover, and the skirts of two dresses over her for coverings. The rest of the furniture consisted of two boxes with lids, used as seats and table, and an open box filled with miscellaneous clothing. There was also the inevitable cook-stove. In the back porch stood the water barrel, partly filled with ice, a barrel of salt fish, one of salt pork, a partly empty flour barrel, and a bucket of molasses. On the floor were a large kettle, the garbage pail and slop bucket, also the water pail and can of kerosene. Nearby was the pan for baking bread, evidently used also as a dust pan to take dirt and ashes from the floor. Ragged hooked mats clogged with dirt were frozen on the floor. There was much to be done to restore order and cleanliness.

It seemed really necessary to plan special food for our Christmas dinner if possible. I had in mind a roast from the bull that was killed some time ago, but when Mr. Jones,[82] the man in charge of supplies, led me over packing boxes and coils of rope to where several carcasses were frozen in, the bull was at the bottom of the heap and so stiff with frost it could not be moved. We think it cannot be taken out until it thaws in the spring, and then it will be spoiled. I abandoned my visions of sirloin steaks and asked Mr. Jones to give me a dozen chops from a sheep I saw hanging there and two of the four chickens someone had given Dr. Little. When the chops appeared for lunch with fried potatoes around the platter's edge, Mr. Lindsay, who attributes all his ailments to lack of fresh meat, was electrified. His expression was unbelieving but joyful.

The *Portia*'s smoke appeared over the point in mid-afternoon, and Miss Keese and I took our cameras and rushed to the wharf. People were coming from all directions, and on reaching the shore we found the boys harnessing dogs to komatiks to go out. There was great excitement. The dogs, barking and eager to be off, were getting entangled in their harnesses and stopping to roll in their glee before starting across the ice to the steamer, where people were already walking away from her side instead of landing in boats. The English cook for the hospital arrived and was brought ashore on a komatik, a novel experience for her and a fitting introduction to her new manner of life. Dr. Grenfell did not come, but the steamer brought several patients and two strange young men, representatives of the Harmsworth people of Grand Falls, Newfoundland, who have bought fifty of the reindeer. The men are to wait here to conduct their part of the herd to Lewisporte, and one of them (a Norwegian) will act as interpreter. Mr. Lindsay is relieved of the Lewisporte part of the responsibility but wishes Dr. Grenfell were here.

Our Christmas dinner was a grand success. Among my gifts from home I found a little Christmas tree with decorations, which we used as a centerpiece for the table. There were also paper cap bonbons for each place, and the table was gay. It was really a six-course dinner: soup, broiled chicken garnished with sausage meat (Mr. Lindsay's suggestion), potatoes, turnips and onions, cabbage and beet salad, canned plum pudding, nuts, raisins and sweet chocolate, followed by coffee. And the men smoked real cigars instead of pipes to celebrate the day. We had a very jolly meal and much fun over the small gifts and jokes on the tree. It has been a happy day and in some ways more genuine than any Christmas I have known. Moreover, it was without fatigue.

December 26th Dr. Little has made his rounds and reports Mrs. Green much better. In fact, everyone is recovering. We may soon consider the grippe a closed incident and return to normal.

Mr. Lindsay asked today if he could invite one of the reindeer men to luncheon. He is an Englishman named Cole,[83] who has been in this country about three years. He had a great deal to say about the reindeer and more about the timber his company is cutting in the center of Newfoundland. The company supplies pulp for British newspapers. They do not leave a stick standing and denude the country. I wanted to argue the subject but as hostess could only touch upon it. He evidently enjoyed ladies' society, for instead of returning to the study after luncheon for a smoke with the older men he stayed on, and I finally left him with Miss Keese while I took one of the little Christmas trees to Rhody, the small scurvy patient at the hospital. It was the first time I had seen Rhody smile. She sat still, not saying a word. The other children were delighted.

December 27th Ever since hearing there were to be real desks in Miss Keese's schoolhouse, I have wanted to photograph the room before they are installed, and I went this morning. The snow is deep, but the path is packed down so that one can walk on top of the drifts. The desks are there, but now the minister, who is supposed to constitute the school committee, objects to them because they interfere with the use of the room for the Methodist prayer meetings. It does seem (contrary to this worthy young man's point of view) that the children's education and comfort come first.

Mr. Pitman, a fellow passenger on the *Portia*, spoke of the unsatisfactory educational system on the coast. There are schools in nearly all settlements, but their number is disproportionate to the size of the population they serve, as they are all denominational, and instead of a general educational fund to support one school, each denomination has its own: Methodist, Church of England, Roman Catholic and Salvation Army. The Salvation Army is not generally represented, and in some communities there are not enough Roman Catholics to support a school, but Methodists and Church of England are found nearly everywhere. The income of a fisherman's community, however, is usually too precarious to guarantee the expense of a year-round teacher of any denomination, and sometimes her services are divided among two or three villages during the winter. In some places, there is a teacher only once in two or three years. As a consequence, the educational standards are low and the curriculum elementary.

At St. Anthony, the village is on two sides of the harbor. The Mission itself is non-sectarian, but it is located on the Methodist side. On the other is the Church of England. The Methodist side does not send its children across the harbor because the Catechism is taught in the school. The Church of England side will not send its children to the Methodist side because the Catechism is not taught, and there it is! Dr. Grenfell is trying to establish a union school[84] on the Mission side and hopes for success, but it takes time to overcome prejudice, tradition and ignorance.

After luncheon, the weather was so wonderful and the shadows on the snow so blue I took my pastels and sat outside in a sheltered spot to paint, although the mercury registered only nine above zero. The snow was so dry I sat on it with comfort, but my hands nearly froze, and I moved indoors to paint the same view from Miss Keese's window. The shadows on Fishing Point were an exquisite, transparent blue and purple, the ice at its base a curious nile-green [pale greenish blue]. I am continually impressed by the beautiful colors.

December 28th This morning I found Dr. Little sitting in a stiff chair with newspapers under it and a towel over his shoulders while Mr. Lindsay, in the character of barber, was cutting his curling locks. Even the wavy front lock was gone. The doctor's hair was beginning to turn up at his neck and he looked like a German professor, but now he looks like a shorn lamb, and I shall have to get acquainted with him all over again. Mr. Lindsay stood very straight, with the shears in his hand, his pipe in his mouth, smoking vigorously, with a serious expression as if the gravity of the occasion impressed him. Mr. Lindsay also will need the services of a barber.

He has been wild for fresh meat. Notwithstanding the chops he had on Christmas Day and the chickens, he still clamors for more. I ordered some chops again, but Mr. Jones, who acts as butcher during Reuben's illness, is not expert, and the hunks of meat would do no honor to the profession. Mr. Lindsay was sure he could do better than that, so he cut a hind quarter of the sheep, and we had roast mutton today for the first time in two months.

Miss Keese and I did a dreadful thing this evening: we played solitaire, and it was Canfield![85] It is considered a heinous sin by the local Methodists to touch cards, and at the end of the game we put them away hurriedly with a guilty feeling in the depths of the wood closet, where they lie near a box of cigars, making the place a den of iniquity. To strengthen the evil association, the closet also contains matches and wood. Nothing could be more suggestive of the penalties to come. Miss Keese and I were so exhilarated, for some unknown reason, that we started the phonograph, playing a lively two-step, and actually danced in the narrow confines of the front hall. It was a real Saturday night spree – and "we are all miserable sinners"!!

I have tried the wide Labrador snowshoes for the first time and find it easy to walk over the tops of drifts without going through. One does not have to "spread" so much as I thought, though the gait does somewhat resemble a waddle. Mr. Lindsay tried skis today and had so much difficulty he is planning to take lessons from the Norwegian waiting here for the reindeer.

December 29th This has not seemed like Sunday – that is, a Newfoundland Sunday, usually associated with church-going and psalm-singing. It snowed all day, a drifting snow with keen wind. I did not go to church but washed articles we never entrust to the local wash-lady on the principle of cleanliness next to godliness.

This morning when I went in to see Dr. Little, who is still quite an invalid, I had to push his door and climb over a snow-drift to enter the room. The fierce wind had blown fine snow through cracks around the window-frames during the night. A light blanket of snow covered his bed, and articles on his bureau were small mounds on a white field. It took some time to dig him out. The sound of the snow shovel reminded us of winter streets at home. After dinner, Mr. Cole appeared and I invited him to supper. He accepted, remarking that he did not know if he could find his way back through the drifts, so of course I asked him to spend the afternoon. He came to see the men, so we did not have him on our minds, and Miss Keese joined me when I went to the hospital to read to Mrs. Burt and Rhody.

Mrs. Burt wanted me to read her last mail to her, and I struggled with bad English, lack of punctuation and some words and phrases whose meaning I did not know. The letters were very loving and sympathetic. I felt that I was drawing the curtain from something sacred, but it was the only way the poor woman could know their contents, for she is blind. Her husband is to bring her little daughter in the next *Portia* to stay at the orphanage. He is anxious to work here this winter in order to be near his wife, who must stay at the hospital. She has already the patient expression of the blind.

We plowed through drifts back to the Guest House to find Mr. Cole still here, and I humored the family by giving them canned lobster and griddle cakes for supper. At nine o'clock, our visitor said good-night and left for the study for a final word with the men. He was still there when the evening cocoa came in, and it was some time before Dr. Little ushered him out of the door. Truly, he is a "stayer." He will be useful when we decorate the hall for the Christmas celebration, and he seems only too glad to have something to do.

December 30th Today has been so exhilarating I could have thought myself back at the tender age of fifteen. Miss Keese, Mr. Lee, Mr. Cole and I started off for greens to decorate the [Loyal Orange] Lodge for the Christmas party. The snow was deep, and we threw snowballs and chased each other like small children. Drifts covered the trees, and we had to dig down to cut branches, then thaw them before they could be used. Alf[86] passed us on his way to the hill for the tree. The solitary figure climbing upward on the snow reminded me of something. Later, I realized it was "Excelsior" – he needed only the banner.[87]

The hall is decorated and looks very nice with the platform dressed in green and the motto made by Miss Storr: "Glory to God in the Highest, and on earth peace, and good-will to men." Mr. Pelley is to be Father Christmas and master of ceremonies. He seemed disappointed when I suggested trimming the tree, so I asked him if I might come down and help him decorate tomorrow, and that made it all right.

On the way home, Miss Keese and I had our first real komatik ride. After Alf left the tree, we got on the komatik and were drawn by six dogs, Alf running alongside shouting the curious sounds – half word, half cry – that the dogs seem to understand, and we went flying down the Gully Bank, clinging for dear life, almost running over the dogs, who tore ahead to keep out of the way. All very exciting and great fun.

December 31st Yesterday was a weather breeder. It could not be perfect like that two days running. I woke this morning to find it snowing, a sorry day for the belated Christmas celebration which, it became apparent, would have to be postponed. I had invited the members of the staff to a late supper with a request that they produce literary matter in some form for the "Chronicle," which was to be read aloud while we waited for the Old Year to die, and everyone appeared at supper with a contribution. We had a gay meal and became quite demoralized, even the usually sedate Dr. Little beating on his plate with his knife and fork when he wanted griddle cakes and yelling for food. I was prepared for this, for before the gong sounded I was greeted at the door of the study by a shout of "I want food" from five hungry men. The contributions to the "Chronicle" were very interesting. Dr. Little calls it the Black and Blue Club, and this seems quite appropriate, for no one is spared. We watched the Old Year out in the study. I wonder what the New Year will bring and feel that some change is impending. May it be a happy one.

January 1st, 1908 This has been another happy day. It was so cold and windy this morning that the festivities almost had to be postponed again, but it calmed at noon. The day was confused as to meals, but we had a fine dinner at the hospital – real roast beef, the first we had tasted for months. It is remarkable how much stress one lays on food down here. I never knew it to be such a general topic of conversation, and when there is a roast or any form of real meat we sit in silence and watch the carving knife as it gives us our portion. All this enthusiasm over such meat as we have is almost pitiful. All the children were to receive sweets on the tree, and after dinner we filled the bags, 140 of them. There are over 200 inhabitants in St. Anthony, most of them children. I did not know there were so many.

At six, we started for the hall, half a mile away. Two men carried the dismembered gramophone and its records, in spite of doubts as to their safe arrival. We had no lantern and stumbled along the hazardous path quite blindly, for there was nothing to distinguish it from the white country around us. Finally, we reached the Lodge and found a full house, even a row of men in the gallery. Mr. Pelley had stretched a cloth across the room in front of the tree, and the people waited expectantly.

It was an interesting sight. In the center of the room stood a large stove with a roaring fire, and on benches around it were apparently the entire population of the village. Mr. Pelley had told us (confidentially) that something new was expected each year to provide added interest, though he did not put it exactly that way. This year he had planned to have a supposed son of Father Christmas come along too, and to make him very unusual, his face was to be black. What relationship was supposed to exist between Father Christmas and a black boy I do not know, but it was a grand success.[88] Old Santa, in a fur coat, with reddened face and long white beard, entered the hall, followed by the black boy in a sheep-skin jacket and a straw hat. To some of the children they were really unknown creatures, and one little girl even cried from fright until told her kind old Santa had come to bring her something nice.

When the curtain was lowered and the glittering tree appeared, it was fun to watch the reaction of the children. There was no outcry, not a shout or squeal. They sat with wide-open eyes, dumb with amazement, and when called to take their gifts from Father Christmas, some of them were still half frightened and all solemnly excited. Several men in the gallery provided an uproar by refusing to come down when their names were called to receive their gifts, which resulted in those on the floor rushing the gallery and pulling the others to the platform – a general free-for-all with confusion, horse-play and applause from the whole party. Dr. Little sprang to rescue the kerosene lamp that hung in the line of battle and stayed nearby to guard it. The evening ended with gramophone music and a performance by Mr. Cole (impersonation of a crying baby) which brought down the house. Then we stumbled home in the darkness along the rough, icy road. The gramophone miraculously survived the trip.

7

Landing of the Reindeer[89]

2–10 January 1908

January 2nd We have been busy all day preparing for Dr. Grenfell, who arrives next week. Although much remains to be done, I went this afternoon with Miss Keese to the schoolhouse to assist in rehearsing the "concert." There was no fire, and it took some time to start one. Meanwhile, we swept up the snow, blown in through chinks of the building. The children spoke their "pieces" in the way familiar to anyone who has visited a country district school, but when "Little Red Riding Hood" was rehearsed and real action required, they quite rose to the occasion. Ephraim [Dyson], as the wolf, was as realistic as a boy on all fours could be, and the scene between Wroughton, as woodsman, and the wolf became a fracas.

It was bitterly cold on the way home, with the wind in our faces. The mercury registered 20° below zero. Mr. Lindsay said he had frozen his nose again during the day. It was quite white, and Reuben held it in his hand to thaw it. He should have a nose-muff – another job for me.

January 4th The reindeer have not come yet, and Mr. Lindsay begins to fear something serious has happened, since they are so long overdue. If there is more delay, it is possible Dr. Grenfell will be here to land them after all. He must be on the *Portia* now, on his way from St. John's. Mr. Lindsay also thinks of going home to England if the reindeer are lost and coming out again in the spring if a new herd can be shipped. Everything seems doubtful, and unrest is in the air.

January 5th The reindeer have come! We are so excited everyone talks at once. We had just returned from church and were at the dinner table when we heard Reuben's excited shout that a strange whistle had been heard, and then, over Fishing Point, black smoke was clearly visible. From all directions men started out across the ice, and a group gathered on the

Saami herder, 1908
(Grenfell Historical Society [GHS])

hill at the harbor's entrance. For some reason everyone suddenly became silent and serious. Scarcely anyone spoke, and as soon as we rose from the table Mr. Lindsay, Mr. Lee and Reuben, with snowshoes on their backs, started for Crémaillière, where the steamer was supposed to be.

Mr. Lindsay walked ahead with the air of a conquering hero; Mr. Lee looked like an explorer in his sealskin suit, and Reuben followed with the appearance of having the time of his life. Miss Keese and I, who wanted to go but were afraid of being in the way, were overjoyed when Reuben said it was only two miles, and he was just going for fun and would stand by us. Soon after leaving the village the path was practically unbroken. We had no snowshoes, and we sometimes sank through to our waists in the soft snow, but struggled on until, on reaching a hill-top above the sea, the little harbor of Crémaillière lay below us: and there, fast in the ice, far from shore, was the steamer, dark against the broad expanse of white. We half slid, half ran down what seemed almost a precipice and in a few moments reached the tiny settlement of half a dozen wretched little huts where a few rough, unkempt people and a large pack of dogs were moving about.

The steamer was surrounded by "slob" [slushy ice], but the men said it was safe to walk on, and we started for the ship, all of us using poles to test the strength of the ice. When it seemed weak, we jumped from

Saami herder and his wife, 1908 (GHS)

pan to pan and walked around weak places if the opening was too wide. As we neared the ship we saw the Laplanders standing in a group, looking over the bow, their bulky fur garments and queer caps with four pointed horns suggesting the Vikings of early history. They seemed as much interested in us as we in them, and we probably satisfied any curiosity they may have had concerning inhabitants of the strange land they were about to visit.

I never expected to walk to a steamer's side and climb up a rope ladder, slippery with ice, swaying from side to side and out with the weight of my body, but that was what I did. The Laplanders gathered around us when we reached the deck and were much interested when I changed the film in my camera. I signified to one of the women that I wanted to photograph her, whereupon she struck an attitude and looked solemn, which was not what I wanted. They are very short and when walking give the impression of being bow-legged or bent-kneed. They had on skin clothes: two garments, one with the fur inside, the other outside, apparently made of odd bits put together with gay cloth. The men and women both wore skirted coats, short and very full, with a belt below the waist, suggesting a little girl at dancing school.

The women wear close caps of bright colored cloth and embroidery with side pieces tied under the chin, and the men's cloth caps – black with gay embroidered head band – have four pointed horns that flap in any direction and are used as pockets to carry a pipe, etc. Deerskin gaiters cover their legs to the knee, and deerskin shoes (fur outside) are bound around the ankles with inch-wide woven strips to keep out the snow. The toes of the boots are pointed and upturned to fit under the straps of their skis. It is said they wear no stockings but stuff hay in their shoes for warmth. One of the Lapp women had injured her knee during the voyage and lay on a pile of skins on the lower deck. She, with her family and two dogs, were eating their noon-day meal together, and we discovered the men carry a sheath knife attached to a belt under their outer fur garment, which is used for a variety of purposes from cleaning shoes to cutting bread and butter.

I lowered myself down another slippery ladder to the deck where the reindeer were penned and found them so tame they ate from my hand. They are lovely creatures and in excellent condition in spite of the rough voyage. Their pens were clean, and they show evidence of the careful preparation made for the hazardous voyage. After weeks of anxious waiting, they scarcely seemed real. They are to be landed tomorrow morning, and we are planning to go over early to photograph them.

January 6th This has been another day of excitement. The whole harbor is demoralized, and everyone who could possibly get off has been to Crémaillière to see the reindeer landing. Mr. Lee started off early this morning without waiting for breakfast, going on snowshoes to Crémaillière, where the steamer was yesterday, only to find an empty harbor. The *Anita* had vanished, and he came back with the news to prevent our starting. Then followed a series of reports. "The steamer had put to sea in the night during a gale that swept the coast." "She was still in the harbor but around the point, wrecked on the rocks, and the reindeer would be landed at once as they were practically on shore, fast in the ice." In a few minutes a later report: "They would be landed this afternoon."

This seemed definite, and we planned to go out at noon. I decided not to wait for the others and started with snowshoes and camera, joining two of our boys on the way. As we reached the path to Goose Cove, we met two men on komatiks who told us the deer were already ashore. They were "all over the country" and would soon come over the hill. It was a great disappointment to miss the dramatic landing, but there was no help for it, and we sat on our snowshoes to wait for them.

The main part of the herd was on shore, but the rest were still being landed from a large port in the steamer's side. Men and boys formed lines beside the landing stage from the ship, which lay on a rock, fast in the ice. As the deer emerged one at a time from the port, they ran toward the shore, where on the top of a bank one of the Lapps, with two tame deer beside him, stood clanging a cow bell and calling something unintelligible that the deer seemed to understand. After the first wild gallop on leaving the ship, they sauntered single file up the bank past the Lapp with his bell and passed on to the hills beyond. In spite of every care, five of the deer broke through the line of men and ran off across the ice to the sea, where waves were churning the edge into a foamy, treacherous mass. We could see three of them, rising and falling with the waves above the ice level. Poor things! To think of surviving the perilous ocean voyage only to perish like that! The others that strayed made their way across the harbor and will doubtless join the herd.

After the deer were all landed, several people who had appeared on shore joined me and we went out to the ship. The ice was far more hazardous than yesterday. The wind had driven the pans together, but spaces between were filled with slob ice, and in jumping across, one sometimes went through. Several times I sank to my knees. It was a grey, lowering day with heavy, purplish sky above the white hills, a day typical of the season and of Newfoundland. The little harbor and the stranded ship seemed more lonely and the whole scene more dramatic in that somber light. We went on board and watched the bales of hay and moss and the queer *pulkas* (sleds) being hauled up from the hold. Sometimes the Lapp women took a ride up from the lower hold on a bale of hay, to everyone's amusement. It reminded me of a witch on a broomstick. The most exciting moment came when the Lapp woman with a broken knee was lowered over the ship's side in one of the pulkas. I stood on the ice below, waiting with one remaining exposure in my camera, and seeing a pulka descending with a huge bundle in it supposed it was she, only to discover, after turning the slide, that what I thought to be the woman was tents, bedding, etc.

In the first excitement of arrival on the scene, I snapped my camera wildly at every deer in a new position and found myself with an empty camera when the woman was taken across the ice by the harbor men and transferred to Dr. Grenfell's komatik, with dogs waiting to drag it to the hospital. It was a sight I shall never see again. The woman was pathetic yet funny. At least a dozen men helped drag her from the ship in her pulka, but when they reached the komatik and tried to transfer her to it for the rest of the journey, the sight of all those strange men speaking a strange tongue, trying to take her to a strange place, was too much for her and she refused to budge. The men all stood looking helpless. Suddenly she saw me, and the presence of a woman seemed to reassure her. We exchanged smiles, and when I pointed toward the komatik and arranged the skins for her, she threw back her own coverings, disclosing a frying pan which she clutched as if it were her one treasure, and consented to be tucked in and taken away behind dogs for the first time in her life.

On reaching the village, we found that Reuben had been followed by the woman's husband and son, who refused to be left behind with the rest of the Lapps, and he was about to take them to his house to see what effect the sight of such strange creatures would have on the children. On the way to the house, some men and some of the orphanage children saw them, and it was quite a procession that filed in through the low shed leading to Rube's door. I wanted to see what happened and, on entering soon afterward, found them all seated along the wall in dead silence, the boys with awed expressions, their eyes and mouths open, and the Lapps huddled in a corner.

It was very funny. The silence was unbroken until Mrs. Simms[90] brought out the loaf and teapot and invited the visitors to come to the table. They accepted and approached with their queer waddling gait while the dumb, open-mouthed audience watched their slightest movement. Reuben passed granulated sugar for their tea, but they did not seem to know what to do with it. He put the spoon in the boy's hand and motioned him to take some. He responded at once and put the sugar in his mouth, licking the spoon before returning it to the bowl.

It is such a pity Dr. Grenfell could not be here for this exciting event. He was in communication with the *Anita*'s captain when part way up the coast, so we know he is on the *Portia* en route to St. Anthony. I can imagine his impatience. The next excitement will be his arrival. The *Portia*'s coming is always an event – at least it has been thus far. This time it is uncertain if the steamer can reach St. Anthony on account of the ice. She may be unable to come beyond Conche. As many of our winter supplies and feed for the cattle are on board, the uncertainty is alarming. There are so many "ifs" one becomes dizzy, then philosophical.

The Laplanders are to live in tents located near the herds, some distance away, but the camp is not yet ready. Meanwhile, they are sheltered in the bunk-house and a small, unoccupied building near by. For them, a house with walls instead of a tent involves changes in living conditions, and the Lapp women, who are very friendly, often visit the Guest House kitchen, evidently much interested in what goes on. Naturally, the domestic routine is at a standstill during their visit, which may be regarded as an entertainment, although sign language takes the place of audible conversation, perhaps more exciting on that account.

The woman whose knee was injured during the voyage is still in the hospital and sits in bed, weaving on a tiny primitive loom about six inches wide and a foot long, like a wooden comb with alternate slits and holes through which the woolen warp is threaded. One end of the warp is tied to the bed's foot-rail, the other knotted around her waist, and she can tighten or loosen it by leaning backward or forward as she weaves a narrow strip an inch wide and about a yard long which is used to bind around the top of a deer-skin boot to keep out the snow. They are very pretty, and I have ordered one to wear as a belt.

The principal topic of conversation has been the fate of the *Anita*. The men said the gale had driven her on a rock, but the high tide had lifted her from it and she now lay free but leaking badly. Only the right wind could save her, for she lay between a shoal and the rock. Reuben and the men shook their heads, agreeing that with the wind now blowing, the ice would break up, and not only would the ship be lost during the night but the only hope for the men was to come ashore before nightfall. Otherwise, they would "all be dead men." They said it with a kind of gruesome satisfaction – not that they are really pessimists, but sometimes one enjoys a feeling of cold shivers down one's back. Everyone is sorry for the captain, who after bringing the ship with its important cargo so successfully through such a terrible voyage may yet lose it.

Later. Mr. Cole has just been to the ship and reports the ice badly broken, and if the wind does not change the ship may break up, though the men will probably come ashore. The moss landed on the ice for the deer has been washed or blown away. They will have to depend on spruce tops and what moss can be found under the "glitter" [coating of ice].

We were all in a mood for relaxation after nerve strain, and tonight the long anticipated "concert" provided it. We had made bets as to the ability of any of the children to remember their "pieces," Mr. Lindsay predicting not one would perform creditably. Fortunately for him the bets were small, for with two exceptions everyone did remarkably well. Mr. Cole was a trump. He not only helped Miss Keese all the afternoon but when evening came offered to fill gaps in the program by performing himself. His stunt was to play clown. He improvised a ragged, two-colored wig, and I made a gaudy collar for him to intensify the ridiculous effect. He tumbled about all over the stage and did a cry-baby act that took the audience by storm. They shouted and squealed with joy at the horseplay.

The program was lengthened by the unexpected offer of some of the men in the audience who signified to Miss Keese that they were willing to contribute to the entertainment, and I was amazed by their ability to memorize so much. Bob Newell recited a long poem. Eli Curtis contributed something long and moral, with assurance and fluency that filled me with admiration. He even made gestures and skipped about the stage. As he is sexton of the church and I have seen him only in his professional capacity and judged him by his unobtrusive manner, it is no wonder I was struck dumb.

The most exciting part of the evening was when Mr. Lee and Mr. Sundine[91] (the Swedish interpreter) ushered in four of the Lapps. The hall was crowded and their arrival produced a gasp, then dead silence. The strange looking creatures and their gay clothing in the midst of that commonplace audience was enough to create a sensation anywhere. The women had on gala attire: gowns to the knee, very full and made of gay red and white striped stuff with a ruffle at the bottom, the inevitable fur leggings, and shapeless boots and a little cap close on their heads.

The performance went on after the first breathless pause, though the people, especially the children, who were afraid of anything so weird, still looked furtively at the Lapps. The climax of the evening arrived when three of them were persuaded to mount the platform and do a song and dance, or rather make the attempt, for the boy was so frightened he could not utter a sound, and the others after a short effort gave it up. But the audience howled with glee.

It was a "grand"[92] concert.

8

Dr Grenfell Arrives

11–25 January 1908

January 11th I feel a trifle incoherent. The *Portia* has just come with Dr. Grenfell on board, and he, Mr. Lindsay and Mr. Cole have been here talking of plans and projects. They have just gone out. It is after midnight, but I must write a word to Miss White at the Boston office about the reindeer, and it must go when the steamer returns for the mail at 5:00 a.m.

Dr. Grenfell is full of plans of all kinds with sometimes little regard for details. He approves of what has been done to the house but is so absorbed by the reindeer that he has had time to talk of little else. He has not yet seen the workshop. I hope he will be pleased by what has been accomplished in spite of many interruptions, such as the grippe epidemic and reindeer excitement.

January 19th Our winter life grows daily more interesting. We begin with breakfast at eight by lamplight, the knives often so cold we hold them with our napkins. Dr. Grenfell sits opposite me at the table, laughing, making jokes, teasing. We have marmalade and toast, with bacon in a silver dish. After breakfast we have prayers, and the maids file in, in English fashion.

A week ago, Dr. Grenfell had his four trunks taken to his room and asked me to help him unpack them. I wonder what the people who crowd to hear and see him – who even flock to hear me talk because I talk of him and his work – would have said if they could have seen me on the floor sorting his clothes, packing away his summer underwear, his silk handkerchiefs, frock coat (which, he explained, was made expressly for his audience with the King)[93] and all his other good clothes. He pranced around the room in his khaki overalls, flannel shirt and skin boots, wearing his silk hat to show me how he looked in such an article. I came to the beautiful doctor's gown, symbolizing the honorary degree conferred on him at Oxford,[94] a gorgeous black silk affair with the scarlet cloth hood and the mortar board worn with it.

He made me dress up in it and put on the mortar board, then led me to the mirror to see how I looked. He had tales to tell about the souvenirs I came across, menus and invitations for grand affairs, and his remarks were most amusing and characteristic. It is interesting to have him consult me about household matters, and he seems to enjoy the whole atmosphere of the house. I see on his face what I call the "purring" expression which means content. He comes in and takes off his boots by the fire, puts on his felt slippers, which he keeps behind the stove, while he tells about patients he has been to see or about some plan he has in mind. He leaves things all over the house, and I meekly go around or send the maids to pick up after him without a murmur. That was only at first, however. Now that he has settled down, I find he likes an orderly house and manner of living as well as the rest of us, and if things are disorderly he notices it. He even changes his coat for dinner at night. He has given me a pair of lovely snowshoes with little colored wool balls around the edge, and I am delighted with them.

While in Montreal he bought Canadian blanket suits for himself, Dr. Little and Dr. Stewart. They are bright blue with dull red stripes at the edge, large pointed hoods and a red sash. Miss Keese and I have begun to make similar suits for ourselves of bright red blankets like those on our beds. They will have black stripes on the edge and be very fetching, with full bloomers of the blanketing instead of a skirt. The coat will reach nearly to the knees. Miss Storr and Miss Kennedy are making white suits with light blue stripes, so when we all go out together we will be like a living American or English flag waving over the country.

My present occupation (aside from teaching weaving, planning food, scolding maids, etc.) is making nose and ear muffs. The nose muff is my own invention, and Mr. Lindsay says he would perish without it. He is just starting for the reindeer camp, and this is my parting gift. He has frozen his nose several times. Everyone is freezing cheeks, noses, chins, etc., and it is not uncommon to see someone suddenly clutch his neighbor's face and warm it with his hands. I frosted my fingers the other day and today my toes a little. I keep my nose covered. Nothing is seen of my face but my eyes.

Sunday was frightfully cold with a bitter wind. We went to church in the morning and I was not going again at night, but Dr. Grenfell urged me to. He said it always seems more cozy to come into the house after being out in the cold, so several of us went after all. Coming home, the wind was so fierce it nearly blew us over, and we were glad to find shelter. Today was another bitter day, but the loom-room was warm. Dr. Grenfell has had the windows fixed, and the wind does not now blow in around the sash. The pupils are doing well, and we have some beautiful weaving finished, but clay cannot be used in this extreme weather, so I am giving the boys wood-carving instead.

January 21st Dr. Grenfell has ordered two more looms to be made in the local work-shop, and they are to be begun tomorrow. I have the harnesses and reeds for them. At present there are more workers than looms, and I shall teach some of them dyeing and preparation of the warp. Mark, my special boy, who has been making pottery, is carving wood, and there are nine boys in the evening class. They come four nights in the week and are as happy and enthusiastic as any group of boys could be. I am busy all the afternoon at the loom-room and rarely reach home until 5:15.

The evening class is at 7:15. When I reach home at night, the maids have cocoa ready, and the men all come in to sit around and smoke. Dr. Little has his banjo, Dr. Grenfell tells stories, and we make merry sometimes until midnight. That is the time I do the family darning, which does not interfere with the use of my tongue. Tonight I am trying to write letters that Dr. Grenfell will take when he starts tomorrow morning on a trip to the saw-mill. He hopes to intercept the *Portia* or meet the mail-man. The *Portia* will not be here again this winter. We shall not see her or have any regular mail for five months. It may be June before she can break through the ice. The real winter is just beginning, and it seems as if it were nearly over. The mercury goes lower and lower, and we have frequent snowstorms. The snow is three feet deep on a level and in some places much deeper, with only the tips of trees appearing above it.

Mr. Lindsay started out yesterday with a Laplander, six deer and pulkas and three men with all their camp luggage, but it was so late they could not make camp before dark and came back to spend the night, bringing a tale of travel under difficulties. In one place they had to climb beside a high waterfall. It was very steep, and the deer could not pull the pulka over the rocks. One of them sank quite out of sight in the snow and had to be hauled out. In the struggle he became mixed up with Mr. Lindsay, stepped on his toe, knocked him down, and Mr. Lindsay became entangled in the harness. Altogether it must have been very exciting. They went on again this morning but were back at cocoa time, tired, dishevelled and hungry. I feed hungry men at all hours.

When Dr. Grenfell first came he often forgot about coming to meals or took them at the hospital without notifying us. I never waited for him, for we are prompt and Dr. Grenfell begged me not to because he hated to feel hurried. Lately, however, he has been on time – sometimes even ahead. He has a way of piling up his woolen mittens, his heavy wool vamps [thick socks] and inner soles behind the dining-room stove to dry, and tonight he came in with wet feet, took off his skin boots and wool socks and discovered enormous holes in his fine thin stockings. He left them behind the stove to dry, so I have pinned them on the mantel like Christmas Eve, with an announcement on the wall that we like Christmas so well we have decided to continue it indefinitely by hanging up stockings at intervals, such as Tuesdays and Fridays.

This afternoon, when I came in from the shop, Dr. Grenfell called me to the gun-room. The carpenter was there, and they were planning a settle like an inglenook around a corner where the stove is to be put up. There is also a plan for curtains. This is the room the men have used exclusively for photography, gun cleaning, oil skin clothes, and all their sporting outfit. I was supposed to have no part in it. When I told Mr. Lindsay, he was disgusted and said Dr. Grenfell was fast deteriorating and growing too domestic. I did not dare tell him about the curtains.

On Saturday I took a trip to Goose Cove, where Dr. Grenfell had arranged to have a Christmas tree [community social]. It is a little settlement about seven miles away, which in summer we reach by boat or over a terribly rough trail and in winter by komatik across frozen bays and ponds. Dr. Grenfell's secretary, Mr. Jones, and one of the men went over in the morning and carried the gifts. The temperature was about zero but still and clear. Miss Keese and I walked on snow shoes part of the way, and then Dr. Grenfell overtook us and I went on with him in the komatik with nine dogs. Alf, his driver, and Miss Keese came in another with seven dogs, and Mr. Lee followed on snowshoes.

I cannot fully describe that ride through the beautiful country. We went flying down the steep hills, the dogs racing to keep ahead of the komatik, and across the bay where the reindeer were landed. Occasionally the dogs met another team. Then the men ran ahead with their whips to prevent a scrap – very exciting! Great rocky hills rose directly from the ice that once was open water, and the trail ran around them. We reached the scattered group of houses just as the sun was setting, and the color across the ice was something indescribable. The Christmas party was to be in a small building used as a school-house or chapel, and a kindly woman asked us in to get warm before going there to decorate the tree and to return for tea, a simple meal of bread and butter, tea and prunes.

At dusk we went to the school-house, where Dr. Grenfell distributed the presents, introduced Father Christmas and was the spirit of the party. I loved to see him take the children in his arms, laugh and play with them, and soothe them when they were a little afraid of Father Christmas with his long white beard, which disguised a well known friendly neighbor.

When we left the house the whole world was flooded with moonlight. The white glory of that night was something unforgettable. We were facing the moon all the way, and in places it shone like silver on the tops of great ice-covered crags where the snow had blown off. There was no sound but the slight scratch of sled runners and soft pad of dogs' feet in the snow. It was so cold my eyelashes froze over my eyes several times.

We are wearing more and more clothes. Dr. Little is assuming enormous proportions and Dr. Grenfell looks quite like a giant. When I go out, I wear two pairs of golf stockings, wool stockings and bed-shoes and a pair of vamps inside my skin boots, a sweater under my fur-lined coat, a knitted helmet under Father's sealskin cap, and a pair of wool mittens with skin mittens over them. Even then I am not overwarm.

January 22nd It was soon after Dr. Grenfell's arrival that members of the family were for the first time referred to otherwise than formally. It began by the Doctor mentioning that young Lee reminded him of "The Young [King] David," and "David" was his family name forthwith. Mr. Lindsay, his first Mission assignment having been director of peat excavation, is dubbed "Pete." Miss Keese, whose formal name seems inappropriate for such a gay, lovable personality, is referred to as "L.M.K." (Little Miss Keese). Dr. Grenfell is known as "the Doctor" – *the* Doctor who treats physical and mental ills, stimulates and vitalizes the community, animates and encourages the coastal people. His title is used with loving respect by everyone. Dr. Little is "Dr. Little." There are no obvious traits in his fine human personality to take liberties with. As to myself, I think I am sometimes referred to as "the Missus" but never addressed as such.

From the first, the daily life of our small group, notwithstanding differences in nationality, temperament and background, has been happy and harmonious. Now at last the Doctor brings his unusual personality to fill the place we have tried during the past months to prepare for him, and he adds zest and purpose to all our activities. It is inevitable that in our English-Irish-American household there should be varying opinions in small things, including food preferences. There is, for instance, the matter of toast: some of us like it soft and toasted lightly; some prefer it dry, crusty and hot. Now we find the Doctor likes it hard, burnt black and cold. It is served in an open silver rack, which would insure its being cold in any case. We have so little association with roasts and chops that preferences as to "rare" and "well done" are immaterial, although we are sometimes called upon to choose. At one time, the conversation turned to a discussion of what article of food one would choose for a first meal in one's own country, Mr. Lindsay's choice of cold beefsteak pie making more impression on me than any other preference.

January 24th Interest in the reindeer project seems to be international. Miss White has sent from Boston a clipping from a London newspaper (quoted from a letter written by Mr. Wood, secretary of the Royal National Mission to Deep Sea Fishermen) which describes the reasons for their importation: "The reindeer is everything to the Lapp. It clothes him, feeds him and is his beast of burden. The herds of deer are owned by families rather than individuals and some of these common herds number 10,000. Altogether, it is estimated there are 400,000 reindeer in Lapland ... The experiment of introducing this herd of Lapland deer into Labrador is an exceedingly interesting one. Its chief promoter has been Dr. Grenfell and, if successful, it will prove a source of wealth to Labrador and be the means of conferring a boon upon the scattered population of the coast of this outlying portion of our Empire."[95] In this country, especially in the eastern portion of the U.S., interest approaches excitement. The Doctor has received many letters of inquiry accompanied by suggestions, some of them very amusing. Miss White reports that my letter about the deer's landing is being sent to various parts of the country and read at meetings.[96]

The reindeer have naturally been Dr. Grenfell's main interest since his arrival. He was impatient to test their usefulness to the people, and to prove their importance for transportation he left yesterday with Mr. Sundine, the interpreter, by dog-team for the deer camp. From there, they started across country with a deer and pulka and one of the Lapps, who was supposed to know the deer's little ways and the best methods of dealing with its moods and movements. To the Doctor's disgust, instead of flying over the snow in the pulka like Santa Claus in Christmas illustrations, the galloping deer guided by a flowing rein, the Lapp walked ahead on his skis and led the deer while the wholly uncomfortable Doctor sat in the pulka, which was sometimes tipped over by tree-tops, sometimes stuck in the snow. At last, the deer became entangled in the one trace and sank to his neck in the soft snow. The men, in attempting to haul him out, sank through themselves, and all became one floundering mass of animal life. Finally, the Doctor and Mr. Sundine left the Lapp and started off by themselves,

A conference on reindeer management, 1908. *Left to right*: William Lindsay, Saami herder's wife, St Anthony herder, Saami herder Nels Ander (JLP)

attempting to drive instead of lead the deer. This did not please the deer. He walked very slowly, frequently stopped, turned, and (according to Dr. Grenfell) "made faces at them." He also opened his mouth and seemed to gasp after every effort.

They went only as far as Englee, where they spent the night in a vacant house, burglarized with desperation born of necessity, and returned to St. Anthony the next morning after travelling nine miles instead of the sixty they had planned. The Doctor came in disgusted with the conduct of both Lapps and deer, convinced that pulkas are only fit for firewood and that the Lapps know nothing about reindeer. This impression is strengthened by discovering that two of them are village Lapps and have no herding experience. The Doctor plans to keep three deer here while training them to be driven, not led. The Lapp insisted they were too weak to be driven after their voyage and would not be ready for several months. Imagine the Doctor's reaction!

January 25th This evening, Kate (the cook) called me to the hall. She said one of the Lapp women was in the kitchen acting queerly, and I found her lying full-length on the floor near the stove. She monopolized one side of the room. On the other were the maids, Mark and two other boys, all silently gazing at her as if they expected her to die on the spot. I called Dr. Grenfell, who roused her and called Mr. Sundine, the interpreter. She complained of smoke in the bunk-house and made signs of something in her eyes. Dr. Grenfell went out to the house and found they had taken off all the stove covers before building a fire and had not put them back. Of course, woodsmoke filled the room. When living in the open, they have a fire in the middle of the tent, and smoke passes out through a hole in the top. Probably they never used a stove before and thought smoke would go out the same way in a house. They are supposed to leave the house on Monday and make their camp with the other Lapps near Mr. Lindsay on the hills, but they now demand all sorts of things, including a real house. Observe the effects of civilization.

Later. A real excitement! We have just seen Dr. Grenfell trying to break two reindeer for driving. After his experience with the pulka, he decided to try a komatik and there they went – two deer harnessed abreast, the Doctor driving. The deer were trotting quite fast, and all that was needed for the Christmas card effect were horns on the deer (which they shed every year and grow again in the spring) and a white beard on Dr. Grenfell. Mr. Sundine was driving another deer and komatik, while Mr. Lee hovered on the outskirts. We watched them with excitement until they vanished over the hill, a splash of color on the snow, for the Doctor wore a gorgeous costume of white deerskin jacket, blue sash and red cap, grey trousers, high yellow skin boots and black fur mittens. He looked "wonderful" (in local parlance).[97]

9

Work and Play

26 January – 18 April 1908

January 26th This has been a nice day, like most of our Sundays here. The family went to church this morning, Dr. Grenfell preaching a fine sermon which should help these simple people – real meat and drink instead of the dry husks sometimes offered them. On the way home, the Doctor wanted me to go with him and the orphanage boys to feed the foxes, so I borrowed Miss Keese's racquets (snowshoes) and floundered up the hill to the fox farm. It is an awful hill. I don't see how the reindeer came down yesterday without broken legs. The foxes were very lively. "Jack," the "patch" (a cross between silver and red fox), ate from our hands through the wire netting, wagged his brush, and laid his ears flat in an amusing, dog-like way. The red fox is also very tame, but the white foxes are too shy to come near. The silver and black did not appear at all. They have burrowed holes in the snow and keep quite out of sight.

Dr. Grenfell had been asked by Mr. Richards, the clergyman from Flower's Cove (whom I met on the *Portia* last year), to come across the harbor this evening to take the Church of England service, and most of us went to hear the Doctor preach another fine sermon. Each one seems better than the last. Mr. Richards joined us after our return to the Guest House, and Dr. Grenfell showed him around "our little home," as he called it, with great pride. The other men came in later for a comfortable smoke. Dr. Grenfell read Kipling aloud. The evening ended with lively discussions, and so to bed after a happy, peaceful day.

January 29th Dr. Grenfell was in the porch when I came down for breakfast. He said he had been outside and the air was "sweet as milk," like an April day. It seemed a favorable time to walk across the harbor to make purchases at Biles' store[98] and (incidentally) take Dr. Grenfell's skin boots to be mended. The path led to the little church where

Miss Keese is holding her classes, pending repairs of the school-house on this side of the harbor, and I made a short call, finding the room warm and comfortable, a contrast to the miserable condition of the school-house on this side. I am told the government school board has a limited number of grants for teaching. Probably that is why there is a school session only once in three years at the little [St. Anthony] Bight settlement. A teacher's salary is only $250, out of which she pays her living expenses. Apparently, one should be a philanthropist as well as teacher in this region! Mrs. Biles had been to see me and I stopped to return her visit. The house was typical of the better grade of local houses: there were several rooms and an "upstairs," referred to as a "second flat" by someone who had known city life in St. John's. Mr. Biles said his daughter Maude is slowly failing. She has tuberculosis, the scourge of this country.[99]

It is reassuring to know that Dr. Grenfell's preaching in regard to sanitation is beginning to bear fruit. Noah Simms, a village neighbor, told me this morning he is making an effort to have the Orangemen's Lodge[100] members pass a resolution to form a kind of total-spitting-abstinence society, and the men are considering it seriously. There was further evidence of the leaven working when Beatrice Suley (one of the weaving girls) told with disgust of someone from Quirpon having such a dirty house where the man [was] spitting on the floor, what she said to him, how he took it, etc. She also asked for some "Don't Spit" cards and wants to make a mat like Dr. Little's. It is evident she has heard the matter discussed. All this pleased the Doctor greatly.

This afternoon was the beginning of the matting club. Several women came, but evidently with the idea of looking around before committing themselves. It was a busy room: four looms, a spinning wheel and dyeing in operation, one girl making baskets, another preparing material. The rabbit skin blankets (an original product) are also on the way.

It is odd how little interest we have in letters. Dr. Grenfell rejoices that he is beyond the reach of regular mail and would be glad to have no more for a month. I can understand his reaction after being swamped by correspondence all last winter.

Mr. Sundine appeared for luncheon. He reports the deer in fine condition, the new stove burning properly, everyone comfortable and a state of calm at camp.

February 1st It has been a mixed up day. I begin to feel like the brook, "Men may come and men may go, but I go on."[101] There is mild excitement in packing provisions in nunny-bags[102] and tin boxes, seeing that the clothing outfit is intact, clean handkerchiefs tucked away in corners, nose muffs in order and such details, but it is sometimes like a snow flurry and quite blinding. Mr. Lindsay and Mr. Sundine were to start for camp after breakfast, but the Lapps did not appear, and lunch time found Mr. Lindsay standing with hands in pockets, gazing over the hills in the direction from which they and the deer should come. He was outwardly calm, but inwardly ...! It was trying, after packing his week's supplies and hurrying through breakfast, to be ready by nine o'clock.

My latest accomplishment is barbering. Mr. Lindsay's hair is one of his trials: it is rather stiff and over his forehead grows in what is known as a "cow-lick." Since his appeals to Dr. Little for hair surgery were disregarded, in spite of Mr. Lindsay's services to him, he became desperate and operated on himself. The result was a moth-eaten appearance, far from attractive, and while we waited for the Lapps I tried to remedy matters. It was a novel experience for me, and the sight of Mr. Lindsay enveloped in a towel, sitting meek and resigned with an occasional reference to his Irish barber, was almost pathetic. He was quiet during the operation, but having been told he was considered one of Dublin's best groomed men it was doubly moving to hear him murmur, "If my Dublin barber could see me now!" – which may have been a reflection on my efforts.

February 2nd After recovering from a two-day [sic] headache and again joining the family, I was at once taken around to see the "improvements." On the dining room wall was an illustration of the school board controversy, done in Dr. Grenfell's best style, quite descriptive. I had heard hammering in the gun-room during my retirement and found the cozy-corner seat was finished. It was two inches too high and four inches too narrow. The seat is also a locker, and cushions have to be disarranged to get anything from it. There was no material for cushions, so we dyed some unbleached sheeting and used excelsior [wood shavings] found in a barrel for stuffing. One must be ingenious and resourceful in Newfoundland.

February 3rd Dr. Grenfell and Dr. Stewart went to Bréhat last night, the Doctor to take the service and Dr. Stewart for an opportunity to talk with him. There was a meeting before they left to discuss plans for starting a co-operative store. I asked why it had not been done before. They said Mr. M[oore], the local trader,[103] had been friendly to the Mission and, as he lived at St. Anthony, they had made no effort as it would be a competing enterprise. Now, however, when Mr. M. makes his home at St. John's, the time seems to have come to establish something that will be of real benefit to the people. The meeting was very successful, and the village was all in favor of it. Mr. Jones of the Mission is to be in charge and is well fitted for the position. They have taken Rube's Point across the harbor and are to do the necessary carpentry as soon as possible in order to have everything in readiness for supplies as soon as they can be obtained in the spring. The people will be greatly benefited, and I am glad they recognize it.

We went to church as usual this morning and were nearly blown down. One could actually lean on the wind. The church was cold, and the building creaked and groaned like a ship at sea. If I had not known it had probably weathered worse storms, I should have thought it was coming down.

February 3rd [a second entry for the same date] Dr. Grenfell and Dr. Little started today by dog-team for a long professional trip. The weather was perfect: clear, not too cold and everything sparkling with fresh snow. They were to start at two, and from breakfast on I packed supplies, took last stitches, etc. There was a general air of activity. In the middle of the morning, Dr. Little appeared decked out in full regalia: long trousers, flannel shirt, grey sweater, and over that, khaki overalls and Dr. Grenfell's white deerskin jacket, belted with a red sash tied at the side; high skin boots, a black toboggan cap and fur mittens. He looked "grand," as the natives say. To this gaiety of attire was added a jovial expression like that of a boy off on a lark. Even his thin beard looked hopeful. He had just gone to his room to make his will and settle his accounts (quotation) when Dr. Grenfell rushed in demanding luncheon at once, as they were about to start. During the ensuing turmoil in the kitchen, both men discovered the shoulder straps of their overalls needed adjusting, and the resulting delay gave Kate time to prepare a decent meal.

They were to be away only two weeks, but the preparation seemed enough for two months. The two komatiks waited outside, one with sleeping bags, clothing, medicine chest, etc., the other with Dr. Grenfell's long komatik box (resembling a local coffin) packed with provisions. I took a photograph of them, with Dr. Grenfell striking an attitude on one of the komatiks. We hovered near by with our cameras while the Doctor and Alf, the driver, fastened the dogs – eighteen of them – all rolling, kicking, and yelping. When all was ready, the Doctor shouted "Look out!" and we jumped to one side as they started off towards the fox farm and disappeared over the steep hill trail with a farewell wave. Miss Keese and I feel quite widowed, and the house does not seem a house at all, only a shelter.

February 4th This afternoon we returned to routine, and the loom-room was full of workers. Some of Miss Storr's bible class stopped on the way home, the women appearing pleased and interested.

We are trying to see how much can be accomplished before Dr. Grenfell returns, and everyone works with a will. I gave Ted McNeill[104] some designs for his carpentry class, trays to be carved or finished with copper. Dr. Grenfell has given us a corner of the work-shop for the potter's wheel, metal bench, etc., where nothing will be disturbed. We are to find hiding places for the tools, for the carpenters who use the shop during the day take anything in sight to open boxes or remove nails and lack the reverence for carving tools which I am trying to cultivate in the boys.

The night was glorious as I left the house. A young moon and brilliant stars hung over the hill-top, other stars appearing one by one. Across the harbor, dim lights gleamed here and there in the little houses, a reminder of humanity with its joy and pain of living. I thought of our wanderers and wondered if they were also out under the stars or safely sheltered in a red-hot kitchen with tea and bread and butter on the table.

February 5th I might have known yesterday was a weather breeder. Today has been bitterly cold and windy, the snow so blinding one can see only a short distance. I can understand how travellers often freeze in the open. The Lapps were to move to camp today. They will have to wait.

Mr. Ash brought matting frames for the class today, but none of the women appeared. "Too dirty," the girls said. I am often surprised by the attitude of the women, especially the older ones, toward the weather. If it snows, which it does so frequently, not a soul will stir out, and intense cold and wind also tend to keep them by their kitchen fires. It surprises them that we who are strangers to this climate go out in all sorts of weather. When one takes into consideration, however, the difference in clothing, it is not strange. Their underwear is often insufficient, and their jackets are sometimes without a lining. Their skirts are long, and they usually wear heavy, ill-shaped high boots with no overshoes, though some of them have lately adopted native, high skin boots which before our advent were worn only by men and boys. Some of them have fur caps but usually only a cloth cap or tam o' shanter and on Sundays a regulation "hat." Their chief protection is a woolen muffler. Miss Keese went bravely across the harbor to her school this morning, but the weather was so bad she stayed at the Biles' for dinner, and they brought her home by komatik at nine o'clock.

February 6th The Lapps all moved to the camp today, taking their household belongings, the women and the dogs. I was in the loom-room and missed taking a photograph to add to my collection of "starts" by man and beast. The new camp is about six miles from Goose

The reindeer camp, 1908 (GHS)

Cove, and Mr. Lindsay says we may come out for the day when a komatik is available. That means after the Doctor's return.

The evening class boys are working on the mottos Dr. Grenfell wants placed on some of the houses. The first, "Praise the Lord, ye birds of wing," intended for the pigeon house, will be ready before the Doctor returns (Davey refers to it as "Praise de Lord, birds as has wings"), and the second, "All ye children shall be taught of the Lord," is well under way for the school.[105]

It is a brilliant moonlight night with a gorgeous aurora. I long ago exhausted all the adjectives in describing such wonderful phenomena. Therefore, with variations of such splendor I have nothing left but repetition. Tonight, great sweeping swirls of light crowd the heavens, unlike anything I have seen before. There was little color, but the form was wonderful. One realizes more fully than ever what atoms in the universe we are when such forces, almost awful in their beauty, are in action which we cannot understand. One feels the need of silence, as in some great presence, while the glorified heavens bend above a dim and silent earth.

February 7th This has been "blue Friday," surely. Our small family of three has been depressed all day, and the atmosphere is tinged with indigo. The weather was bad, and when Miss Keese came home at noon she was almost in tears, having had a violent fight with pupils in order to enforce discipline. She told a sad tale of a refractory girl who refused to stand in a corner for punishment and rushed home to her mother, who lives conveniently near. The mother came bursting into the school-room bareheaded, with her sleeves rolled up and "saying things." It ended by her taking the offending Rosie home, but not before Mr. Simms, one of the school committee, had been sent for to settle the matter and had spoken firmly to the mother, but not firmly enough to please little Miss Keese, who was in the frame of mind of the good clergyman who, in time of stress, asked a layman to swear fifty cents worth for him. It is unfortunate that Mrs. Jackson lives so near the school. There has been trouble with her before. Miss Keese left for the afternoon session still perturbed but returned much comforted, for the delinquent Rosie appeared at school as if nothing had happened. Mrs. Jackson had evidently thought the matter over in her quiet moments.

No sooner had this wound healed than other calamities followed. David [Cuthbert Lee] rushed into the dining room with great excitement. The one deer remaining at St. Anthony had slipped its halter and departed. David was wild. His comforting supposition was that the wind that afternoon had been from the quarter of the rest of the herd, and the deer had broken loose to join them. It was some comfort that the rope was left behind and the deer not in danger of being caught in the bushes, as they feared was the fate of the others which had escaped. As it was late, there seemed to be nothing for the boy to do but meditate on the situation, which was not cheerful.

The family said my turn would come, and that evening it did. Jack came to conduct me to the class and told me Walter had helped himself to raffia, which he and Tom had twisted into cigarettes and smoked downstairs in the shop, where the floor was covered with shavings and chips. I turned them both out of the class, and the rest were good as gold all the evening. It was well to have a test case, for there are several new boys.

Mark turned his first piece on the potter's wheel and amazed me, it was so good. He is charmed with pottery and would like to do it all the time. It is interesting to see the difference in his expression when he is at work modelling, wood-carving or drawing. It seems to bring out latent qualities, and I wish he could devote more time to it just as a matter of character development if for nothing more. Two men from Conche who have been here about a week on account of relatives at the hospital were greatly impressed by the potter's wheel, which they had never seen before. They have also been in the loom-room and said they should tell Conche people of the great work being done here, an encouraging reaction which may mean increased interest in the work and its possible establishment in other centers, which is just what we want.

February 8th Mr. Lindsay came in from camp this afternoon, weary and faint from hunger, having had nothing to eat that day but oatmeal, a cup of tea and a biscuit. He looked like a tramp and retired to his room with a jug of hot water, appearing at luncheon groomed like the gentleman he is. He entertained us all the evening with tales of his experiences with the Lapps and their attitude towards things in general. There seems to be some friction between the Lapps and the powers that be. They claim that their contract states they are here simply to train, tame and herd deer, nothing else, and one of them flatly refused to do some of the manual work connected with the camp. We looked up the contract and found it just as they stated. Camp duties are not their affair, and anyone can do them who wishes. Moreover, they train only a certain number of deer – six, I think – and those they keep entirely for their own use. Mr. Lindsay, Steve and John can train deer for themselves under their supervision but may not use those the Lapps have broken. They claim, moreover, that the deer should not be used for another month, as they are not yet recovered from the voyage or sufficiently acclimated. They are also indignant because three deer were brought in to St. Anthony to stay and have kept one of them in camp.

This being their frame of mind, they must be almost pleased that the two remaining deer have escaped. They are very gentle with them and fulfill their part of the contract to the letter. Mr. Lindsay told how all four men wandered about all day in search of one deer that had strayed from the herd and did not abandon the search until it was found and brought back. They start out from camp on skis to where the herd was grazing the day before and follow their tracks to their new location, then keep them in sight during the day. The snow is very deep – five or six feet on a level except on some barrens where the wind has blown it off.

If snow were the only covering of the ground there would be little difficulty, but it has rained at the freezing point, making a "glitter" that in some places is several inches thick over the moss and hard as heavy glass. When such places are found, the herd goes on to a more promising feeding ground and often eats the moss hanging from the trees (like our southern moss), called locally "moldow."

The Lapps are by no means the primitives we thought them at first. They have much intelligence and more business ability and system

William Lindsay tends the reindeer herd, 1908 (GHS)

than an average person. They read and write, keep accounts and insist on paying and being paid for everything. The Lapps insist that the pulkas and harnesses are theirs and should be paid for if used by the Mission, that the only things paid for are the little "Lapp dogs," as we call them, which are used to herd the deer. The women are clever with a needle, and the injured woman in the hospital is not only weaving on a tiny loom but knits as well and at present is making one of the tents in which they are to live – all this while sitting up in bed with a broken knee.

The nurse has given her a bright pink jacket, which pleases her color sense, and her husband comes often to sit by her bed, speaking their language, which to our ears sounds like grunts. She is very fond of singing and often sings in the night, a performance that seems to entertain the other patients so much they do not mind being kept awake by it.

Owing to some misunderstanding, the Lapps were told when leaving home that it was necessary to sell everything they had and, on reaching the ship, found they were to provide their own pulkas, harness and tents, which they were forced to buy at ruinous prices. They are now making some of the tents to live in before they can leave the bunk-house. David and L.M.K. have tried to make some of the harnesses, and more are now being made in the workshop. I watched them yesterday. The men had been in the woods to cut birch trees under Reuben's direction, choosing curved trunks or large branches and shaping them like the collars of the harnesses brought over by the Lapps. The harness is made of cowhide, but in broad bands that will not injure the deer.

Mr. Lindsay described the tent, which he shares with "Sunbeam" (Sundine) and Johnny. They had to dig to the ground where the snow is six feet deep. Only the peak of the tent must be visible. The guy ropes have to be attached to tree branches in some mysterious fashion. The tent is satisfactory in snowy weather but will leak during the rainy season, and something will have to be done before that time comes. There is a stove with a pipe running through a hole in the tent-top, which heats a circle about four feet in diameter. They draw their sleeping bags as near the stove as possible and say they sleep warmly.

Herding is often a test of endurance. Mr. Lindsay had to be brought in one day, thoroughly exhausted after tramping through snow from nine in the morning until six at night. He had tied himself with a rope to one of the Lapps to avoid being lost in the storm and thought his last day had come. He says such ordeals are spoiling his temper.

February 9th Yesterday someone reported hearing the *Portia* whistle – only imagination, of course, as she was given up for the season long ago and could not possibly get through the ice. We wish it were true, for we have had no mail for over a month. The mail-man came yesterday by komatik and dog-team but only with a few local letters, and none for us. We are growing very keen for letters.

February 12th It was half past one when I went to bed last night after an orgy of writing in the quiet house, unwise perhaps in view of next day's duties but an opportunity hard to resist.

The boys have nearly finished the text for the birdhouse. It is to be in place when Dr. Grenfell returns and will please him, for he plans to decorate the sides of several buildings with biblical texts carved and painted in the workshop. The boys are interested and were very good in class tonight. Perhaps the affair of Walter and Tom was a happy chance after all. Tom came tonight, smiling as if nothing had happened, but looked somewhat crestfallen when I told him I meant what I said the other night. It was funny to hear the other boys moralize about it afterwards. All of them are so enthusiastic they want to come every evening. Mark works persistently at the potter's wheel, hoping to evolve the form which he says he has "in his hands," waiting to come out. I think that boy will make something of himself.

Today Ted McNeill showed me the picture he had framed for Dr. Grenfell. It is a lovely print of waves breaking on rocks, full of action and quiet color. Ted thought it beautiful and could not understand why some of the men could "see nothing in it." I asked him what Mark thought about it and was not surprised to find he admired it too. I have hung it over the sideboard in the dining room where Dr. Grenfell can see it from his place at table.

Today Julia, one of the loom-room girls, mastered the intricacies of making a warp in one morning, much to my surprise; but joy was tempered by the discovery that she has little knowledge of reading and writing and none whatever of mathematics. For weaving, one must know something of the four elements [rudiments of learning], for much calculation is necessary. I suggested that Miss Keese could teach her in the evening with Phoebe and Eunice, and she is to begin next week. She has never been to school. It is pitiful to think of minds lying dormant and latent abilities undeveloped.

This was the afternoon reserved for matting (hooking mats), and it was a blow when only Mrs. Pelley appeared. When I asked where the others were, she said they all went to Miss Storr one afternoon a week for bible class and thought that was enough for engagements. It troubled me at first, but later I discovered a matting class had met last year at the hospital, and Sister Williams,[106] the nurse, had given them a cup of tea. I suspect that in my zeal for work I have not had a social function in mind, so when the girls have made several really lovely rugs, the women will be invited to see them in the solarium when it is warmer, and we will have tea there. We will see what happens.

February 16th At nine o'clock tonight I heard dogs yapping and a man's voice shouting the unmistakable komatik-dog language. It sounded like Dr. Grenfell. I could not believe it, for he was not expected until next week. Then I heard other voices, tramping of feet and the Doctor's cheery voice in the hall. He is home again. The next morning, I reported all that had happened while he was away.

He approves of the plan for the women's matting class and the encouragement of Mark, and he likes the picture where it is hung.

He took many photographs on the trip, which he developed last night. Unluckily, he had followed someone's suggestion to use alcohol to hasten the drying. Instead of trying the effect on one, he had trustfully dipped them all into the alcohol, and they turned green and shrivelled to about half their size. I appeared during the period of groaning and gnashing of teeth. It was a pity, for some of his films were among the best he has ever made. After dinner, we all helped to make prints from the good films. The dining room table was cleared: Dr. Grenfell was at one end with the bath, L.M.K. and I at the other with the lamp and an improvised screen consisting of a chair with a cloth over it. Between us sat David and Dr. Stewart, preparing the printing frames. Some of the first prints were failures, and no one knew the reason until Dr. Stewart discovered there was no bromide in the developer. It was a wonder our efforts were finally crowned by success.

I woke in the night hearing dripping from the eaves, and this morning a balmy breeze was blowing from the south and the snow rapidly melting. It was also raining, and poor Mr. Lindsay is in fear it will suddenly freeze and spoil the moss on the barrens where the deer are now feeding. It is quite dreadful that so much depends on weather.

We have a new centerpiece for the table. Dr. Grenfell brought with him three little bisque figures of the wise monkeys – "See no evil, hear no evil, speak no evil" – and during his absence we had used them on the table at meals. Yesterday he asked me for sealing wax. I could not supply it and wondered why he wanted it but found out this morning when he appeared at breakfast with the monkeys fastened with cement on a bit of wood – "On a rail," he said, and that in turn fastened on a piece of birch firewood with the bark on it. They were on the table ready for dinner when the Doctor took them away again and brought them back with a bit of ground-pine added, the decorations being a part of a dinner favor given him during his visit in the States. The "See no evil" and "Hear no evil" face one way, "Speak no evil" the other. We plan to turn that one towards anyone with gossiping tendencies.

The Doctor is troubled because congregational singing at church is so lifeless. The tunes still drag but are brisk compared to a year ago. Tonight, we chose Moody and Sankey hymns which the Doctor thought might have more "go." We need a leader, and the Doctor is no vocalist.[107]

When we left the church, the beautiful white world lay before us, a glorious moon over the frozen sea and great white hills. It gleamed like silver on places where the ice was bare of snow. The air was balmy as early spring, and nothing could have been lovelier. Dr. Grenfell walked with me, and both of us sank deep in soft snow when we accidentally left the hard path. The Doctor remarked that to fall when forsaking the narrow, crooked path after leaving church was an inconsistent thing to do.

The cheerful event of the day was Dr. Little's return. He came running beside the komatik in late afternoon, tired and a trifle thinner but in good spirits. He had many tales to tell of his three-hundred-mile trip. A trying part of it was a run of thirty miles along the shore, when there was a choice of going over broken ice, piled like great boulders with crevasses between them, caused by the rise and fall of the tides, or running over the tuckamore, covered with snow six feet deep, where komatik and dogs broke through frequently, dog traces became tangled and everything was generally mixed up. We shall look for an increase of patience and forbearance on the part of Dr. Little and Alf after such an ordeal. What is the use of a constant state of exasperation and repression for two days if it does not affect one's character?

Another qualified joy of the trip was sleeping accommodations. Fatigue had made Dr. Little indifferent to bare boards, little disguised by two pairs of blankets and his sleeping bag, which had at least a familiar odor all his own; but there was a time when it was necessary to sleep in a bed, and there was but one bed for him and Alf. It had an unfamiliar and untraceable odor, and this combined with close proximity was too much for the fastidious resident of Marlborough Street, Boston, who would not even allow his temporary family to tend him during his illness. In recounting this part of his adventures, we noticed a thoughtful expression on his face rarely seen there.

He was able to relieve many ills on the way and urged several patients to come to the hospital for treatment, as Dr. Grenfell had done on his part of the trip, but the people dread the long komatik ride across country in the cold. Dr. Grenfell said he thought perhaps it would give them courage to see him go on his way alone with only seven dogs, and perhaps it will, for to travel alone is hazardous. The weather is very uncertain, and in a blinding snowstorm it is impossible to see the path if there is one. Men are often lost. Other things may also happen: a fall or a broken leg would mean loss of life.

Dr. Little came through the reindeer camp and took tea with Mr. Lindsay, who still has a wounded deer as an inmate of the tent. This needs explanation. I forgot to say that Dr. Stewart, Miss Storr and Miss Kennedy visited the camp on Wednesday, taking "Jack" and "Jill," Dr. Grenfell's pet house dogs. The dogs, excited by the deer, stampeded the herd, and one of the deer was quite badly torn. They brought him into Mr. Lindsay's tent for care, an inconvenient intrusion. Moreover, the newcomer demands a light kept burning at night to enable him to partake of mossy refreshment when he so desires. In the dark, he might knock over things, become frightened and do himself an injury, to say nothing of injury to other inmates of the tent. Imagine having a nightmare come true and a whole live animal suddenly descend on one's chest, and think of the stove going over and utensils clattering about. It is certainly better to have a light.

February 19th A change of weather from gloomy snow clouds to an evening of beautiful color and still, cold air. This morning, we saw the arrival of a new patient who followed Dr. Little's advice to come to the hospital. He was brought over a hundred miles by dog-team across the bleak, snowy country. He lay in a box filled with blankets, fastened on the komatik. That is the third new patient received lately, and the hospital is nearly full.

Miss Keese had a party for her birthday with her girl pupils and the boys of my class as guests and the gramophone for entertainment. By eight o'clock our living room and hall were crammed, and it was fun to watch their enjoyment of the music, not to mention cocoa and cakes. L.M.K. was gorgeous in a pink necktie and many brooches, the gifts of her girls. When one considers how little money these people have, it is almost pathetic to receive gifts from them, and the articles represent a lovely spirit. I still cherish a celluloid box of brightest pink given me by one of the women.

This is a digression from today's events, which culminated this afternoon in attending Maude Biles' funeral and coasting down hill afterward – the second to counteract the first, for the funeral was awful. It was a bright day, but the wind was very keen, and the women were muffled in all the warm clothing they possessed. The procession started from the Biles' house across the harbor and trailed off in the direction of the church on the opposite shore. The coffin was a rude wooden box made by local carpenters. It was covered with bright pink cloth trimmed with narrow black spangled galloon [braid]. On the coffin were a few wreaths and bunches of bright paper flowers. Five young men drew it on a komatik to the cemetery. In the midst of beautiful gleaming whiteness on hill and bay, the procession was the only evidence of human life and the soft crunching of crisp snow under many feet the only sound to break the winter stillness. If we had turned back at the church door, the funeral would have left only an impression of peaceful solemnity, but the ceremony was depressing, filled with warnings and exhortations by the parson and his assistant from Griquet. They sang all the verses of two mournful hymns, and the gloom was increased by constant moaning, wailing and rocking to and fro of some women in the mourners' seats.

After the service, during which the bearers' hats rested on the coffin beside the paper flowers, the congregation walked to the cemetery, where deep snow had been shovelled to prepare the grave. In spite of bitter cold, another long service followed. Some of the more heroic men bared their heads, the others expressing their reverence by tilting their caps to windward. Thoroughly chilled and depressed, we turned away, feeling ready for anything to divert our thoughts, and when we saw Dr. Grenfell start out to shoot target ducks we followed, then coasted and tobogganed with Archie and Jimmie Green the rest of the afternoon. The fine air and exercise blew all the cobwebs away. While waiting for the komatik, I sat in the snow and watched the sunset light grow rosy on Old Man's Neck and violet shadows creep across the harbor. It reminded me of something vaguely delightful: an elusive association of pleasure, deeper than the realization of beautiful coloring which in itself sinks into one's soul and gives one peace.

February 20th The mail came yesterday and with it my article about the reindeer in the *Boston [Evening] Transcript.*[108] Dr. Grenfell asked for a copy. It is to be used in a pamphlet to be sent to all subscribers to the Reindeer Fund, and he wants a photograph of the ship and one of the herd to print with the pamphlet.

February 21st This has been an afternoon to remember. The weather was perfect. After luncheon, Dr. Grenfell said I needed air and exercise to cure the headaches I have had lately, and he intended to take me for a walk. We started off, he on skis and I on snowshoes. We walked directly across country towards the "Head." It is not often one can fix one's eyes on a definite objective and travel directly towards it. (If we could only do that in real life without meeting so many obstacles and cross-roads without sign-posts!) The snowdrifts are soft around the tops of small trees, and we floundered in up to our waists. In other places, the "glitter" was swept clear by the wind. When we reached the summit of the hill the sea lay before us, a gleaming expanse of ice delicately colored in the afternoon sun with the Grey Islands a violet silhouette in the distance.

Taking off our snow-shoes and skis, we walked cautiously on patches of rock bared by the wind to the verge of the precipice. It was a wonderful sight. Four hundred feet below us a wide belt of open water lay near the rocks absolutely still and black, and in it floated small blocks of ice that from the height looked like lily pads. Snow and ice trailed across the steep sides of the cliff and softened its grim surface, while from a depression in the hill-top a stream had flowed and frozen in great greenish-yellow masses to the water's edge.

Behind us stretched the great white hills and the scattered village that meant home for us temporarily. The sun was warm and the air so soft it did not seem possible it could be mid-February in northern Newfoundland. On the way down, the Doctor left me far behind, for he was on skis and went flying over the snowdrift, sometimes with a tumble at the foot, where he waited for me while I clumped after him. Afterwards, we climbed the hill above the orphanage where the Doctor wants to build a teahouse. It is a natural lookout, and from it one can see far across wild country and the icy sea. Dr. Grenfell read his article on "Faith"[109] to L.M.K. and me this evening, a happy ending to a memorable afternoon. I wish more days were like this. The Doctor says they do one more good than medicine, and I am happy to take such sugar-coated pills.

On reaching home, we were amazed to find Mr. Cole. He has been on the road from Grand Falls since the 25th of January, covering 250 miles mostly on foot as the dogs had to drag the luggage. He looks well and is already tanned by the spring sun. He starts back with the deer and two Lapp families on Wednesday, taking Mr. Sundine.

February 22nd This has been, if possible, a happier day than yesterday. I shall be spoiled if it continues. At breakfast, Dr. Little announced that my physicians had again decided I needed fresh air and exercise and that at twelve o'clock Dr. Grenfell and I on one komatik and Dr. Little and Miss Kennedy on another would start for Goose Cove for the afternoon. The Doctor decided at the last moment that we would take lunch, and I rushed to gather supplies and have them ready before the dogs were put in.

The trail to Goose Cove was very beautiful, winding in and out among great spurs of rock, up hill and down, with glimpses of frozen sea and the far country and hills. It was great fun to fairly fly down steep places, the dogs galloping to keep ahead of the komatik and Dr. Grenfell sometimes trailing his legs in the snow to act as a brake, while I simply "hung on." At one place we missed the trail, which was covered by a deep fresh snowfall, and even the dogs went through. It looked for a while as if we were stalled in the tuckamore.

At Goose Cove, we went to the Noels' house to eat our lunch and have a cup of tea. The Noels are kindly, hospitable people and received us with simple courtesy. There was a fish net under way, and we learned how to make the knot, marvelling at the rapidity of Seth Noel's achievement in that line. The men had been sealing, one of the native occupations at this season, and they took us to the fish house to see the rows of dead seals, some of them ghastly and gory in their skinless state. There were also a number of little unborn seals, which are known as "whitecoats." They are uncanny little things, and their white fur is much more attractive than their "ensemble." Dr. Grenfell bought four of them. They were lashed to the komatik, and he used them as a seat until they began to thaw and he was unpleasantly conscious of dampness in his nether garments. We took another road home across the ice, around the point. It was a lovely afternoon.

February 24th A glorious day again. After a busy morning, Dr. Grenfell, Mr. Cole and I started on snow-shoes to find the wood path the Doctor wants to use for hauling logs to build the cabin on the hillside nearby. We wandered about until we found snowshoe tracks and followed them to the top of a hill, where a scraggly tree stood alone, too wretched to tempt the woodsman's axe. Dr. Grenfell said we should plant a banner on the height and produced his red cotton handkerchief, which he hoisted to the top after striking off the small branches to increase the impression of a flagpole. Then he and Mr. Cole struck attitudes while the banner, twelve inches square, was flung to the breeze. We felt and acted like small children. On the way home we flushed a grouse or ptarmigan, and we heard a chickadee sing. We found the trodden path and followed it to the harbor. In our gay mood I photographed Dr. Grenfell and Mr. Cole as the "St. Anthony Minstrels," the Doctor playing on his staff for a flute and Mr. Cole on his snow-shoe for a banjo, with St. Anthony and the distant sea for a background.

February 26th This was the day appointed for the doctors to take the ladies to see and photograph the official round-up of the reindeer, but through some misunderstanding Alf had gone to the woods for logs and taken the dogs, so Dr. Grenfell told me to start and he would overtake me. He soon appeared with a deer and the small komatik, and I discovered that the thrill of riding behind a deer is not always associated with speed. The deer ambled now and then when urged but took a lukewarm interest in travelling faster than a slow walk. This indifference gave Dr. Grenfell, accustomed to a rushing start with the dogs, a painful surprise: that an animal born into the world a deer should so mistake his use in life.

We tried to help by pushing along in the soft snow with our feet, as if paddling or rowing a boat. I was reminded of Frank Stockton's story[110] of Mrs. Lecks and Mrs. Aleshine as they "swept along" with a broom through the water after their famous shipwreck. We tried every known call and gesture, and some unknown. The Doctor used all the unintelligible words for driving dogs and invented others, even resorting to sounds used in driving horses. He waved his arms like a windmill and ran alongside the komatik, shaking the one rein.

This had some effect, but the deer resumed his slow pace when he felt two people on the komatik again.

The deer had other mannerisms. He occasionally stopped, turned and faced us, opened his mouth and gasped. He put a hind leg over the one trace, which is fastened under the belly and should be between the hind legs. When the Doctor tried to adjust it, the deer resented interference and struck with his fore-feet. Only once did he voluntarily quicken his pace, when he spied a patch of moss laid bare by the wind. He made a dash for it, and there we sat and waited while he feasted. The climax arrived when the Doctor allowed his feelings to overcome him and struck the deer sharply with the leading rope. The deer sprang forward, snapping the slight harness. While it was being repaired, the cause of the trouble stood with eyes closed, apparently sleeping. After this, we resigned ourselves to slow movement. The Doctor evinced remarkable patience, and with his characteristic optimism and allowance for others' shortcomings remarked that the deer "was only a little fellow after all," and the whole situation convulsed us with laughter.

We found Mr. Lindsay clad in his new Lapp garments of deerskin, gay with red and yellow trimming. He even wore Lapp boots tied with the colored woven bands, but his cap was not in keeping – it lacked the four horns. The camp is situated on high land with hills rising on two sides and to the south the ice-covered sea. The Lapp tents are made like Indian wigwams. Material something like a Navajo blanket, very soft and thick, is stretched around poles, leaving an opening in the top through which smoke ascends from the wood fire in the center of the tent. I looked inside and found the beds were piles of deerskins on the ground, and boxes of household and personal belongings served as seats and tables. It was a very warm, snug place, shared by the little herding dogs belonging to the household. Outside, one of the men had brought in the carcass of a seal which he was chopping with a hatchet, and on a frame of poles other pieces of seals' anatomy were drying in the sun. Beside it, several articles of gay clothing fluttered on the family clothesline. Around the camp, the three hundred reindeer were grazing on moss and the moldow found on trees in the locality. Presently, they began to mount the hill to the barrens above the camp. I shall always remember the sight of those delicate creatures filing singly or in small groups up the trail to the top, where they stood outlined against the sky.

Our return trip was the same as far as speed was concerned, but we were more resigned, perhaps because we ceased to expect the impossible. It was fun, and I would not have missed it for anything.

February 27th Mr. Cole and Mr. Sundine came in last night for a final glance at St. Anthony. They both left us today and started on their journey with the deer.[111] It is an awful day, a "mild" of the worst kind. Travelling will be difficult if the snow melts, for it is six feet deep on a level and drifts are deeper. It is hard to realize we shall probably never see them again. Time has little to do with the growth of acquaintance and friendship in this isolated place.

February 28th I have again turned barber. Dr. Little has bowed to the inevitable and the wishes of the family, and today I trimmed his beard. This was in preparation for his impersonation of King Edward [VII] at Dr. Grenfell's birthday dinner tonight. It occurred to us that it might be amusing to bring together at the party some of the personages he had met or might meet. We were a strange and motley crew. Dr. Little, gorgeous and extremely good-looking with a cardboard crown on his head, bejewelled with gilt ornaments from curtain fixtures. His chamois jacket was turned inside out with a ribbon stretched across its red lining, decked with everything in the nature of orders we could find. My fur-lined coat, with fur side out, fell from his shoulders, and he was otherwise unfamiliar in a white starched shirt, high collar and white cuffs with gold cuff buttons. They held a suggestion of the gay world we used to know. I could scarcely take my eyes from them.

Dr. Stewart appeared as Sir William MacGregor, distinguished from ordinary citizens by a steamer rug draped over one shoulder and also disguised by white collar and cuffs and a "boiled shirt." David was Booker Washington[112] – that is, his face was black in streaks and his fair hair covered with a black stocking, but it was a Booker Washington who might have been end man for a minstrel show. L.M.K. was our old friend Miss G., with her ready notebook and camera, in quest of information, and Miss Kennedy, who was too busy to come in costume, was labelled "Miss Macomber"[113] to conceal her identity. I was Mother Eddy,[114] disguised in the one good gown I brought with me, and Miss Storr was the star of the evening, impersonating Mrs. Dane of Grady, Labrador, who is wont to come on board the *Strathcona* during its stop in her neighborhood and pester Dr. Grenfell's life. I have never had the doubtful pleasure of meeting the original but judge from Miss Storr's impersonation that her ailments are many and varied, that her many children are also sufferers, and that she demands speedy cures in the form of medicine for all of them – also that she has but a hazy idea of the laws of thine and mine. She must be a bulky lady, for Miss Storr's ample proportions were increased by two sweaters, another coat and a long cloak with baggy sleeves. She used them as receptacles for medicine bottles to be filled, family silver (purloined from the table) and bits of food "for the children," while her remarks brought down the house. She kept up the characterization until coffee time.

Dr. Grenfell was completely surprised: so astonished that he was quite dumb at first and I was afraid he thought it silly, but there was no doubt of his enjoyment and appreciation. We had the best dinner obtainable, a six-course dinner with real roast beef. Dr. Little wrote a poem and sang it to the tune of "Mr. Dooley"[115] with banjo accompaniment. I add it here:

1

The day had been a long one, the hour was getting late.
Sleepiness stole o'er him. Thought he, "I'll sleep till eight,"
When in the dark a step was heard. Just as he got in bed,
A husky Labradorman appeared below, and said,

"Oh! Dr. Grenfell, my wife, she ain't been well.
The children's got the scarletina too.
My sister's kids is bad: such stomachs they have had!
You'll only have to walk a mile or two."

2

There were forty-two bad stomachs and twenty-six bad heads,
And thirty-seven children in two or three small beds.
There were swellings that had "pitched down" and backs they seemed to "find,"[116]
And variegated ailments of non-recogniz'ed kind.

3

He tackles tumors, tears out teeth, tests stomachs – undertakes –
And treats all human ailments at the very cheapest rates.
When he's not cutting people up, he cuts all sorts of capers
And spends his leisure moments in writing for the papers.

Oh! Dr. Grenfell, Oh! Dr. Grenfell,
Why don't you take a little rest?
At the rate you're going, there surely is no knowing
If flesh and blood can always stand the test.

4

A place that's called St. Anthony was fading from the map.
But that was sixteen years ago: St. Anthony's come back.

There's storehouses and club houses and Guest Houses, I hear.
And everything is going fast except the new reindeer.

Oh! Dr. Grenfell, Oh! Dr. Grenfell,
In honor of your birthday is this feast.
You are an inspiration to the population,
And we hope you'll live a hundred years at least.

February 29th The first excitement of the day was the appearance of Dr. Little at breakfast with a clean shave. Every hair on his face had vanished. I am reminded of his remark some time after his arrival that under his disguise he was a nice Boston gentleman. He looks like quite another person, and I feel shy, as if he were a stranger and I need to make his acquaintance all over again.

Another excitement developed today, an unpleasant one. Walter, one of the orphanage children, was found to have taken money from Mr. Jones at the co-operative store, over forty dollars. Everyone has been dreadfully upset, especially because Elisha, a bright, apparently frank little fellow brought from Labrador by the Doctor last summer and a favorite with everyone, was also implicated. It seemed unbelievable.

Dr. Grenfell called a number of witnesses, and a trial was held during the evening. On returning from my class, I found the dining room still a court-room and went to the study. Questions and answers could be heard. Then there was a general movement, and Dr. Grenfell came in looking very stern and said he was going for Elisha, who was proven guilty. Another suspense, then the Doctor reappeared looking pained and angry. He was followed by Elisha, looking as subdued and uncertain of himself as Elisha could look. L.M.K. had meanwhile come in, and the Doctor called Dr. Little and David, saying that if the ladies would retire he would administer justice. We fled, almost as scared as Elisha must have been, expecting to hear sounds, but nothing broke the stillness, and we do not yet know whether something was preventing the caning or Elisha was a stoic. It is all very confusing and baffling. Everyone is much upset but especially Dr. Grenfell, who is inclined to think of the affair as a failure for which he is somehow responsible. We are especially troubled about Elisha. Our rating of Walter has never been high.

March 1st We have had a death in the family. The Doctor's little black dog Jill, the companion of Jack, died yesterday. A few days ago the Doctor appeared with her in his arms and took her to the hospital, but he could not save her. We all feel dreadfully, Dr. Grenfell as if he had lost a member of the family and a faithful friend. She was a great pet and one of the most intelligent and affectionate dogs I ever saw.

March 2nd The affair of the thieving boys is still unsettled, and the duties of a magistrate bear heavily on the Doctor. He is appointed magistrate for the coast by the Governor of Newfoundland, and there are no other officials with whom to share his responsibilities. He is authorized to call meetings of the people for discussions of local affairs and controversies, hear testimony in cases of wrong-doing, act as judge as well as justice of the peace and exact penalties according to his judgement. He is baffled by conflicting evidence in this case, but it seems that Walter was the real culprit. Elisha, younger and more guileless, became involved through ignorance, but he seemed unable to tell the truth and Dr. Grenfell, after trying to make him give a reasonable statement, at last was actually amused by him. The Doctor asked me to be present at the second trial, and I wish I could remember all the arguments, questions and answers. The child apparently has no moral sense, perhaps the result of influence and environment before Dr. Grenfell brought him to the orphanage. He seemed utterly unconscious of his fault. It ended by Elisha's being returned to the kindly supervision of Miss Storr. Walter was given a flogging and will be sent to camp to make himself generally useful under Mr. Lindsay's supervision.

March 6th Dr. Grenfell labored with Walter until late last night without satisfactory result. This morning, he asked Dr. Little and me to take him to the deer camp. Walter carried his own nunny-bag on his back, Dr. Little having tested its weight to be sure it was not too heavy for the boy.

Mr. Lindsay's tent is located in a snow bank among small trees at the foot of a hill. A stream flows by, but the only indication of it is a hole about four feet deep where water can be seen gurgling at the bottom. The rest of the stream is many feet under ice and snow. This is the water supply.

Alf had a narrow escape a few days ago. He came from across the bight with a load of wood on the "cat" [catamaran] (sled for hauling) and to avoid climbing Old Man's Neck went on the flat ice around the point. When he was nearly at the mouth of the harbor the ice gave way, and he and the dogs went through, but not the komatik. I do not know by what means they escaped. No one seems to think of the adventure as especially remarkable. Alf himself accepts the mishap as a part of the day's work. Death is never very far off down here.

L.M.K., Archie Ash and I went on an all-day tramp and skating expedition today. As we neared the pond, we saw a "cat" waiting with a log on it and heard someone call. It was David, waiting for the two boys who had gone to the camp for logs, which they bring down the narrow path and leave on the edge of the frozen pond to be transported by deer to St. Anthony. The woods are about eight miles from the hospital, and a gang of woodsmen are there cutting. They have a small camp and stay through the week. The boys drag logs from the camp to the pond by hand through the narrow wood path; then two deer help to haul them across the pond to the hospital. They have much wood to get out before snow melts. Everyone is hurrying to gather his year's supply, and one sees komatiks and dog-teams running in all directions. The woods around St. Anthony are denuded of large trees. Only small spruces, alders and birches are left.

The Newfoundland woodpile is a different affair from ours. Instead of the logs being cut uniformly and piled neatly along a stone wall, as we see them in New England, the trunks are trimmed of small branches and stacked together, small end up. The pile resembles a wigwam. In this way, they shed water and when needed can be pulled from the deep snow that drifts around them and would bury them if they were piled horizontally. They looked very strange to me at first, but like all the curious ways of this place have become so familiar I seem to have always known them.

March 8th St. Anthony is susceptible to epidemics. This time it is a happy one and takes the form of conversion. We heard last night that Mr. Pelley is conducting a revival. Two meetings have been held, and several young people have come forward to declare their wish to lead a better life. Among them was Kate, our cook, and this morning her expression was very calm and sweet. Dr. Grenfell spoke of the revival to the maids at prayers and when he preached this morning mentioned it to the congregation, suggesting ways of applying their new resolution to everyday life. Mr. Pelley is such a sincere, fine old man that a step of this kind on his part is of more value than if a stranger had made the appeal or taken the lead.

March 9th I have given the Doctor two new red handkerchiefs for flags. He announced at lunch that they needed planting and invited us to walk with him to distant hills for the ceremony, but the loom-room was full and there were other pressing duties. The flag-raising will have to wait.

Among urgent matters, badges for the sports headed the list. Mr. Pelley said he wanted something on them besides lettering, some emblem different from last year, and this being a reindeer year, I suggested a reindeer head. He was charmed when I drew the design, but

Dr. Grenfell did not like my reindeer horns, remarking that they looked like bushes, and he proceeded to draw the right kind, Mr. Lindsay assisting as a connoisseur.

It is a terrible day. The storm that began during the night by noon became a blizzard. We did not think L.M.K. would be able to return from school, but while we were at luncheon she appeared with a wild tale of crossing the harbor after sending Davey for Mark to come with the komatik to help her bring the children home. She protected them with rugs as much as possible, and none of them had frostbites, but Mark froze both ears. People speak quite casually of having a nose, ear or cheek "touched." The victim does not realize he is frozen, for the part affected simply turns white, and there is no sensation until afterward, when there is a great deal. In severe weather, everyone watches his neighbor's face and reports the appearance of tell-tale spots.

March 11th A howling blizzard and no one has been outside the house, not even the men. The day has brought together more intimately our small, congenial family surrounded by its peaceful, homelike atmosphere. Toward evening the storm abated, allowing the hospital and orphanage staffs to join us for our weekly dinner at the Guest House. Mr. Lindsay was here from camp, still storm-bound, and this was the first time our entire staff of nine had been free to meet. It was a merry meal, everyone in high spirits, one cause of our gaiety being a list of the staff's characteristic utterances posted on the bulletin corner of the dining room wall. I add them for reference:

Dr. Grenfell: "Come on. Let's be off!"
Dr. Little: "Where is my pipe?"
Mr. Lindsay: "It's really most extraordinary!" (Expressing amazement at certain manners, customs and limitations of people and things.)
Miss Kennedy: "I must go and take the temperatures."
Miss Keese: "That boy is *such* a dear!"
Mr. Lee: "I think I'll go and begin to gather moss." (Indicating the ruminant character of the youth and how long he considers a step before actually taking it, also how often thinking about it is the main issue.)
Miss Storr: "Give me air!" (Suggestive of shivering days at the orphanage.)
Dr. Stewart: "Yes?!!" (His manner of saying "What is your business with me? Hurry up and state it!")

The kindly, tolerant spirit and sense of humor of our little family perhaps explains our happy winter, so free from friction and misunderstanding. Everyone contributes to the notices of protest or appeal posted on the dining-room wall and everyone responds. Our "Chronicle" usually contains subject matter of a highly personal character: one's foibles and idiosyncrasies are laid bare, and replies in kind are expected and enjoyed. These activities may be called the lighter side of missionary life, and monotony is unknown.

March 12th The first day of the annual sports. The two doctors, L.M.K. and I donned our gay blanket suits and sallied forth at nine o'clock. It was bright and sunny but bitterly cold and the wind blowing a gale. Yesterday's storm had delayed preparations, and we found Dr. Grenfell preparing the firing line. There was a "running deer" for a target. I found Dr. Grenfell on the floor painting the head a few days ago, and the men of the shop provided the body. The legs looked as if the deer were stiff with rheumatism, and the members of the herd would have disowned him. When some small antlers were bound on the head, we recognized him as a deer. Otherwise, we might have thought him a horse with trousers. The "running" was managed by fastening him to a small komatik which men on either side of the range pulled back and forth by a line. The wind was so fierce the firing line retreated to a sheltering bank and fired across the ice.

The Doctor had told the men casually that those who came from a distance would be fed, and we planned to use the bunk-house, but the response to the invitation was so overwhelming that we moved to the Guest House piazza, where the men sat on everything available: boxes, chairs, window sills, on the floor. I stumbled over their feet on trips between the kitchen and their plates, which were always empty. When the guests were "full up," to use a local phrase, we realized our own lunch had been overlooked, and Kate prepared it under difficulties, for men still filled the kitchen, silently watching the five maids while they tried to perform their regular duties.

During our meal, Dr. Grenfell saw people entering the loom-room and said someone should be there to show them around. I hurried out to conduct a silent group while explaining and enlarging on the benefits to be derived from learning and practising crafts in hope of stimulating interest in those from other settlements.

There were other races in the afternoon but few participated, for the wind was bitterly cold. After more tea, the men stood and sat about, not knowing what to do with themselves, and Dr. Grenfell entertained them by performing wonderful stunts with Indian clubs. He then demonstrated what was possible in the use of boxing gloves, and after he had received a black eye from Dr. Little and had given

Four maids employed by the Grenfell Mission at St Anthony, August 1908 (JLP)

Dr. Little a sprained thumb, to live up to the sporting character of the day, the party broke up and we had our own dinner with Father Thibault,[117] the Catholic priest from Conche, and the Reverend Mr. Richards from Flower's Cove as guests.

March 13th Blinding snow and a gale of wind this morning prevented everyone from venturing out for the second day of the sports, but at noon the storm abated and people began to gather for the remainder of the program. There was a sack fight. Six men tied in sacks butted each other with their shoulders until each in turn was knocked down. The contest was confined to men and large boys, but one sporty small boy entered and hopped about like a sparrow, butting the men, who ignored him as too small to be noticed. The sack fight culminated in a contest of doctors against clergy, and the sight of Parson B.[118] and Father Thibault as fighters was worth the show. Dr. Grenfell and Dr. Little were funny enough, but one could more readily imagine their participation than the parsons, who put up a sturdy fight. Dr. Stewart carried the day for the medicos, all the others being literally laid out by his superior avoirdupois. The sports ended with a tug of war – the Mission against the natives – the Mission having no chance against the hauling ability of native fishermen.

By late afternoon everyone had gone home, one komatik after another starting off across the ice, the excited dogs eager to be off. Quiet has descended on St. Anthony, and only our house guests, Mr. Richards and Father Thibault, are still with us. There was a Church of England service across the harbor this evening, Mr. Richards being here to officiate, and several of our household walked over. It was a very simple service, more impressive under such primitive conditions in so small a place. I think it gives one a deeper realization of the actual fellowship of Christians than a larger and more formal service and congregation usually does. Perhaps it accentuates the awareness of life's realities, a feeling that tinges all one's experiences here.

March 16th The Doctor has been busy preparing manuscript before starting with Dr. Little tomorrow on their last medical trip of the season. I have helped him most of the day, and he has read aloud extracts from his "Faith" articles and his "Log."[119] He writes for a while, then stops to play with Jack or stuff a puppy skin, then back to work again.

Apropos of puppy skins, there is sometimes a surplus of puppies and those unwanted are dispatched as soon as born, and the Doctor has been stuffing the skins to make a team for a small model komatik. He has already stuffed several, assisted by Reuben, who also practises taxidermy. They have made harnesses for the pups decorated with bits of gay yarn, and the komatik team when finished will be an interesting addition to the curios in the solarium.

March 17th The doctors left this morning, supplied with pork buns, coffee in thermos bottles, and a few luxuries. ("Pork buns" may need a word of explanation. Pork is used as shortening and prevents the bread from freezing, as it would if only water were used to mix the ingredients. With the addition of molasses, they are very good and nourishing.)

The doctors went alone without Alf, each with a komatik and seven dogs. We are now so accustomed to "starts" that this one did not cause the excitement of their first trip several weeks ago.

Mr. Lindsay came in this afternoon and will stay a few days to nurse a cold. Dr. Stewart has dosed him, and I have tried to make him happy with hot-water bottles, lemonade and other comforts. It is fun to take care of people.

March 18th Dr. Grenfell left unfinished work on his desk and asked my help during his absence. Today, I have proofed his "Faith" manuscript and answered letters that needed attention. These are busy days. Ted [McNeill] has just finished the new loom. It is a fine piece of work.

It seems quite as good as those sent from the States, which he used as models, and more interesting because in the absence of iron fittings, he has made them of whalebone or hard-wood, quite appropriate in this environment. It is encouraging to know looms can now be made locally to supply the hoped-for demand for extension of the industry.

March 23rd Mr. Lindsay is still here, for which we are thankful, for it is lonely without our doctors. The deer have been a source of anxiety this week. They have hauled logs every day, and Mr. Lindsay is sure they are not fit to do it. Some new boys have been taken on as apprentices. They do not understand the deer and cannot harness the wild ones without throwing them. Mr. Lindsay was wild when he found two boys sitting on a deer's head and explained to them that different animals required different treatment. The worst thing one could do to deer was to sit on their heads because it choked them. The boys seemed chastened, but today a deer with an injured head was brought in and taken to the stable to be cared for. The boys said, without further explanation, that something had struck it. Another problem is the proper tethering of deer brought in from camp for use. The apprentices have little judgement, and it is impossible to supervise them continually. Altogether, we wish the doctors would come home.

In the shop, things are more encouraging. The text for the bird-house is finished, and the boys are really beginning to carve small articles. They grew quite excited over my book of designs based on objects familiar to them: fish, birds, crabs, flowers, boats, waves, etc. Their recognition of the motifs indicated imagination, and the local names for some of them were so queer I made a note of them. I will try to get a blackboard and encourage them to make their own designs.

From the shop this afternoon I saw the goats wandering down from their little hut on the hill. Reuben says they come down every day for their food. I said to Archie that I thought it remarkable the dogs had left them unmolested, and the words had scarcely passed my lips when there was a great uproar outside, and the men rushed out to find the dogs making an attack. It was the goats' supper time. Their food was being cooked in the shop, and they were all about the door waiting for it. There was no fatality, but it was a narrow escape and one feels insecure about the poor beasts.

March 24th Mr. Lindsay left us today, having as he said no shadow of an excuse for staying longer. He had searched the house for odd jobs or things to mend and put in order. He started after luncheon and had been gone only a half-hour when the doctors returned, to everyone's surprise. They had had wonderful weather, fine hard roads, and made good time. It was fortunate Mr. Lindsay helped me put the text on the birdhouse yesterday. The Doctor saw it and asked when the next one would be ready. I told him "soon." It is sometimes the thing not done that he notices.

March 25th Several weeks ago, Dr. Stewart announced that he would like to go home to Scotland to be married. It was quite startling, although we knew he has had matrimony in mind. We have noticed with interest the photograph on his desk. His plan is to go as soon as possible in order to be back by June first, when Dr. Grenfell and Dr. Little have to go to St. John's for the *Strathcona*. After their return, Dr. Stewart and his wife will go to Indian Harbour on the Labrador for the summer and Dr. Wakefield, an English friend of Dr. Grenfell's, will be at the Guest House as resident. It seems odd to be making plans for the summer as if the winter were over. Every mail brings announcements of people who are coming to help the Mission during the summer. According to the latest statistics, I think we shall have to arrange a number of stout pegs in the solarium for visitors' use. I fail to see where there will be even floor space for them all. Bowring Bros. Steamship Company should turn over a tidy commission to the Mission for the increased sale of passenger fares.

The first to arrive will be a nurse from Johns Hopkins hospital. A Mr. Kohl is due to help make drains, build roads and turn the tuckamore into a fertile garden and pastures. The Doctor's cousin from Montreal and Mr. Kohl had planned to arrive on snowshoes the last of the winter but may not come until June. Sometime in May Dr. Wakefield will arrive. He will stay at this house. The first of July a Mrs. [Martha] Sayre of Toronto [Bethlehem, Pennsylvania] and her two sons, the elder a professor [divinity student] at Williams College [Union Theological Seminary], are coming. Added to these will be squads from Harvard, Yale, Princeton, Columbia and various boys' schools. There are visits from Governor and Lady MacGregor and others from St. John's in prospect. Our dear family will be scattered, Miss Keese and I only remaining to face new problems which are sure to arise. I fancy myself singing about the middle of July,

> The house is running over; we do not like them all.
> We sometimes feel we're surely in a fix.
> Oh! give us – do give us – those joyous days of yore
> When this family of ours numbered six!

I have made a mental disposition of Dr. Grenfell's guests as to beds. The rest will have to bring tents. This country will surely blossom like the rose. I can see vegetable gardens, green pastures for cows, sheep and goats. There may even be a lawn – even a pergola! Who knows? The agricultural possibilities are great. Privately, if lettuce, spinach, radishes and potatoes can be persuaded to grow, I shall feel happy and actually joyful if there can be a few plants with blossoms. The bulbs I brought from St. John's and kept in a drawer felt the urge of spring and even dared put forth tiny shoots, but froze when planted in the solarium. I shall have to wait until it is warmer before giving the others a chance to follow their natural instincts.

March 27th As I left for the shop this evening, a great yellow banner waved overhead in the deep blue sky. It was the beginning of an aurora, again unlike what I have seen. Sunset light still lingered behind the western hills, and the wide white surface of the country was tinted with pink and violet. Later, Dr. Stewart came for me to see the sky. It was a mass of quivering, shifting light, sometimes concentrating in the arc with long brilliant streamers to the zenith, which we associate with arctic regions, then a great mass of color intensified in another part of the heavens quivered and changed to other forms. At bed-time it was still streaming upward, taking the form of illuminated smoke as from some great fire beyond the hills.

March 28th There are cases of serious illness at the hospital, and Dr. Little is greatly depressed. He had a bad operative case this morning. I telephoned him at Dr. Grenfell's suggestion to ask him to take Miss Keese and me out for a komatik run this afternoon. We planned to visit the logging camp, and he was to meet us on foot. Alf had the best dogs, leaving us a scrub team with the exception of Brin, the leader. Some of them were lame, and we felt we must get off and help them all we could. It was a comfort to turn back. When we stopped to rest, most of the dogs lay down, and when we were about to start Brin seemed to know it and sprang at each dog with a short snap and bark, as much as to say, "Come on, wake up and bestir yourself." We dreaded the long steep hill on the way back, not only steep but crooked, with trees to be avoided. But I knew Dr. Little would take us down safely if anyone could and moreover would be careful of the dogs, which some people are not.

The treatment of dogs is the only cloud in my sunny sky. They work so hard, are poorly fed and frequently beaten. Sometimes they deserve it, but often it is because they are not pulling the load fast enough. The poor creatures are driven away from the back door when they come to find scraps of food or drag empty tins from the refuse barrel in hope of finding a bit of fat. When slops from the house are emptied in a deep hole in the snow some distance from the house, the dogs gather from every direction. They almost disappear into the hole in an effort to chase the refuse to the bottom, and one sees only a waving tail and a bit of hind-quarters appearing above the snow, while others look on with envy, awaiting their turn. Some time ago, they broke into the back porch, even breaking a window, and we now have bars on the outside to prevent further burglary.

March 29th After lunch, Dr. Grenfell announced we were going on snowshoes up the hill to feed rats to the foxes. The Doctor pays Beatty Simms five cents for every rat he catches on this side of the harbor. Beatty has become quite professional. This morning he brought four. On the way, we stopped to feed one of the rats to "Pete" the eagle. He is very funny when he chortles, screams and flaps his wings at the sight of food. I had not seen him eat before and wanted to know how he did it, but his manner of attacking the rat's head after having it firmly in his claws made me uncomfortable, though the rat was stone dead.

We went on up the hill to the flag and found it in good condition considering the gales since its raising. The Doctor wanted to select sites for possible houses, but it was difficult to choose. There are so many with lovely views across the harbor, all of them presenting the problem of approach up a steep hill. We decided upon one nearly behind the orphanage which had the view, the shelter of the hill from the north, full sunlight and an easier approach than the others. I wonder whose house will be built there?[120] Dr. Stewart's is to be on the way to the peat bog, behind the Guest House.

March 30th Dr. Stewart left us today. We wanted to photograph the staff yesterday but refrained out of respect for local prejudice against such indulgence on Sunday.

Dr. Little brought some fishermen's boots to throw after the prospective bride-groom as he started off, and we all threw rice. Alf is to take him with the Doctor's komatik to Port Saunders, on the West Coast, and he is to find his way from there as best he can. I believe this ninety miles of the trip is the most uncertain, for there are no settlements by the way where he can find food and dog-teams. He will travel 250 miles to the railroad, thence to St. John's, unless he finds better time can be made by sailing from Boston or New York on a faster ship. With good connections, he will be about three weeks on the way and have two or three weeks in Scotland. That girl should appreciate such a journey for her sake. She probably does.

Mr. Pelley came to me in the shop today and I knew he wanted something. It was only "a little piece of work," he said, about thirty stars for the boys to wear on their caps when they parade in the Orangemen's Walk[121] on Thursday. We decided on silver stars: i.e., pasteboard covered with tinfoil. Mrs. Jackson also approached me with a request for our maids' services in helping wait on table, and Reuben asked for food which everyone is to donate for the entertainment. I have promised these and also a cake and feel quite popular. Everyone seems to feel the spirit of festivity. Think of being in a place where one celebration upsets the social equilibrium of the whole village for a week. My boys were so restless at class tonight they were practically useless, and L.M.K. says the school's attendance is small, owing to assistance in home baking or some activity connected with the great occasion, for it is the great occasion of the year. We are all going, of course.

March 31st A glorious, cloudless day without wind. The prevailing restlessness made concentration difficult this morning, and after lunch I started off alone for a walk, taking the path to the logging camp. The day was so lovely I went on and on but after walking nearly the length of Long Pond felt tired and lay flat on the ice, outstretched with my face to the sky. The stillness was wonderful – no sound except now and then a faint murmur of spruce branches on the bank. The sun was so warm it was hard to realize I was lying on a snow- and ice-covered pond.

The boys have been told there will be no more classes until Monday, when I hope they will be calmer. L.M.K. has no school tomorrow. The schoolhouse has to be cleaned for service during the Orangemen's Walk. Dr. Grenfell is planning an all-day picnic tomorrow and we hope for fine weather. There will be few more opportunities.

April 1st No picnic. It is a wretched day, but in spite of it the Doctor went to the woods and cut an enormous birch tree, one he marked the other day. He bore it home in triumph, drawn by fourteen dogs on a komatik that wobbled from the weight.

April 2nd A perfect morning for the Orangemen's Walk. We decorated Jack's tail in honor of the day, put on orange badges, and I dyed some white cotton orange for flags. L.M.K. and I donned our blanket suits to add gaiety and were as good as two more flags.

The procession started from the Orangemen's Lodge and marched to the harbor, then to the hospital, where we took photographs. It was a colorful sight, the long line of men with orange scarves on their dark coats, the standardbearers in red jackets. The squad Mr. Pelley has been drilling every Saturday evening wore white shirts and gay scarves. There were banners waving and an English flag, of course. The band consisted of a drum and several concertinas, each one playing a different tune – a trifle weird, but gay. After passing the hospital they walked to the church on the other side, where Mr. Richards held service.

We were all invited to have luncheon with them at the Lodge at one o'clock. The hall was filled with women and food. I never saw so much cake in my life. Long tables followed the wall on each side of the room, and on the platform was another table for the elect, which we were told included us. All over the room and under some of the tables were boxes of more food, the women sitting on them until their contents were needed. Then they rose, drew out pies and cakes and sat down again. The cake was of every description, some covered with the lurid pink frosting I had noticed at the church picnics. We had provided cornbread, thinking this might be an opportunity to demonstrate the value and attractiveness of cornmeal used in this form. A woman bustling about with an important air proved to be the famous Mrs. Jackson, L.M.K.'s one-time antagonist. To our amusement, she was very attentive to us, especially so to L.M.K. Evidently, she wishes to bury the hatchet.

The women made merry with us and dropped all sorts of things into the hanging hoods of our blanket suits. We removed tins of cocoa, pieces of bread and cake, knives and forks, etc. It takes very little to amuse them. When the Orangemen came up the hill we scarcely recognized some of the men, whom we had rarely seen with such clean shaves, white collars and well fitting clothes. Even some expressions seemed changed, and Reuben carried himself with a poise and dignity I had not noticed before.

The day culminated with a "concert" at night. The weather had turned "dirty,"[122] and a heavy wind and cutting snow met us when we started out to brave the uncertain path, but weather evidently mattered little on Orangemen's Day for the hall was crowded when we arrived. Dr. Little acted as chairman, and the "concert" began. There were recitations, songs and one duet, "As the Train Rolled Onward,"[123] the most remarkable of its kind I ever heard. The Doctor's address was worth hearing, but the drill by Mr. Pelley's boys was the star performance. Dear old Mr. Pelley had made it up by himself. He had never seen soldiers, had no idea of military affairs or tactics except from what he had read, and his pride in the boys' quick attention and performance was beautiful to see as well as his dignity and bearing in leading them. This ended the entertainment.

We left the hall to face a blizzard so blinding we could scarcely find our way. Dr. Grenfell piloted some of the little boys, and all the children had someone with them, but it was hard for those who had to cross the harbor. Mr. Richards came home with us for the night. We are all tired, but this day of typical festivity in St. Anthony has been something to see.

April 3rd Still bad weather and no picnic. I have spent much of the day correcting manuscript for the Doctor, who went out to blaze a trail to Mr. Lindsay's new camp. He came back with the sad news of having found portions of a deer that had been almost entirely devoured by dogs. Something will have to be done to make the people realize they must make more effort to restrain them. The regulations, drawn up by Dr. Grenfell in his capacity of magistrate and agreed to by the men require all dogs to be kept tied, muzzled or clogged or with their masters. Any dog found at large will be shot. At the Doctor's request, I have spent the evening lettering large notices to be displayed conspicuously in all the settlements. Each has a sketch of a reindeer head at the top, the Doctor's idea and design.

These measures following the introduction of deer to partly replace the dogs are necessarily painful. The people cannot get on without dogs in this climate, where any other animal is impracticable for general use, hauling through deep snow, bringing wood and water – for everything, in fact, that horses do for us. Deer do not answer the same purpose in the winter, for they sink almost out of sight in deep snow early in the season. Without dogs, many patients would have been unable to reach the hospital, for they came many miles over country where deer could not travel. But the dogs are unquestionably a menace to deer.

April 6th At the Orangemen's concert, Dr. Grenfell mentioned in his address that he had many logs to be brought from the woods before the snow broke up and would greatly appreciate any help that could be given him. The harbor men immediately responded, each offering to haul at least one load. Today was the first haul, and men from the Bight and Goose Cove were on hand early. They were back with thirty logs at noon and had lunch in the kitchen.

The snow on the harbor is heavy, and a new plan of the Doctor's is to mark a road by the best route to the log pile as a guide for all komatiks to go the same way and beat a hard path. I went out with him to help gather armfuls of branches which we stuck in the snow all the way to the end of the harbor where the woodpath enters. We felt very proud of our work, the Doctor more than once referring to the good we were doing, which was generous of him since the idea was his. He gathered most of the branches and told me where to place my share. As we gathered our last lot, six teams of dogs with loads of wood came down the hill one after the other. I wished for my camera, but no picture can give a true idea of the fascination of dog-teams, whether one is riding on the komatik or watching men guiding a load of wood down a hill or across the open. The working of the dogs, the calling of men with their strange cries – "Look-at-the-bird," "Look-at-the-man," "Cra-r-r-r" – and anything that occurs to them, all shouted in a high key that adds to the excitement. I still feel the thrill of my first sight of a dog-team starting out in the December snow.

April 7th All the St. Anthony men hauled today. Thirty teams returned together at noon, left their logs at the pile and came to the Guest House for dinner. There were dogs of every size and kind – big, little, brown, black, white, yellow and combination – all barking, howling, getting tangled in their own traces and those of other teams. There were fights, and the men often sprang into the midst of the yelping, snarling mass, shouting and swinging their long whips. The teams lay beside their komatiks, overturned to prevent the dogs from running off with them. They looked very peaceful, but when I approached one the men warned me to keep away as they might attack a stranger. I had been told it was not safe to approach a dog when in harness. They apparently have a strongly developed instinct of guardianship which, I am sure, is a praiseworthy quality. Occasionally, a dog belonging to one team rose with hair bristling to growl defiance at one of another team, and their exchange of incivilities was amusing. The men brought in sixty-five logs and the Doctor is gleeful.

April 9th Disturbing news came yesterday that the store at Goose Cove had been entered and goods stolen. Dr. Grenfell sent Reuben and Mr. Jackson as constables to search the houses and make arrests. They returned tonight through a heavy storm with two prisoners in custody, Reuben with an air of much official importance. One of the prisoners was from Crémaillière, the other from Goose Cove. The constables had done real Pinkerton work,[124] tracking them in the snow, fitting snowshoes into foot tracks, etc. One of the men finally confessed, and the constables returned in triumph. The men are on the piazza in semi-darkness and have had a meal of bread and water. I stumbled over them on my way to class. They will be lodged in the bunk-house with Mr. Pelley as jailer until the trial on Monday.

It is a bitter night. I went to the shop and had a hard time returning after the class. The wind was so fierce I could scarcely breathe, the snow a blank white wall. I could not see the Guest House and found myself wandering near the turf piles. I can easily imagine one perishing in the open on such a night.

April 10th The blizzard is worse than ever. The oldest inhabitant acknowledges this is typically "dirty weather." Even the doctors were storm-bound. Most of us did odd jobs while the storm howled outside our quiet room, and when we gathered around the tea-tray before the purring fire in the little stove, Dr. Little brought his banjo and played softly in the twilight.

April 15th We have symptoms of spring fever and I am growing restless. This was another morning of glory and I longed to be out in it. It was cold. The water froze in Dr. Grenfell's bath, and he said he had to be careful not to cut himself with thin ice. There was heavy frost on the windows. Much of my day has been spent in the loom-room, and I have avoided looking outside at the inviting country where I would like to be. It is weather advantageous to the hospital patients, who have been out on the porch nearly all day. They are becoming tanned in the sunlight, restored in body and spirit.

We took the piece of pattern weaving off the loom this morning, also some lovely small pieces. The girls are encouraged. Tom Boyd's carved tray is much admired – a real triumph considering it is made from a lowly pine soap box.

Our cherished electric table bell can never be depended on, so lately I have substituted a large Swiss cow bell brought over by Dr. Grenfell. A fair-sized rope is attached to it, and it has been used on the table, rope and all. Dr. Grenfell calls it the "domestic tinkler," which seems a light term to apply to such a heavy article.

The Doctor announced this morning he was going after luncheon to cut another birch tree and asked Dr. Little to go too. There was a search for dogs, for Alf had taken most of them. I wish I had a photograph of the Doctor as he appeared across the snow, carrying a great dog in his arms and clad in his remarkable costume consisting of grey shirt with rolled-up sleeves, blue running shorts and gay striped stockings. He wore also a red toboggan cap, dark glasses, and a broad grin. The dog, which had been sick, was called Jessie (my own name), and it gave me quite a shock to hear her referred to by the doctors as "that brute Jess." I suggested that if I annoyed them they could relieve their feelings by calling me all sorts of names by proxy.

They returned at dinner time, having had a tussle with a great twisted birch tree which they cut after digging twelve feet of snow from its base for space to swing an axe. Dr. Grenfell should be able to write two sections of his treatise on whales[125] after relieving his mental pressure to such an extent.

April 16th This morning, just before my start for the loom-room, Dr. Grenfell appeared with a strange young man whom he introduced as Mr. Burton, a Methodist minister from the strait who was to spend a day and night with us. He is the large-faced, bespectacled type with a pleasant smile and a twinkle in his eye. I knew Dr. Grenfell was still struggling with his article on whales so took Mr. Burton to the loom-room and other places of interest. Later, I turned him over to David and met the Doctor for a conference concerning household matters, loom-room and shop affairs in need of adjustment. A general clearing-house of plans and arrangements is occasionally necessary, we find, and I hope the wage scales and other matters are now satisfactorily arranged for the season.

We sat in the study for a while after coffee while Dr. Grenfell read his latest chapter to us; then I dragged our guest to the other room, where we discussed sectarianism, the right and wrong of theaters, dancing and card-playing, and he gave counsel regarding smoking in general and on request his own case in particular. I know my broad views shocked him, and I tried to handle the subject with care. He is a nice, earnest young fellow and apparently enjoys his visit. He is, I think, one of the travelling ministers without a definite parish who go from settlement to settlement, relying on any means of transportation available and on the hospitality of native homes, which is always offered gladly. They are friends of all the people.

Dr. Grenfell is groaning because the hole around the birch tree that he and Dr. Little dug with such difficulty will be snowed in today and have to be dug out again.

April 17th No Bartlett's Brook today. One of the coldest days of the season, with fierce wind and snow blowing in clouds. Visions of lying on the cold pond, looking into a hole in the ice and waiting for fish to bite, were not alluring to anyone, so the Doctor suggested that Miss Keese and I should go in with him to the birch tree. Everyone told us to dress warmly, and I never realized it was possible to wear so many clothes.

We had twenty-one dogs and three komatiks. There were enormous drifts, some across the path, and travelling was very heavy. The country was beautiful with its fresh covering of snow and soft veil of wind-blown drifts over the high hills. In the woods, the trees were laden. It was bitterly cold, and Dr. Grenfell turned to ask me if his chin was frozen. It was difficult to realize that a few days ago I was walking along the same trail on snowshoes without coat and mittens.

The hole around the tree must have been nearly twelve feet deep and not filled with snow as the Doctor feared. I had perfect faith in his

L.M.K. wields the axe at a birch tree, April 1908 (JLP)

ability to get the tree out but must confess I did not see how. It was done by cutting the trunk in sections, L.M.K. and I being allowed to wield the axe one minute each, and then the great twisted trunk was hauled out by using the komatiks with pieces of small branches for rollers and the motive power supplied partly by us. We must have pulled the weight of at least twenty pounds. Alf's dogs, the best and strongest, brought the log in. It took all their strength, and it was interesting to watch them haul. They walked sideways to bring their whole bodies against the harness, digging their claws into the snow, heads down, tongues out and every trace taut. Dr. Grenfell had a hard time with his load of smaller sticks. It tipped over twice and we all assisted in righting it. The other dogs mixed up with our dogs. The leader ran around the wrong side of a stump and stopped the komatik abruptly. All those things are liable to happen on any komatik ride and always prevent "a cruise [excursion]," as the natives call it, from being monotonous. We had a great time and came home glowing, though a little weary.

The barber shop was open again this morning and Mr. Lindsay was the customer. He says it is a "lovely cut" I gave him, and I really begin to feel like a professional. I have not yet attempted a "pineapple" [close clip] but imagine that would be easy, as one would not have to consider delicate gradations of length.

April 18th Cold and windy and no day for the trouting expedition. They say trout will not bite at such a temperature, and we refuse to freeze for nothing. I spent the morning in prosaic darning. Dr. Grenfell is writing another chapter, but his urge to counteract the great expenditure of grey matter by using all his muscles was too much for him, and he and Dr. Little started off at 12:30 to attack a brother birch tree, Dr. Little with rather subdued enthusiasm but a willing spirit. L.M.K. and I promised to walk in to meet them and started at 3:00 o'clock. We walked to the flag and found it still flying but faded, and only half of it still clung to the pole. Alf met us with the dogs hauling some big logs and stopped to let "Brin" give us his usual greeting. He put his paws on our shoulders and smiled. He is a remarkable dog, one of the most intelligent I ever knew. During the two doctors' recent trip together they lost their way, and after what seemed aimless wandering Brin started in a direction Dr. Grenfell thought was wrong. The Doctor tried to make him go the other way. Brin turned to face him and sat down. The other dogs also stopped. After the Doctor had tried again without result, Dr. Little said, "Why not let him go his way and see what happens?" Within five minutes they found the right trail. The dog had been through those woods only once – two years before in a snowstorm and at night.

We found a tree covered with moldow and stretched ourselves out on the branches while waiting for the doctors. Blue shadows passed over the White Hills, and the silence was broken only by the sound of wind through spruce branches as we waited in that peaceful spot. After a while the doctors appeared, walking beside the dogs. Dr. Little did not look very happy. The loads were difficult to steer downhill, often sliding to one side. When we reached the last rise near the house, we helped haul. I am also becoming expert at jumping off and on a komatik in motion.

Dr. Grenfell and I have been invited by the Noels to spend the night with them at Goose Cove, where the Doctor expects to preach tomorrow. I have been ready since dinner for Seth Noel to come for us, dressed up in my brown silk blouse and yellow butterfly pin. The Doctor has dignified his grey flannel shirt with a white silk tie, also his dark coat. We sat around in the study for a while after dinner and, when nothing happened, settled to reading and writing, the Doctor finally going out to visit a patient. The moon has arisen over the Head, shining gloriously across the white country, and I can see it through long icicles hanging from the piazza roof, turning them all aglitter.

It is eleven o'clock. The Doctor has just come in. He reports there is a special meeting at the Orangemen's Lodge tonight, which may account for Seth Noel's non-appearance. I suppose he did not realize it is well to notify guests of a change of plans. We conclude the lodge meeting held too late and he did not think of notifying us, or perhaps there really was not enough room in his house for us both (which seems likely) and to ignore the invitation was a happy solution of the difficulty.

10

To the Gates of Death

19–25 April 1908

April 19th Easter Sunday! What would the straw-hatted public at home think of such an Easter Day as this? Snow over fences and parts of buildings, frosted window panes. The weather is lowering and disagreeable.

The Doctor was late at breakfast. We had nearly finished our oatmeal when the door opened abruptly and he strode in without a smile, looking very stern but particularly well groomed. He had even waxed his moustache. On reaching his place opposite me, he swung one leg high over the chair-back, revealing a gay red, yellow and black striped Richmond Club stocking below dark blue serge shorts, then slid into his seat, looked around the table and grinned. It was a gay meal.

Mother had put an Easter gift in my trunk, which proved to be a bright paper egg containing a tiny fluffy chicken. I had put the egg at Mr. Lindsay's place at breakfast, as he frequently clamors for such an impossible article of food. Dr. Grenfell promptly added the chicken to our wise monkey centerpiece. Some time ago, he found a tiny American flag and planted it beside the admonitory monkeys. Now the chicken, tilted at a rakish angle, adorns the tip of the flag. All sorts of liberties have been taken with our centerpiece.

Everyone went to church this morning, the Doctor taking the service. A part of the lesson was that used for the burial of the dead: "Lo, though I walk through the Valley of the Shadow of Death, I shall fear no evil"; "The years of our life are threescore years and ten …"[126] For some reason the familiar words sounded ominous. On our way home from church, we saw a komatik and nine dogs crossing the ice near the hospital. These people do not take their dogs out on Sunday unless from absolute necessity, so we wondered why they came. They proved to be a team from Canada Bay, calling the Doctor for an emergency: a patient he had operated during his last visit, a case of acute osteomyelitis. The family, through ignorance, had not treated the wound properly, and the boy was in danger of blood poisoning. He needed assistance at once. Dr. Grenfell started directly after dinner and insisted upon travelling alone to make better time, although at this season, when ice on harbors and bays is breaking up and brooks bursting through, few men travel without a companion. He took his seven best dogs with Brin as leader and Jack, his little black cocker spaniel, for company.

We watched him climb Fox Farm Hill and turn near the top for a final wave of the hand. Jack ran on ahead, wild with joy at being allowed to go. He will make Lock's Cove tonight, stay with friends, then go on to his bungalow on Hare Bay, where he plans to meet the men who came for him, and continue with them to the village where the patient lives.

April 20th The lower harbor ice is breaking up gradually. Open water has almost reached Biles' wharf, which has been frozen in all winter. It rained all night, and objects we had forgotten are appearing above the snow. We realize we have a fence. We are aware also of barrels, piles of boards and rubbish which we wish were still buried from sight. Rain has dripped through the piazza roof and soaked the puppy-dog team and various articles of clothing hung on the walls for convenience. No dog-teams or deer went out for logs today. If they are not brought in this week, it may be impossible to haul them at all.

During the day, there has been much discussion among the coastwise men as to the chance of the Doctor being able to "get on." "Th' Doctor can't get on today," said they. "Th' road 'e's wonderfu' 'eavy. 'E won't be able t' stir out." The general impression being that it was impossible for him to leave Lock's Cove this morning if he had reached it at all last night in the storm, and we rejoiced when the snow hardened sufficiently for him to travel without difficulty. Every man we met had a word to say about it. "'Tis civil weather[127] for th' Doctor after all. 'E'll 'ave a fine run this evening" (*evening* meaning afternoon).

Part of the route, as I have already said, lies across the end of Hare Bay, an arm of the sea about twenty miles wide. Dr. Grenfell had crossed it on an earlier trip during the winter, when the ice was hard and smooth and the komatik ran almost of itself, but with warmer weather it will become treacherous and, as in the case of ice in all bays, is liable to break up at any time with a change of wind and bring in the first sea of the year. By crossing Hare Bay, one can shorten the journey by about fifteen miles and avoid travel over very rough country, but we had begged the Doctor to go around in any case, and he said he would, so with his promise we feel reassured.

Today, Reuben brought in the head of a goat that was killed by dogs, all stuffed and mounted. I added it to the solarium museum, which is beginning to be interesting. There are three caribou heads, a fox head and brush, besides the goat's head. Three long coiled dog-whips hang under one of the caribou heads. Numerous Indian clubs swing on part of a bamboo fishing pole. There are dumb-bells, boxing gloves, skates, skis and snowshoes. The English flag is draped over the dining-room door, English and American flags at the far end of the piazza. The furniture is miscellaneous. A large curio case with glass doors, an invalid's chair, a leather-covered examining chair, packing boxes (painted green) used as benches, a table (usually littered with odds and ends) and an old trunk filled with rope and reindeer tackle. On the floor is the Doctor's long komatik box, which reminds us of a coffin. There is a huge pile of antlers in a corner, and it is difficult to clean behind them. In another corner, unfortunately near the entrance door, is an accumulation of what looks like rubbish, but it may be valuable, and I do not dare touch it. There are bits of cow-hide, deer and seal-skin, pieces of walrus ivory, ends of rope, blocks of wood, bottles of unidentified liquid. During the past few days, a large piece of seal fat has dangled from a large hook fastened on the wall. It is not pretty, but its excuse for being there is its use for oiling the many pairs of foot gear that are hanging all over the place. There are long skin boots, black ones with white feet, yellow ones (soled and unsoled) and occasionally moccasins or slippers, also caps, mittens, sashes, belts, oilskins and overalls, sometimes a shovel and frequently a gun or two. When I arrived, this collection of rubbish and general piazza accessories would have sorely wounded my housekeeper's instinct. Now, after six months, I survey the mess with placidity and meekly pick up only what I know may be legitimately removed.

This has been a day of odd jobs: laundry lists, household affairs, loom work, afternoon and evening classes. I am ripping David's old trousers apart, preparatory to using them as a pattern for new ones made of native homespun. I did not want to do it, but my heart softened when I was told that without them he would soon be reduced to wearing a barrel. So for the reputation of the Mission, I am about to embark on my latest form of industry, men's tailoring. Dr. Grenfell also wants a pair made from some serge he has on hand, which brought a protest from Dr. Little, who says the Doctor has trousers galore – even for distribution.[128]

We rejoiced when the snow hardened with a change of wind. The mercury fell rapidly this afternoon and it is still blowing hard, but the sky is clear and the night bright with stars. Archie came in during the evening and said the harbor ice is breaking up near Biles' wharf. It may be free by morning.

April 21st I little thought last night, when I finished my log at 12:30, that I should write what I do tonight. It has been a day of horror and thanksgiving. The most terrible thing that could happen nearly happened.

I awoke this morning at 5:30 with the roar of the sea in my ears, a sound so unusual that for a moment I could not think what it was, and on going to the window and hearing the waves could still scarcely believe it possible, for all winter the ice on the shore had held the waves, and the silence had been absolute. It seemed as if some great and awful creature had suddenly awakened from its sleep and roared hungrily. The morning was perfect, the air sweet and frosty. There was no sound but the song of a few birds and the roar of the sea. A soft haze of opalescent color hung over the hills, and the early morning shadows on the shore were a delicate violet – a morning so full of beauty that it almost gave one pain.

I went back to bed but a few moments afterwards heard hurried footsteps on the snow. The side door was flung open (our doors are never locked), and someone rushed up the stairs calling Mr. Lindsay. I recognized the voice of Steve Pelley, the reindeer apprentice, though it was hoarse with excitement and feeling. "Th' Doctor's adrift on a pan o' ice out in th' bay," he gasped.

In a few moments, all those connected with the Mission and a number of men besides had gathered at the corner of the hospital piazza, a crowd of quiet people with sad, anxious faces and lowered voices. When I reached the group, they were discussing the possibility of its being some other man, each one trying to foster a little hope in his heart. The news was first brought by a man from Lock's Cove who had come through in the early morning and stopped en route at the reindeer camp, where he told Steve Pelley that the Doctor had been seen the evening before, just at sundown, on a pan of ice with his dogs around him, drifting out to sea. He was far from land, and before a crew could launch a boat it was too dark to see him. The man had left Lock's Cove before daylight and could tell no more. Steve Pelley had come directly to St. Anthony with the news. There was so little anyone could do.

Mr. Lindsay and some of the men of the harbor went with glasses to the high cliff above the sea, thinking that by some chance he might have drifted in that direction. The boats were all hauled up for the winter and frozen in under ice and snow. The suspense was terrible. Some of the men, however, were optimistic. The Doctor's own man, Reuben, who is always light-hearted, assured me that "Th' Doctor could live a week on a pan o' ice, Miss, an' 'e 'as 'is bread-bag with un, an' someun's sure to pick un up 'fore long. Like 's not, 'e's ashore a'ready, ef 'twar th' Doctor."

While we stood helpless and uncertain, two men from Goose Cove stopped on their way to Griquet. Reuben met them and came back more optimistic. "'Twarn't th' Doctor 't all. 'Twa' someun else comin' from th' other way. Th' Doctor 'ad George Reid's b'y with un, an' 'e'd never take chances like dat with a b'y in 'is care. An' dis man was alone an' 'e 'ad only five dogs an' th' Doctor 'ad seven, an' anyway 'e was on a bit o' ice no bigger dan th' floor o' a kitchen, an' 'e couldn't 'a lived on dat tro' th' night. It would 'a bin all abroad in th' sea. 'E's perished before dis, whoever 'e is, poor feller. But, sure, 'twarn't th' Doctor, an' 'e said 'e'd go 'round, too."

The hour for work passed and the breakfast hour. We knew the Doctor would want the work to go on, and some of the men went to the woods for logs. Everyone tried to believe it was the other poor, unfortunate traveller who had been seen, but I think everyone felt in his heart it could be no other than the Doctor. Something had to be definitely known, and Mr. Ash quietly started for Lock's Cove with a komatik team to find the truth. Then we waited. We tried to be busy. It was what the Doctor would have liked us to be, but it was hard work. Men stood about in groups, and everyone spoke with lowered voices as in the presence of the dead. One thought was in every mind, one dread in every heart.

"Ef th' Doctor's gone," said one of the older residents, "dat's th' end o' th' French Shore an' St. Anthony. 'Twill all go down wi'out th' Doctor. 'E's th' one dat keeps us all goin'."

At about eleven o'clock, Reuben came to the house and called Mr. Lindsay and Dr. Little. I heard them talk in low voices, but it was not until later that I knew what was said. A man had just come from Lock's Cove with further news. George Reid, a fisherman at whose house Dr. Grenfell had passed Sunday night, had heard a man was seen on the ice and, on taking his glass (the only one on that part of the coast), had seen a dark object on the wide white expanse and recognized the Doctor with dogs around him. He was on a small pan of ice more than three miles away and drifting down the coast towards the open sea. It was impossible to reach him that night, for it was then after sundown, and they feared the pan would not hold until morning or even be within reach of the shore if it were still afloat. To those who heard this, there seemed to be no possible chance of any rescue.

The morning wore on, and we counted the hours that must pass before we could possibly hear any tidings. At one o'clock, we were at the luncheon table, trying to eat, when Archie Ash burst into the room with the words half choking in his throat: "The Doctor's come!"

Gradually, we learned the details of his rescue. It was a mistake that a boy had started from Lock's Cove with him. He had gone alone, expecting to meet the men who had come for him at the bungalow on the other side of Hare Bay. When he was seen on the ice, it was impossible to go for him. It was then after sundown and the sea so high a boat could not live among the great blocks of ice in the surf. George Reid and other men walked the cliff all night, noting the direction and velocity of the wind that they might know in the morning where to look for him. Few men of Lock's Cove slept that night, and many women also lay awake with anxious hearts. There was sorrow in every household. With the dawn, five men were ready and eager to go out,[129] though with a sea running it was extremely hazardous. High seas made it difficult to launch the boat, and as they rowed through the slob ice great blocks, surging together, threatened to crush them. At times, large pans of ice blocked their way and they climbed over them, dragging the boat to launch on the other side. They took kettles of tea and food for several days, for they might be caught in an ice-pack and carried to sea or obliged to abandon the boat and either await possible rescue or try to walk ashore if the pack were blown into the bay. They did not hope to find Dr. Grenfell alive. It seemed impossible that anyone so lightly protected could survive a night of such exposure and bitter cold. They only hoped the pan of ice had held together during the night and they could bring back his body. Even to find the pan at all was a mere chance.

After rowing and struggling a long time, they saw afar off in the early morning light a dark speck above the level surface of snow and ice. They saw it was a man, a man standing and waving a signal, and they redoubled their efforts, fearing that even with help so near he might succumb before they could reach him. When they took him and his remaining dogs into the boat, they were nearly nine miles from Lock's Cove, and nine miles of perilous return still lay before them. Unless they were questioned, very little was said by the hardy men who brought him in of that battle with ice and sea. It was loving service they were glad to give for the Doctor, and the dangers and uncertainties of a fisherman's life are considered inevitable.

They took him to George Reid's house, which he had left twenty-four hours before, and gave him clothing and food. He was, of course, unable to travel on foot, and they urged him to wait, but he knew we had been told he was lost and insisted upon going on. He did not know his feet were frozen until heat in the fisherman's house brought the painful realization, and before the borrowed komatik and dogs were ready it was quite impossible for him to walk without effort. To his chagrin, he was obliged to be drawn all the way. The news spread rapidly along the coast. As he passed the little villages, people came out to shake his hand and tell him of their joy. Mr. Ash met them on the trail half way to Lock's Cove and told them how the news had reached St. Anthony.

At the hospital, the men who brought him stood talking with a group outside the door when Ruth Keese and I entered. We found the Doctor inside. He was so changed one would scarcely know him. His eyes were bloodshot, his face a curious dark red and his hands so swollen with frost he could barely use them. His voice was strange and weak as he told us of his terrible experience. It seemed nothing less than a miracle that he was still living.

This is the outline of his story. He had tried to cross Hare Bay in spite of the danger and was less than 100 yards from shore when the wind veered suddenly, the ice broke away and began to move out toward the sea. He found he was really on "slob" ice, like heavy snow packed closely by the wind, but so soft he could push a stick through it. When he discovered there was no possible retreat, he tried to reach shore by urging the dogs forward, but they became frightened, hesitated, and he and his eight dogs sank through the soft mass into the water.

The father of the boy the Doctor was on his way to see had been drowned earlier in the season when he fell through the ice and became entangled in the dogs' traces. With this in mind, Dr. Grenfell's first act was to cut all the traces free with his clasp-knife and climb with the dogs on the nearest pan of ice. This, however, was so small it could not hold them long, and when a larger pan drifted by he tried to reach it. With little Jack in the lead, he pushed the dogs off toward it and, as they climbed up, drew himself on it by the trace of the leader, Brin. But this pan also was too soft and small, and it was only after three attempts and immersions that he and the dogs succeeded in climbing on a pan strong enough to hold them. On this they drifted toward the open sea.

Between him and the shore was soft ice formed by large pans grinding together. It was impossible to swim through or he might have reached shore. His only hope now lay in the possibility of being seen by someone from Lock's Cove if the pan held together. Once outside the bay in the open sea, there could be no help from any quarter.

The Doctor had freed himself of outer clothing when he first sank through the ice and in the intense cold had only his underwear, a flannel shirt and sweater, running shorts and two pairs of stockings (one of them the gay Richmond Club stripes). His high skin boots were taken off when they filled with water, but he had tied them to the harness of one of the dogs and his clasp-knife to another, hoping to save them. Nearby on the small pan was the komatik with everything he needed –

dry clothes, sleeping bag, food and hot coffee in a thermos bottle, also his axe and rifle. He tried to harpoon the pan with his knife tied to a dog-trace but gave it up as he feared to lose his knife, and finally the komatik sank through the slob pan and disappeared. He said he felt he had lost his last friend.

As the hours passed, he realized something must be done to prevent freezing to death, for his clothes were already frozen as well as his stockings, which he now began to change, one over the other and back again. He then cut off the tops of his high skin boots, lacing them together with dog-traces to make a jacket. This tempered the icy wind only a little. There seemed to be only one thing to do: kill some of the dogs and use their skins for protection. The seven he had with him, besides little Jack, were the best of the pack and it was terribly hard to do it. He chose Watch, Spy and dear old Moody. He had to kill them with his clasp-knife. Afterward, he tied the skins together and wrapped them around him like a cloak, with the hair inside. Finally, he ripped the rope harnesses apart using the strips of flannel covering for puttees to wrap around his legs. The frayed rope he used to stuff inside his trouser legs and sweater to add a little warmth.

It was now late afternoon and he had drifted a long way but was still in sight of the cliff beyond Lock's Cove. Although he was far from land and there were no settlements near, there was a bare chance that if he had a flag someone might see him. He said he had not a shadow of hope but intended to die hard anyway. He needed a pole for his flag, and the skinned bodies of the dogs with their outstretched legs gave him an idea. He dismembered them and tied the legs together. When they froze stiff he took off his flannel shirt, tied it to the ghastly pole and waved it until dark. He did not dare sit down, for even a few feet of elevation might make a difference in visibility at sea.

When it was really dark, he made a windbreak of the dogs' bodies and laid his head on them, then called Doc, Brin and Jack to him and they crouched close around him and kept him warm while he lay very still, hoping the pan would last through the night. After a while, he slept a little and dreamed he was on shore, while at times throughout the night he seemed to hear the old familiar hymn, "God, My Father, while I stray ... Thy will be done."[130] At daybreak, he was again on his feet waving his flag. After a while, he thought he saw a glint of oars in the distance, but he had been deceived the night before when such a glint had proved to be only light on a piece of ice. He had now drifted about fifteen miles and was nearly opposite Goose Cove. Soon he would be in the open sea, and the pan was only half its original size. Only death lay before him. Then he saw again the glint of light rhythmically repeated, with a dark spot that indicated a boat, and he knew he was saved.

We left the hospital with the Doctor after he told us his story. Outside, near the door, a group of people had gathered, a group of almost silent men and women who felt more than they could utter. Perhaps in no more convincing way could the love they bore the Doctor have found expression than in a silent grasp of the hand and tears in the eyes of more than one sturdy man of the harbor as he stood before them, almost as one risen from the dead. Although his feet were badly swollen, he presently wanted to show us how he used his dog-skin cloak and his flag. The ghastly flagstaff made of dogs' legs lay on the porch where the men had left it. He had put on his shirt, which he had used as a flag, but now took it off again and tied it on the pole to show us how it looked. Everyone tried to treat the matter lightly, and we helped him put on the jacket made of boot-tops and adjust the dog-skin cloak and wrap the puttees around his legs. As he stood outside the door waving the flag, everyone tried to laugh at the queer object he made and someone suggested photographing him. Then he hobbled home and collapsed.

Only a man of great vitality as well as resourcefulness could have survived such an experience, and no one, however strong, could fail to feel keenly the mental and physical reaction when need of effort for mere existence had ceased. We have always seen the Doctor vigorous and active, but he now moves like a man eighty years old. He cannot get warm and has sat all afternoon in his big chair near the stove in the study, half sleeping. He says he is sure heaven is not a cold place anyway. The hymn that was with him during all his hours on the ice is still in his brain, and at times he begins to sing it. He talks about the dogs and imagines he is again sinking through the awful "slob" ice. His feet and hands pain him more, and there are two wounds from bites given him by the dying dogs. I have put him to bed with a hot-water bottle, plenty of coverings and an improvised cradle to raise the bed-clothes from his feet. He does not know the strength of the sedative Dr. Little has given him.[131]

The house is very still.

April 22nd I have played nurse all day, trying to make the Doctor as comfortable as a man can be who loathes his bed when he is awake. I have even brushed his hair and prepared him for visitors who are, however, confined to the family and his secretary, Mr. Jones, to whom he has dictated the story of his experience while the details are still fresh in his memory. His brain is so active it is well for him to make use of what is surging through it to relieve the tension.

After luncheon he insisted on getting up – against the orders of Dr. Little, whom he abuses for giving him that large dose of morphine.

He says his brain is still clouded and he cannot think connectedly. For amusement he began to stuff a puppy-skin for a komatik team, which is nearly completed, while I sat by him with my darning basket. The Doctor goes over and over again the events of yesterday. They are constantly in his mind – and no wonder!

April 23rd The poor komatik team of puppies! On the piazza this morning, David found two dogs curled up asleep, the outer door having blown open during the night, and just outside the door was the little komatik and remains of the puppy-team. The dogs had left nothing but a few scraps of skin, the wire frames and oakum stuffing. The Doctor in his nervous state was much upset and wanted to shoot every dog in sight. The komatik team was his pride and joy. Reaction from his experience is developing and he is very blue.

Unfortunately, there are several things to bother him. Mr. Lindsay came in with a wild tale of dogs having bitten one of the Lapps, and there are many logs for the new barn not yet out of the woods. This is serious because the Doctor is sure the trail will soon be too soft to bring them out. It was troubling him so much that after luncheon I went from house to house through the village and across the harbor to ask the men to give the Doctor a haul. All of them responded at once, and they will go in tomorrow if the good weather holds.

I went home quite breathless with news to find the Doctor in the study with his hat on, looking irritated, bored, and everything that indicated physical weakness and mental discouragement. Some diversion seemed necessary, and I suggested stuffing one of the remaining puppy-skins to start a new dog-team. This stimulated a mild interest and he came to the table, but had scarcely begun to work when the sight of a dog at the refuse hole aroused his ire and he went for his shot-gun. I said, "You are not going to shoot him, are you?" He replied, "Yes, I am!" and left the piazza.

I fled to a remote part of the house, for I could not bear to hear the dog's cries. There was not a sound. After a while, the Doctor called from the foot of the stairs. He was laughing and said he fired and missed (I suspect intentionally), so I went down and we continued to work on the puppies during the afternoon. After that he was much more cheerful. I think it relieved his mind to do something desperate. He has now gone to bed, still using the cradle for his feet, but has discarded the hot-water bottle.

April 24th A beautiful day, and at sunrise the silence was broken by shouts of men as their dog-teams raced across the harbor ice towards the woods. At nine o'clock they were all back, each with a heavy load which they left at the wood-pile. The logs are all out and the Doctor is happy.

It was a gay sight, all those sturdy, sun-browned men in their bright jerseys, caps and mufflers against a background of pure whiteness, above them the deep blue sky and over all a flood of brilliant spring sunshine. They came in gaily, laughing and joking with each other, and after leaving the yelping, howling packs of dogs beside the overturned komatiks they all came to the glassed piazza, where we had breakfast ready. It was the first time many of the men had seen the Doctor since his return from the ice, and I noticed on almost every face, as they shook his hand, an expression of awe and wonder, as in the presence of someone who had knowledge of things beyond their ken – of the borderland of a far, far country. It was strange to them to see the strong Doctor limping with a cane, but after a few moments they crowded around him, each one telling him how and when he first heard the news, what he thought about it and what he did or did not do. There was silence while the Doctor told the story in his simple, direct way, and I noticed tears in the eyes of more than one of his listeners. It now seems like an awful nightmare.

We were busy with the puppy-team again this afternoon. Reuben is charmed with the new pups. He even thinks combination dogs can be made by patching together those which are mutilated. It does not matter if a brown dog has a black tail.

The Doctor's long-planned trouting expedition is on for tomorrow if the day is fine. Boys have been catching fifteen dozen at a time and have sent many to him. Of course, he cannot go but insists that we do, though I hate to leave him alone all day. I am to rouse the household at 5:00. Breakfast is at 5:15.

April 25th This has been a great day to remember. It began at 4:45, when I was awakened by the Doctor calling, "It's time to get up. You should be on the hills by this time." I sprang from bed and was nearly dressed before discovering it was not yet five o'clock.

The morning was clear and warm, yet cold enough to make the trail hard and smooth. We started off briskly a few minutes after six with four komatiks and thirty dogs. Out on the harbor ice where the travelling was good, we were able to ride most of the way, but the trail sometimes led across country where dog-traces caught around tuckamore. Dog-teams that came alongside other dog-teams began to fight, and there was barking, snarling, yelping and shouting before they were untangled and off again.

Part of the trail was along the bed of Bartlett's Brook, where the blue shadows of overhanging trees lay across the snow. Only the soft sound of komatik runners on ice and snow broke the stillness, for the dogs' feet make no sound, and they never bark when in action. When we reached the "steady,"[132] or pond, boys were already fishing through holes cut in the thick ice about a foot square with a smaller hole in the bottom to drop a line through. The men cut balsam boughs to sit on, and the air was sweet with their fragrance in the warm sun. The mysterious calm and silence of the wilderness was broken once by a strange cry off in the hills. Dr. Little said it was a lynx, adding that they are very shy and run at the sight of man, but I have no desire for a test meeting. It is well none of us were ardent fishermen, for the efforts of the entire party resulted in seven small trout. But nobody cared. It was a joy just to be there. After building a fire on the ice for our "mug-up" and lunch, we subsided into a state of inertia, half dozing on our balsam boughs while Doc and Brin lay curled up near us.

The warm sun had wrought havoc with the snow when we returned, and travelling was heavy on the trail. Komatiks sank over their runners, and the dogs pulled with difficulty. Bartlett's Brook, frozen from bank to bank all winter, making a hard, level road for travel, is really a turbulent stream, and in places it showed its true character. The level surface had sunk, and steep, slanting, icy sides led to large holes where black water rushed beneath, ugly places where I held my breath while the dogs drew us as far from the edge as possible. The ice was out of the Bight. Only a few small pans remained. They gleamed white and beautiful against the blue sea, but the sight of a small block of ice in the water anywhere now makes me shudder. We found Dr. Grenfell in the solarium, writing a chapter of his book.[133] He said he had asked Johnny and one of the hospital patients to lunch with him for company, and there was a newly stuffed puppy skin on the table.

11

The Abode of Peace

26 April – 29 May 1908

April 26th A lovely, quiet Sunday. The Doctor is better. We began the day with a short service at home, singing the hymn that haunted him during that day and night. Then he went to church to take the service. At its close, all the men waited to tell him how glad and thankful they were for his wonderful escape. He still hobbles about with his cane but walks much more easily and is in better spirits. He even talks of going to St. John's by the first *Portia*, which may arrive soon.

April 27th The Mission komatiks were drawn up today and laid on the loom-room rafters for the summer. It was a sad sight for me, as I realize I shall probably never ride on one again. In the harbor, the launch *Daryl* is being made ready and everything speaks of spring, although snow fell heavily last night and the country is beautifully white again. Steve came in from camp today with a deer, and I had my first ride in a pulka – that is, I sat in the pulka and held the rein while Steve walked ahead, the deer following him. It took a few steps, then suddenly stopped, turned around and stared at me in a way they have, which is really disconcerting. They are queer beasts.

May 1st We talked of a May Day party today, but enthusiasm was lukewarm with snow four or five feet deep and not a sprig of green in sight. We compromised by walking two miles across country to the new deer camp, sometimes falling through to our waists in the soft snow around branches.

We could see tents on the hillside and on a rocky prominence found the entire herd taking an afternoon rest. They looked, at first glance, like bare rocks. New horns had begun to grow on several of them, the ugly, soft black bunches protruding from their foreheads seeming to have no connection with delicate branched antlers. The horns are shed every year and new ones develop rapidly. Mr. Lindsay says the entire antler grows in only two months. No wonder the process is a drain on their systems, and they should be in full condition before it begins. While we watched them, two or three rose and moved among the others as if to intimate that it was time to be up and feeding again.

On reaching home, we found Dr. Grenfell had just returned from a nearby village, bringing a friend to spend the night. Guests often appear without notice, and I love the sensation of entertaining wayfarers on their journey. It is one of the interesting adjuncts of our northern life, and experiences of the conventional hostess with a house party are sometimes tame compared with it.

May 2nd The two neglected girls from Crémaillière came this morning for a sewing lesson, and we cut out and basted a dress from old material turned wrong-side out to avoid faded parts. It was a close pattern requiring planning and patching – an object lesson in the use of scant material. The girls worked well and intelligently. One can already see their improvement. I only hope they can be helped to appreciate better standards of living and morality. At present, I doubt if they have any.

May 3rd No one went to church this evening. Instead, after our weekly sing for the hospital patients we gathered in the study at cocoa time and listened to the Doctor while he read poetry aloud. Such evenings are popular with our little family. Sometimes the Doctor reads chapters of his book or articles, and there are often lively discussions. One evening he read from Laura E. Richards's two books of allegories, *The Silver Crown* and *The Golden Windows.*[134] One of these, "The Grumpy Saint," we considered especially applicable to Mr. Lindsay. It is the story of a man outwardly a trifle at odds with the world, his sometimes censorious and pessimistic attitude cloaking a tender heart and consideration for his fellow men. When life ended and he approached the Pearly Gates, he was surprised to find them open and Saint Peter waiting to greet him. We alluded to the similarity of characteristics and hope Mr. Lindsay was pleased, for we feel no greater expression of our appreciation could be given him.

May 4th It is pouring rain. We would welcome sunshine after more than a week of dampness. The snow is fast vanishing, and there is a general stirring of spring indoors as well as out: plans for household and loom-room, lists for summer supplies, wage adjustments, house cleaning, which Mrs. Ryan and Lizzie began today. They just love it! It is amusing to find anyone taking that view of something most maids dislike.

May 6th We are longing for fresh food, the only "fresh" for a long time being periwinkles, a snail-like shellfish. To gather them, a hole is cut in the harbor ice, a piece of salt pork on a burlap bag lowered through it to the water and left over night, pegged to the ice. The periwinkles attach themselves to the under side of the bag and in the morning are scraped off and collected. We have cooked them in several ways and the Doctor is very fond of them. His interest is also developing in the use of corn-meal, formerly despised by the British as "dog-food." He finds that many favorite articles of food on the family menu have corn-meal as an ingredient and, convinced of its nutritious value, he suggests inviting the mothers of households to a dinner on the piazza, where they will be offered corn-bread, indian pudding, johnny-cakes and also periwinkles cooked in various ways, with recipes for their preparation. We have already sent corn-bread and the rule for making it to several settlements, which resulted in someone coming fifteen miles to get the rule for "that cake the lady at St. Anthony makes."

Who will say this is not a practical form of missionary work?

May 13th We woke this morning to see a real New England snowstorm: big soft flakes that melted as they fell or clung to trees for a brief return to winter. The Doctor is much better, but effects of nerve strain still linger. He is restless and wants frequent change of occupation. There are many affairs needing his attention: the spring work to plan, building to superintend, the new co-operative store to set in motion, drains, clearing of land, the pottery kiln and new laundry, besides numerous articles to write and a book on Labrador. I am now helping him with illustrations for the latter, which he is making from rough sketches of the northern coast.

Last Saturday, we painted most of the morning until the Doctor, growing restless, announced he was going fishing and wanted L.M.K. and me to join him. We acquiesced as usual and started after luncheon. It had rained and the trail was awful. Ponds were covered with water and slush; the streams had broken out. We waded through ponds and brooks, frequently sinking in the deep snow. I once had to be dug out. Our catch was only three small fish, but it was a jolly walk. The Doctor loved it. I think obstacles often only stimulate him.

For some time, he has planned to go duck shooting as soon as ice leaves the harbor, and the men have been preparing the launch *Daryl*, which has been laid up all winter. She was ready a few days ago, and yesterday men dragged her with long ropes to the water's edge, a repetition in reverse of hauling her on shore six months ago. I packed the Doctor's equipment and enough food for possible emergencies, such as being ice-bound for a few days, but the party returned in mid-afternoon, having found Hare Bay still blocked with wind-packed ice. They will have to wait until it blows offshore again.

May 14th This morning, I heard Dr. Grenfell stirring about and asking the maids if breakfast was ready half an hour ahead of time. That meant something afoot, and on going downstairs I found him nearly ready for duck shooting. Ted McNeill and David (who were going too) had been ready since 3:30 a.m. I poured their coffee and saw them off. They have not returned. Evidently the ice is out of the bay. There was a shortage of guns, and they took any they could lay hands on, to the consternation of Dr. Little and Mr. Lindsay, who have "feelings" about the use of their personal property. Fortunately, it was Dr. Grenfell, not David, who appropriated Dr. Little's. It is a raw, cold day, but perhaps that makes no difference to ducks.

Most of my day has been spent in the loom-room, with occasional stops to make illustrations for the Doctor's book. Nearly all the boys were at class tonight, and I do not see how they crossed the harbor, for the ice is full of holes. They are very philosophical about it and say if one falls in the others will pull him out. Steve Pelley came very near his end today in the big brook. He fell through one of the awful holes we found so menacing and was nearly dragged under the ice by the current.

We have another industry. Several months ago, someone sent the Mission two old knitting machines somewhat out of order. Mark has lately been experimenting and is now successfully running them after cleverly adjusting the parts. Dr. Grenfell will be pleased, for he has wanted them to have a function. Orders for stockings are already coming in from the staff, and this may prove to be quite a valuable addition to our industries.

May 15th We saw smoke over Fishing Point this morning, and as no steamer other than the *Portia* has any business in these parts as early as this we infer she has gone on to Griquet while there is clear water, as a change of wind would drive the ice back to land and block her for several days. She will call here on her return south.

It is interesting to watch the country gradually emerge from its white blanket as it was in the fall to see it disappear and become a blank whiteness. The hills, instead of pure white with a bit of tree appearing here and there, are now covered with young spruce, which is quite green where snow has covered it all winter. Mr. Lindsay says L.M.K. and I may go with him tomorrow to look for deer where they are wandering. He and Dr. Little have lately been out to the hills every day.

May 16th It froze last night and a hard crust covers the snow. We started promptly at eight o'clock. No one knows just where to look for the deer, but on the way to Fishing Point we soon found a part of the herd mounting the hill. A Lapp was following them, wearing a little shawl over his skin clothes, an amusing object! We climbed to the top of a high hill near the sea, and there below us was the main herd, scattered about in groups, the Lapps and their little dogs with them. The does with their fawns were apart from the others, and with glasses we could see them plainly. Two herders were lassoing a few of the stags to bring them "to town" for use with the pulkas, and it was amusing to watch the process. A herder caught one by the leg, and from our distance its struggles suggested a fly trying to kick itself free from sticky fly-paper. The animals are very graceful, and their movements over the snow reminded one of military manoeuvres as they walked in files, pairs and phalanxes. Mr. Lindsay told us he saw one day a doe swim a small lake, and her fawn followed her though it was only two days old.

From the hill-top, where we lay on soft moss and watched the deer, we suddenly saw the launch *Daryl* returning from Hare Bay, and I ran to the cliff edge to wave a greeting. The deer had moved in my direction, and on turning back I found myself surrounded by them. Mr. Lindsay had warned us so many times against startling them, especially with their fawns, that I was terrified and crept by, breathing a sigh of relief when the last one was passed. But they did not notice me, except one that barked strangely. Mr. Lindsay said afterward it was a doe with her fawn. Only anxiety for their young causes them to make any sound whatever.

Dr. Grenfell said he had had beastly luck, including bad weather, and he had only two birds. But he is less nervous, although still restless and eager to be doing something. This afternoon, his activities centered on selecting the site for Dr. Stewart's house, an interesting proceeding which involved merely walking over the land and choosing a location anywhere without regard to zones, building tracts or abutting neighbors. The proposed site is under the hill, not far from the hospital. The house will be built during the summer to be ready for Dr. and Mrs. Stewart on their return in the fall.

May 17th This has been a lovely, peaceful day. To church this morning and the staff here for dinner to partake of the Doctor's ducks. The master of the feast was absent, having gone with the "reindeer king" after service to look at the deer, but the rest of us enjoyed the ducks. Mr. Lindsay appeared later with his temper slightly frayed, reporting he had left the Doctor sitting on a hill-top talking with a man and irritating the deer. That may be, but I sometimes feel those deer are wrapped in cotton wool!

The *Portia* has not come. When she arrives, we plan to don our most spectacular winter costumes for the benefit of Captain Kean and go

The Abode of Peace: Emma White and Jessie Luther on the piazza of the Guest House, July 1908 (GHS)

out to the steamer clad in blanket suits, cossacks,[135] skin boots, etc. Mr. Lindsay will wear his Lapland costume, which is unquestionably wild and woolly. He will also wear his monocle. We think we may create a sensation among the passengers and give Captain Kean a subject for conversation all summer. Dr. Grenfell's cousin from Montreal and a friend are expected on the steamer, also the new nurse. This will be the beginning of the summer colony and many changes. I cannot yet fully realize that within a few days, with the arrival of the *Portia*, the happy family life we have enjoyed in this little house during the past few months will be over and never come again. Dr. Grenfell has named it the Abode of Peace, and so we feel it to be.

The Doctor plans to go to St. John's by the *Portia* on business pertaining to the seamen's institute, a project in which he is much interested. The orders for supplies from St. John's are made out, and I am ready for the *Portia*'s mail, the first of the season. It is well, for if the Doctor does go by this boat there will be quick packing to do.

May 18th The weather has been perfect today, clear with little wind and cool enough to make movement a pleasure. The morning was spent in the solarium with the Doctor, who was working on his chart of the Labrador coast. It worried him, for he feared it was not sufficiently accurate to be of real value. I do not see how he made it at all from notes and drawings on scraps of paper patched together. I am developing a profound respect for anyone who can make a chart.

Mr. Lindsay returned today for lunch, reporting thirty fawns and the herd in practically the same place, near Crémaillère. It is odd they should return to the locality where they landed. It is said they always do. They could not have chosen a better environment for fawning, where there are no brooks or tuckamore which the herders feared as hazards, and they can be plainly seen from distant hills.

The two girls from Crémaillère were in the loom-room this afternoon, ready for their lesson in the three Rs. It seems strange to teach grown girls the elements of education. It is also interesting to watch their encouraging reaction, but they are still untidy. It takes time to eradicate the influence of old habits and environment. The girls would like to stay here and work, but it is necessary for them to be at home during the fishing season as their father has no one to help him, and his living for the year depends on his catch. They may return at the close of the season, but that is when I return home.

My latest new job is coloring artificial eyes for a stuffed head of an angora goat! I was not sure of the coloring and sent one of the orphans to fetch a goat for observation at close range. After waiting a long time, I started to investigate and met Reuben on his way from the stable, leading the shorn billy-goat, protesting all the way and stopping at intervals to plant his forefeet, perhaps thinking himself on the way to execution. He was followed at a safe distance by a distracted nanny-goat, bleating piteously. I found a goat's eye more extraordinary than I thought. It is yellow and the pupil is a horizontal slit. I have certainly been unobservant of goats.

It is so warm we lunched and dined on the piazza in the sunlight, almost giving us the feeling of having moved to a summer residence. The view is lovely – the hills on one side, the harbor on the other. Subconsciously, we are listening for the *Portia*'s whistle. We may hear it at any moment.

May 19th A perfect day, warm, bright and still. The Doctor and I spent it on the piazza, he finishing his chart, I correcting manuscript.[136] He has paused from time to time, as usual, to photograph someone, then back to the chart again. We stopped only when the piazza became a dining-room, and we grudgingly relinquished the table.

The harbor ice is fast breaking up. Doctor says it will go "all abroad" [apart] and float out with the tide. I am interested to see it, having fancied it might go in one great floe. The wharves are gradually emerging on the waterfront, and the high drift that nearly smothered the [Israel] Burts' house is reduced to a minimum. The little trees the Doctor and I planted in the ice to make the komatik road have disappeared. The hills are now dark, and on their sides are great patches of snow that take the form of animals and grotesque figures. Mission men are digging the foundation for the kiln. It is to be in a separate building, attached to the engine house and quite by itself. I can scarcely believe it is really begun.

Reuben came this afternoon to report the imminent arrival of the *Portia*, but she has not materialized. I have looked over the Doctor's clothes in preparation for St. John's, with its formalities. It seems strange to consider boiled shirts and starched collars, which I have suggested it would be well to send to the laundry after arriving at St. John's. The Governor might look askance at shirts finished *à la* St. Anthony.

Mary 20th Tonight was probably the last staff dinner for the winter family. We expect to hear the steamer's whistle at any moment, and I do not like to think of the ensuing changes. The Doctor announced at lunch that he was going out to clear the land for Dr. Stewart's house, and it seemed like old times when he started off with an axe on his shoulder. I went out later to help select the site for the other new house, which will be nearer the water. There will soon be quite a little settlement of Mission buildings.

The Doctor and Reuben have a new project. They are constructing an "icy cave" in the curio cabinet, to be used as a setting for several "whitecoats" (baby seals), recent products of the taxidermy department. The "cave" is made of stiff paper crumpled over boards to imitate a miniature ice-pan. They have smeared this with glue, then thrown lime over it by handfuls, and the consequent effect on one's lungs was rather harsh. Those who entered the piazza stopped short, peered through the fog of lime dust to see what it was all about, then beat a retreat, choking. The Doctor was not satisfied. He not only wanted snow for his cave but glittering ice as well and tried adding pieces of mica he had taken from rocks some time ago, but the pieces were too large to be effective. The Doctor, however, recognizes few obstacles. He had an idea – the coffee grinder! – and he turned the crank while I pressed the mica into the hopper. The effect is quite gorgeous!

There is also another addition to the curio cabinet. A few days ago, I came in to find the Doctor painting grotesque faces on a pair of whale's eardrums. These are heavy bones about four or five inches in size, curiously shaped like a human profile. On one he had painted the face of an old man, a young man on the other. I called them the Young Obadiah and the Old Obadiah,[137] and we made merry over them, the Doctor chuckling in his inimitable way while he added the finishing touches.

The young man, with sleek black hair, jaunty black moustache and bright red cheeks, looked like a smart dry goods clerk, and we imagined him trying to get the better of the white-haired man in a red cap, who was laughing at him and refusing to be taken in – even imagined their actions and conversation. Then the Doctor set them upright in bottles and placed them in the cabinet facing each other. Silly? Childish? Of course it is! But what a drab world this would be without enjoyment of simple things and indulgence in free play of the imagination! These have been our salvation during the winter.

The puppy-dog komatik team is also in the curio case.

May 23rd The Doctor has finished his chart. The sketches are done, and everything is now ready for the Governor's inspection. But one important matter had been neglected – the Doctor's hair. It needed attention before his contact with civilized society, and after luncheon I turned barber. He went to sleep during the shearing process, explaining afterward that the touch of comb and scissors have a soporific effect. He has packed his small belongings and says he is now all ready and a free man. But the *Portia* does not come.

May 25th The *Portia* arrived this morning at four o'clock. I was awake when the whistle blew and heard the Doctor call, "The *Portia*!" Then there was the usual rush and the familiar sight of the steamer out there again in the harbor just after sunrise, with little boats putting out from shore almost as if there had been no interval of winter at all.

The new doctor who was expected did not arrive, but thirty patients did, also an additional nurse who is badly needed at the hospital. Dr. Grenfell ran up for breakfast at 6:45, and I broke the news that he had 124 letters to look over.

Later. Mark has come in to say that a steamer has just passed Fishing Point on her way in, and he thinks it is the *Strathcona.* It is the *Strathcona*! Dr. Grenfell did not expect her to leave St. John's until after the first of June and intended to return on her to St. Anthony. This may change his plans. We look quite populous with the *Portia* still in, two launches unloading freight, several fishing boats afloat (one with colored sails) and at the wharf the Doctor's launch, *Daryl,* which has just been into the bay for wood. Incidentally, we find the *Portia* is not the *Portia* but her sister ship, the *Prospero.* Except for the name, no difference is perceptible.

More excitement! The Doctor has just come in with his cousin, Mr. Spencer,[138] a tall young New Zealander who will be here all summer in charge of gardening, clearing of land, etc.

Dr. Grenfell has decided not to go on the *Prospero* after all but on the *Strathcona.* He can make better time. He has been opening his many letters, reading extracts, asking advice, irritated by some, laughing over others – all of them reminders of a restless outside world.

May 26th The *Prospero* is on her way south without the Doctor. Her next arrival will bring an advance guard of the summer invasion: volunteers, visitors, doctors, a nurse or two. I foresee problems of adjustment.

Captain Kean of the *Prospero* reported open water toward the straits, which stimulated the Doctor to try for a quick run to Battle Harbour on the *Strathcona* before starting for St. John's. He has invited me to go, remembering my disappointment in missing Battle Harbour last year. He is planning to start in the morning.

May 27th We left St. Anthony soon after breakfast. There was sunlight, pale, rather grudging sunlight, and chilly air. Outside in the open sea, a few far-off icebergs gleamed white beyond the expanse of steel-blue water, and as we turned north toward the straits the air grew colder – a penetrating chill – and on the horizon, below the dull grey sky, we saw a line of misty white extending as far as the eye could reach. Field ice. An impression of mysterious menace was in that long white line ahead, a threatening barrier, and soon we were conscious of bits of ice in the water, then more and larger cakes floating nearer, and I heard for the first time the sound of ice bumping and scraping against a ship's side.

The white field was nearer, spaces between the floes smaller, and when an icy fog suddenly and silently enveloped us the *Strathcona* turned and beat a retreat toward St. Anthony. It seemed to be an escape from some great monster or a relentless force, which in reality it was.

May 29th The *Strathcona* left for St. John's this morning with Dr. Grenfell on board. He hopes to transact his business and return in time to greet the first contingent of guests and workers who may arrive by the next *Prospero,* but his itinerary will take time. We at St. Anthony will also have our busy hours. We do not know who or how many may come but must be prepared to accommodate and feed all and sundry. We are planning flexible food, and I am checking sleeping quarters varying from real bedrooms for the Doctor's guests to cots for the boys in a small building now used for storage and dubbed "the bunk-house" (by courtesy). There are, of course, sleeping bags, which might strengthen the impression of some of the students that they are real missionaries undergoing privation. I am quite looking forward to fitting pieces together like a jig-saw puzzle.

12

The Summer Invasion

14 June – 10 September 1908

June 14th Since the Doctor's departure, we have been busy in the clothing store, where garments donated for distribution by the Mission were mixed indiscriminately in bins, making selection difficult. It has taken days of work to sort them, but all is now in order and the articles packed on shelves.

Our latest excitement is burglary – by dogs! Some time ago, they broke a pane of glass in the porch window, another day got into the pantry and ate everything in sight. I went downstairs and drove them out, followed by Dr. Grenfell, who caught and chastised two of them. Two nights ago, I heard them again and found the porch full of them. They ran out when I shouted, but I knew they would return, so L.M.K., who appeared during the commotion, joined me at an upstairs window, where we waited for them with the air rifle (which stings but does not injure). I never before knelt at an open window in the middle of the night with a rifle on the sill, alert for marauders. We felt like real pioneers.

Dr. Little has just telephoned from the hospital that Dr. Grieve,[139] from Battle Harbour, has arrived on a whaling schooner and is coming here for the night. It seems odd to have a visitor materialize in this way "out of the blue" but is characteristic of life on this coast.

The weather is delightful, like the best of our April or early May days, and there is a little green grass. This is the season I love being at home, and I think – perhaps wistfully – of apple, peach and pear blossoms, woods with their misty green veil, the peep of frogs at night, but soon this country will also awake. Flowers will bloom near vanishing snow banks and birches on the hillside be clothed in pale green leaves.

June 27th The last *Prospero* brought several of the summer staff, among them a young man from Providence, R.I., Mr. Paul Matteson,[140] a welcome addition not only because of his congenial personality but his willingness and ability to co-operate in a wide range of activities, including assistance in the hospital and Industrial Department, clerical work and water transportation. Another arrival is Mr. Hause,[141] an engineer sent by the Pratt Institute to install an electric lighting system as a donation to the Mission. He and the local men are already digging post holes, and we feel somewhat dismayed at the prospect of blazing lights and a forest of ugly poles in our primitive community. It seems like an intrusion into our simple life and cherished remoteness from urban glare, confusion and artificiality. We admit, however, the convenience and safety of indoor electric lighting as opposed to the kerosene lamp. Presumably, what is referred to as "progress" will come eventually in some measure, but I am glad to have known the quiet and beauty of this interesting place before the heavy hand of "technology" is laid upon it to diminish its natural and unusual charm.

Still another arrival is Mr. Cushing,[142] who will take charge of St. Anthony's economic affairs during Dr. Grenfell's absence on the *Strathcona*. He will also superintend activities of the many college boys, all eager to work but few with experience. It will be a busy summer for Mr. Cushing with few idle moments.

The whole country is now green, and the days are so long there is really no night. Sunrise glow merges with the sunset twilight, for the light along the horizon does not fade entirely. The mosquitoes, however, have come in swarms. They are the one real annoyance in this beautiful country. There is netting for screens, but we have waited in vain for the men to put them up. They have waited for a rainy day which does not come, and we may soon be eaten alive. I may have to do it myself.

At last the kiln is begun. A small building for it with a concrete floor and a chimney has been added to the workshop. The materials – fire brick, ordinary brick for the outer wall and the iron encircling bands – came last fall. There is much pottery ready for firing, and it is time to begin the kiln, but who is to build it? Two or three men who have built the few chimneys in the community and have some experience are much interested, but they are also wary. The straight wall of a chimney is one thing, the circular wall and curved top of a kiln quite another. No one has ever seen a kiln. With characteristic Newfoundland reserve, the men listened to my explanation, examined the diagram and looked on silently when, realizing something must be done, I tied a cord to a brush, marked the circumference of the kiln on the concrete floor with red paint and began to lay brick – remarking as I did so that I knew so little about it, it might be all wrong – and asking for suggestions from anyone with more experience.

Fortunately, this touched a responsive chord with Mr. Holley,[143] one of the local men, who after watching me for a few moments suddenly interposed. "Miss," said he. "I think it goes this way. I'll do it." And he at once took over, handling the trowel with the deftness of an expert. With gratitude and hope I left him, and all went well for several days. Then there was a halt. All the cement had been taken for the foundation of the new barn, and precious time must pass before more can arrive by the mail boat from St. John's. Meanwhile, Mr. Holley waits.

July 1st Dr. Grenfell has arrived on the *Strathcona*, bringing his friend Dr. Wakefield[144] as well as Dr. Stewart and his bride, who won our hearts at once. It is interesting to speculate on reactions of those making the journey to Labrador for the first time. I begin to think of it as a test of character – certainly of endurance and adaptability – and a fitting prelude to further northern experience.

July 6th The *Prospero* has brought more cement, and the kiln is nearly ready for the arch, which will have to be done carefully. Mr. Holley, though interested – even indicating pride in his accomplishment – is still doubtful about the finishing touch, and assistance from Ted McNeill, who is clever and ingenious, will be welcome.

The vegetable garden started by Mr. Spencer soon after his arrival in early June is at last a reality. The brown earth of the enclosure near the Guest House is striped with rows of green leaves which give promise of fruition. Anticipation of fresh vegetables produced from the earth instead of tin cans fills us with joy. Even a few strawberry plants are hopefully started. Garden cultivation at St. Anthony is not easy. Rocks had to be removed, and the soil, with its substrata of clay and covered with a spongy growth of sparse grass and low plants, required much preparation as well as fertilizing. The latest addition to our livestock, a small horse (the first ever seen in St. Anthony), was brought on the mail boat a few weeks ago, also a plow, which was used after removal of the rocks. A fence with a gate surrounds the "estate" to keep out the goats. Mr. Spencer has done a fine job.

The whole country is changing in color. The hills are now dark green with scrub growth of spruce, alder, and birch. The small areas of pasture are a brighter green, and much of the low vegetation is in bloom. Fishing Point Head is covered with a lovely dull pink mantle. Even the water seems a different color. The sun is up before three o'clock in the morning, and we can work in the shop until nine at night.

Contrary to general impression, there are many flowers in Newfoundland. Most of them are small, growing close to the ground.

Dr Arthur Wakefield, 1908 (GHS)

Some, like the "Labrador tea,"[145] remind me of alpine flora. There are also small fruits: raspberries, cranberries (unlike the New England variety), blueberries on "bushes" four or five inches high, and the bakeapple, which I have never seen or heard of elsewhere than in Newfoundland and Labrador. The bakeapple resembles a low blackberry in form and size but is yellow when ripe and has a pleasant, indescribable flavor. The people cook it for jam and compare it with fruits unknown to them. One woman, soon after my arrival, spoke of it so enthusiastically I asked her what it was like. "Oh! like strawberries, Miss," said she, drawing on her imagination, as she had never tasted them.

July 19th Work and worry have characterized our days lately, and a delightful weekend at Mr. Lindsay's deer camp with L.M.K. and Mr. Matteson was a welcome break. We now feel that almost all details of deer herding are known to us. It was any amount of fun. We wandered, watched the reindeer and were present at milking time to observe the process at close range. A reindeer appears to dislike being milked. It is a lively animal and does not stand still. One Lapp holds the reindeer's head by a rope while another, with a tin can in his hand, awaits an opportunity to milk into it in spite of the deer's sideways movements, jumps and an occasional kick. It seemed an unnatural performance to us, accustomed to a cow's calm, unemotional co-operation and the comparative security of sitting on a stool – and all for about a half pint of rich cream, although when diluted it is equivalent to nearly a quart of milk. Possibly some of the deer are more co-operative, but those we watched provided plenty of exercise for the milker.

The only tent was Mr. Lindsay's own, which he placed at our disposal, but we slept little due to our unusual surroundings and the swarms of insatiable mosquitoes that became so intolerable Mr. Lindsay offered us the use of sleeping bags, an added discomfort during the unusually warm night. The men slept on blankets in the open with netting over them.

July 26th The *Prospero* has come, bringing expected guests – four more than we planned for. The whole house had to be rearranged for their accommodation and places found outside for some of them. Among the arrivals are Miss White, Mrs. Sayre from Toronto [sic], her two sons, Francis[146] and J. Nevin,[147] and Miss Mary L. Dwight,[148] who will take charge at the orphanage during the absence of Miss Storr on vacation. There are other friends of Dr. Grenfell and some college boys fired with enthusiasm and anticipation of adventure after hearing him lecture. Several people from the steamer came ashore and wanted to be shown over the premises. Our reputation for hospitality had to be sustained by feeding them with cocoa, and then Miss White wanted to talk. It was nearly 2 a.m. when I slept.

July 27th The kiln is at another standstill, this time for lack of outside brick. We find to our dismay that the adequate supply sent especially for the kiln last summer has been taken by the local trader to replace what was borrowed by the Mission for chimneys some time ago. There was no other brick to repay the debt. There might be a wait of three months before more could arrive from St. John's. The trader wanted it, so they took ours. Quite a simple arrangement! Meanwhile, Mr. Holley again waits.

Tonight, for the first time, the house is lighted by electricity![149] The glare almost dazzles us. It is truly a comfort to have a light without fear of a draft blowing it out or danger of conflagration. They have wired all sorts of things, even an old iron lantern I rescued from a loft and brought home against a storm of protest. After repairs, it now hangs in the piazza as our main source of illumination, and the opposition meekly admires it. Light will be of inestimable value in the hospital and also in our workshop, where during the winter lighting was a great problem with only kerosene hand lamps on the bench or placed on wall brackets, where they cast shadows on the work. In spite of this appreciation, however, we still resent those thirty-foot poles and wish they were not the necessary accompaniment of our indoor convenience. We are sure Dr. Grenfell will share our reaction, for he even deplored the three little telephone poles installed between the hospital and Guest House last summer.

The new laundry is nearly ready for use. It is one of the most needed additions to the Mission's equipment, but the chimney has been built inside the building, and the boiler occupies much of the space, leaving little for tubs. Miss White's delighted approval of everything is very gratifying because of her personal interest in the Mission and contact with its loyal supporters at home, who will welcome an encouraging report of constructive work under way as well as affirmation of the people's overwhelming need in the medical field, which the doctors are trying to meet. Dr. Wakefield is now in charge at St. Anthony, as Dr. Grenfell and Dr. Little have just started on their long Labrador trip, taking Mr. Matteson with them as secretary. They will be away at least six weeks.

Before they left, I took the Doctor over the Guest House to show him changes made while he was away. He entered our little living room, stopped, looked around, and said slowly, "I think this house is absolutely perfect." If he is satisfied and it seems a real home to him, it is worth all the effort we have put into it. Everyone seems to think it is quite luxurious, yet there is nothing of real value: shelves put up by the boys, painting done by ourselves, rough seats that we have made covered with red blankets, cushions covered with turkey red cotton or bits of hand weaving. There are prints in passe-partout[150] or frames made in the shop, books and curios on the shelves. The bedrooms are pretty, with curtains and bedcovers of 12-cent print, downstairs curtains of 8-cent cheesecloth and a few woven rugs. Perhaps an undefined atmosphere is combined with the visible attractiveness of our simple surroundings, an impression that causes those who enter the house to feel welcome and at home. Anyway, we are happy to recognize the reaction, whatever its cause.

This afternoon George Andrews,[151] one of the men who assisted in Dr. Grenfell's rescue from the ice, came to tell me more about it. His story, perhaps even more moving because told in the local vernacular, was so realistic that I wrote it down as soon as he left me and relived the experience, so terrible for us all.[152] While I wrote, the light across the harbor on Old Man's Neck turned to gold. A violet mist fell over the distant [St Anthony] Bight, and lovely sunset colors cast reflections in the harbor's clear water. The air was very still, and as evening drew near a thin line of blue smoke rose from the Pateys' tilt under the hill. I love to see it at evening and in the early morning, a little sign of homely life in the quiet world.

Dr. Wakefield conducted an open-air service for the patients this afternoon. Even the bed patients were carried down by the men. The little organ was brought from the orphanage, and Dr. Wakefield gave an address. It is the first Mission service since the departure of Dr. Grenfell, who usually conducts one when he is here. We are also to renew our weekly "sing." Miss White is becoming familiar with all our activities.

July 28th We have just heard that the schooner *Blake*, supposed to be on her way with our brick, is not coming after all. The agent is presumably waiting until he has enough cargo to fill her, perhaps with the provisions that were ordered over two months ago. We are short of all sorts of things.

If it were not for the co-operative store and what arrived by special schooner from Boston, we would be living on pork, corn-meal and tea. Dr. Grenfell wired from Battle Harbour to send the brick, and Mr. Cushing and I have written to ship it by the next *Prospero*, but meanwhile we have decided to be independent and make brick from the native clay. Today, Mark and I made a wooden mold and finished a few sample bricks, hoping to have a hundred in a few days. I am more than ever convinced that brick-making would be a valuable native industry to supply the increasing demand and save the expense of importation.

The agricultural department is having difficulties with "Harry" the horse. He objects to drawing peat from the bogs – even the empty cart – and Mr. Spencer finally tied him to a cartwheel and left him in the tuckamore. Later, they went to look for him and found a wrecked cart and harness but no Harry. He is probably enjoying a scramble over the country.

July 30th Harry has returned to civilization. He seems chastened, for he draws without protest. Having had his fling, he is now satisfied to face realities.

Archie came in at breakfast time with a stranger who introduced himself as Mr. Clark.[153] I was surprised to find he was one of the *Pomiuk*'s crew, for she was not in the harbor. A dead calm prevented their entry, and Mr. Clark came ashore in the tender for a pilot and supplies. The *Pomiuk* is a staunch little craft, larger than the *Daryl*, and painted leaden grey like battleships in the Spanish war.[154] A crew of five Yale men brought her safely from Boston along the treacherous coast that has often proved disastrous to larger vessels.[155] They sailed direct to Battle Harbour, then separated. Two men returned south:

DeWitt Clark and Francis B. Sayre working with Harry the horse, July 1908 (JLP)

one joined Dr. Grenfell on the *Strathcona* while Mr. Clark and Mr. Yates[156] came to St. Anthony to serve the Mission in nearby waters. They came for lunch, then started off again, taking Mr. Sayre and Mr. Palmer[157] with them to blow up a wreck near Griquet.

July 31st The *Pomiuk* is still away, probably delayed by foggy weather – a happy chance since it allows time for rearrangement of sleeping quarters. Two of the boys will go to the bunk-house. For the others, we have made two temporary beds in the hall from springs found in the storeroom, while for mattresses – unfortunately lacking – we have substituted old quilts which we hope are sufficiently sterilized after hanging on the fence for weeks in sun and rain. There are nice blankets, but the place where they came from is over the food storehouse, and the odor of last year's frozen meat, hides and salt codfish still lingers, combined with the permanent odor of tarred rope, kerosene and molasses. We hope the boys won't mind.

August 1st The *Prospero* was sighted entering the harbor at midnight, and "Everyone up" was the call. When I reached the porch, the searchlight was flashing all over the place, lighting the hills and picking up the boats already on their way out to the steamer. Her coming at this hour with flashing lights added to the excitement always felt on her arrival. Some of Dr. Grenfell's guests failed to arrive, and there were only eight patients, none of them serious cases. There will, however, be much assistance needed at the hospital during the next few weeks, while Miss Kennedy is vacationing with Miss Storr on the Labrador, for Dr. Wakefield has only the newly arrived doctor and nurse for professional help, aided by the visiting students when called upon. L.M.K.'s present school vacation takes the form of regular assistance at the hospital, and I help when I can. Many patients come on the mail boat, though usually only a few stay to be hospitalized. After treatment, they leave on her return trip. But emergency cases may come in at any time from nearby villages.

Dr. Wakefield is an athlete and a cold water enthusiast who does not allow the icy temperature of St. Anthony harbor to interfere with his morning plunge. He has organized a bathing club after urging (sometimes daring) the students to join him. His hardy example has even induced several village boys to try it. It is not strange that few if any native men or boys have learned to swim. There are no warm sandy beaches, only rocky shores, and the remnants of icebergs (called "growlers") that often drift into the harbor during the summer only add a few degrees of lower temperature to the already icy water.

The ladies, refusing to be outdone by the men, decided to follow their lead with a modified program. On the way to Fishing Point there is a small curving beach, its tiny pebbles gleaming pure and white beneath crystal clear water – an inviting spot with bushes in the background. Yesterday afternoon, we started for the "beach," L.M.K. and I provided with pyjamas for bathing suits. But we had not reckoned with the tide. The pebbly beach was nearly dry. To reach deep water, we would have to clamber over large, partially submerged rocks where swimming was out of the question. But we donned our pyjamas anyway and decided to at least take a dip in the icy water, where we splashed for a few moments and hastily retreated. At least we know how cold the water is and – now that it is over – feel that an experience of bathing in the sub-arctic zone is worth the shock. Meanwhile, Miss White and Mrs. Sayre drew up their skirts and waded decorously in ankle-deep water between the rocks. They reminded us of storks.

It has been really uncomfortably warm, the mercury above sixty degrees, and I think with regret of thin clothes left at home. Reuben's wife was so exhausted by the heat that she had to lie down. The garden is drying, our water supply is very low, and it will soon be necessary to boil drinking water. The last trace of snow has vanished from the White Hills. Even the little rivulets are drying up. We hear of plans for a reservoir on the hill, which is greatly needed, but plans do not fill our cisterns.

Dr Arthur Wakefield dives off a "growler" in St Anthony harbour, August 1908 (GHS)

The outside lights were turned on for the first time last night. They seem like intruders. The indoor lights are most welcome, but no one wants the whole country illuminated, competing with the starlight and moonlight – even the aurora. We hope they can be turned off on clear nights. Nothing further has been heard of Peary[158] on his way to the North Pole. He has evidently gone on without stopping here. The house is very still, but the dogs across the harbor are beginning their unearthly midnight howl.

August 12th We set off in the yawl today, a party of thirteen, to show Miss White the reindeer herd, as so many people will ask about them on her return to Boston. We stopped at camp for Mr. Lindsay to join us, then went on to eat our picnic lunch on a high rock where we could watch the deer. The day was perfect, and we had a grand time in spite of the engine breaking down twice and the kerosene lamp taking fire, incidents that merely added excitement – and anyway, what could one expect of a party of thirteen?

Miss White and Mrs. Sayre have had few dull moments, one excitement being the arrival of Dr. John Bryant and his brother [Owen] with two friends on the Mission schooner *Lorna Doone*,[159] which they had chartered and transformed into a yacht for a cruise to the "farthest north" possible during the short summer. Dr. Bryant was here with Dr. Grenfell on the *Strathcona* during my first summer on the coast. The others were strangers. A round of social activities began with a

John Bryant [?], Dr Norman Stewart, and Cuthbert Lee, July 1908 (JLP)

merry supper at the Guest House, followed the next morning by a fifteen-mile walk to the deer camp and back over rough country. They were with us again in the evening. The next morning, they came to the workshop and bought nearly all our stock of industrial products, then invited us to the schooner.

After travelling about on cluttered motor boats and dirty fishing craft, imagine the luxury of going out in a swift tender with a spray hood to an immaculate schooner yacht with a crew in white duck and a big fresh cabin! Mr. Lindsay, who had come in from camp and joined us, was made especially happy by a bottle of lager, which was something he had not tasted for a year. We inspected the schooner; then they sailed away to the north after what seemed to us a gay whirl of social activity, and we have felt the stimulation of their visit all the week.

The vegetable garden is a reality. We have turnips (the tops used for "greens"), radishes, lettuce and mustard. If the frost holds off, there will be peas, beets, and beans. The general food situation has improved. Of course, our meat supply is limited, but fish is always with us and we prepare it in many ways. We have not yet taken chances on squid, which the men are now catching for bait. It looks unpleasant, but some of the people eat it and tell me it is delicious. "Tastes like oysters," they say. If, however, it resembles oysters no more than bakeapple is like strawberries, I am inclined to be skeptical.

St Anthony hospital with Guest House (*behind left*) 1908
(National Archives of Canada)

Our real luxury is milk. There are now several cows and we even have a little cream. We have built a tiny dairy in a closet under the front stairs. It is lined completely with white oilcloth, which is easily cleaned and prevents dust falling through from the stairs above. Dr. Grenfell saw it before he went away and was delighted with the arrangement as proof of what can be done with limited resources.

Miss White, Mrs. Sayre and her sons leave us on Sunday by the *Prospero* after a visit which I hope has been as happy for them as for us. Francis Sayre and DeWitt Clark, one of the young men from the *Pomiuk*, may walk across country to the west coast to connect with the boat from Battle Harbour. It will be a test of endurance for them but an interesting experience. Miss White is making "last visits" before going home. This afternoon, we walked to Fishing Point Head for her final view of sea, rugged country and mossy barrens from the cliff's edge, also to demonstrate her sporting spirit in following the twisted rocky paths of the steep hillside. Far below us, a group of fishermen in their trap-boats were "jigging"[160] for cod. Their yellow oilskins and the boats, dull green, brown and blue, with sails of red, brown and deep ochre, were a glowing nucleus of color in the clear green water near shore. We stayed until the late afternoon sunlight glowed pink on Old Man's Neck across the harbor.

St Anthony staff, summer 1908. *Left to right*: Dr John Mason Little, Dr Grenfell, Ruth Keese, DeWitt Clark [?], Francis B. Sayre (GHS)

August 24th The long-looked-for schooner *Blake*[161] has arrived at last, bringing needed supplies for the Mission and brick for the kiln, which will soon be finished, I hope. Now where is Mr. Holley? He vanished some time ago.

Schooners are returning from their northern fishing grounds. So many come and go I have lost count of them. Almost every night several are anchored in the harbor. One belonging to the Mission has just been to the bay taking several of the college boys to cut hay for the cattle; another brought lumber from the mill, still another freight from St. John's and Battle Harbour.

A feeling of regret has accompanied any reference to Battle Harbour ever since my abortive attempt to see it last year. There may be an opportunity before I leave, but at present everyone is needed at the hospital. I find my latest occupation is "trained" nursing. To be sure, there have been brief periods of hospital assistance ever since I came, but now they are more frequent and prolonged. Last Saturday, I was there all day and felt quite proud of myself after assisting at three operations but lost my pride yesterday when during quite a simple operation everything turned black. I staggered to the swinging doors and fell through them to the corridor floor, where I sat for a while, waiting for clear vision before returning to the operating room with apologies.

The fishing season is nearly over except with hook and line, leaving the boys free to continue their carving class, which begins next week.

Francis B. Sayre gives the orphanage children a haircut, 1908 (JLP)

Two orphanage boys assist with harvesting peat (JLP)

Francis B. Sayre and DeWitt Clark set out to walk across country, summer 1908 (JLP)

Ruth Keese and Jessie Luther, 1908 (GHS)

So does Miss Keese's school. She is to stay on for another winter as well as several other members of the staff. I only hope Mr. Matteson is one of them.

Mr. Lindsay went to St. John's by the last boat to execute many commissions. He came in from camp expecting, for once in his St. Anthony experience, to be dressed in good clothes, for St. John's is quite a city. Unfortunately, he could not find them and supposes they were taken on board the *Strathcona* by mistake and are probably, by this time, distributed along the coast. He had to set out in his camp clothes: high spiked boots, knickerbockers, flannel shirt with a red handkerchief knotted around his throat, canvas shooting jacket, and bright Kelly green golf stockings, sent with miscellaneous articles to the clothing store and presented to Mr. Lindsay on St. Patrick's Day. He also wore a soft hat with his monocle attached to the brim. He treated the situation as a joke, and I think he really enjoyed making himself as weird as possible to impress Captain Kean of the *Prospero,* who judges him merely from outward appearances and considers him a freak anyway.

Sunday The *Prospero* has come from Battle Harbour and brought news from Dr. Grenfell. The *Strathcona* ran aground somewhere near Hamilton Inlet and broke three screw-blades. They had two new ones and after making a third from some scrap iron are now on their way home.

Ted and Archie have each received a scholarship for a year's course at Pratt Institute and are to go to New York by this boat. Think what it will mean to them and what their first impressions of a great city will be after never having travelled farther afield than Battle Harbour![162]

August 29th The exodus is beginning. The young men are gradually leaving us. Six have already gone. One goes by this boat and two more later. One more visitor is expected, but that is all, and we will be the original family for a week or two before the Doctor and I go away. Mission plans for the winter are already in the making. Dr. Little, Miss Kennedy, and L.M.K. are planning to stay on and Mr. Lindsay will be here. Dr. and Mrs. Stewart are to have this house during the winter. Miss Storr is going home to England for a visit. Plans for the industrial work are still in the air. Mr. Matteson may stay to superintend the shop and direct the boys' classes. One of the older weaving girls is being trained to take charge of the loom-room.

Our cook and waitress are both leaving on Wednesday to be married. A new cook is in training and another girl for waiting, but it will be difficult to replace the fine, loyal girls who have been my reliance since the Guest House was opened. So many things need attention before I go home that I am almost overwhelmed. They must be considered one by one, not *en masse,* for my own peace of mind.

William Lindsay en route to St John's, August 1908 (GHS)

September 7th Mr. Holley (who since the new brick arrived has returned to the kiln) came to me this morning in great trouble, saying that he could not possibly build the dome to follow the sketched pattern. His ruffled feelings were smoothed and he went away, but returned still in doubt, wanting to go over it again. He has since returned for the third time with the lost original plan that Ted had put away so carefully he could not locate it. Perhaps this will end the trouble.

The season is changing. After several cold, cloudy days it is raining heavily tonight, and a "big wind" blowing in from the east is a forerunner of rough weather to come. We must expect it now at any time.

September 8th Those goats! The new family has been in the vegetable garden three times today. I cannot imagine how they got in: no opening can be found in the fence. We are concerned, for there are fresh peas in the garden and we want them for ourselves. Someone must leave the gate open. There are several boxes there, large enough for two to sit on. I have wondered why the engaged maids often come in from that direction.

There has been a funeral today that took the form of an Orangemen's parade.[163] John Simms died at the hospital yesterday and was buried with all the ceremony of the order. The men marched to the hospital from the Lodge in the glory of their full regalia and from there followed the coffin to the cemetery. Since this was only an interval in the daily routine of work, a change of clothing was impossible for most of the men, and there were collarless shirts and working clothes, but the grand orange and gold scarves and gold-trimmed caps made the scene from a distance quite brilliant. Those who went to the grave said the ceremony was impressive, all the Lodge members joining arms around the grave and bending as the coffin was lowered. Fifteen minutes after it was over, the men, shorn of their grandeur, were back at work.

The village was all agog this afternoon when a large white steam yacht came to anchor in the harbor, and we found the owner, Mr. Reid[164] of the Reid Newfoundland Railway, was on board with Governor MacGregor as his guest. They came ashore, visited the hospital, orphanage and shop, then hurried off to get out of the harbor before dark. It was interesting to meet the Governor again, but their brief, unexpected visit was disappointing. It has its amusing side, however, when we contrast such a casual arrival with our elaborate preparations for entertaining him and his Secretary of State as guests two months ago. The Governor brought news of Dr. Grenfell, Miss Storr and Miss Kennedy. Dr. Grenfell was passed at Hopedale going south and (they say) will probably be here this week. We hope it is true, for as usual we need him.

September 10th Charlie [Snow], a young man who came in August from a small settlement down the coast, has joined Mark in the pottery class. His legs are crippled from rheumatism, but he has the use of both hands and came hoping to be taught some trade or occupation while at the same time "gettin' a little learnin'." He has made pottery, shown a definite interest in wood-carving and is struggling with the three Rs. His desperate efforts to express himself in writing and difficulty in spelling even the simplest words are pathetic. I gave him a reading lesson yesterday and told him to write me a letter. Today, he brought a sentimental effusion copied from a book, *The Complete Letter-writer for Ladies and Gentlemen.*[165] I don't know where he could have found it in this place. Mark, my star pottery pupil, has made some lovely bowls and other pieces. Some of them he has turned on the

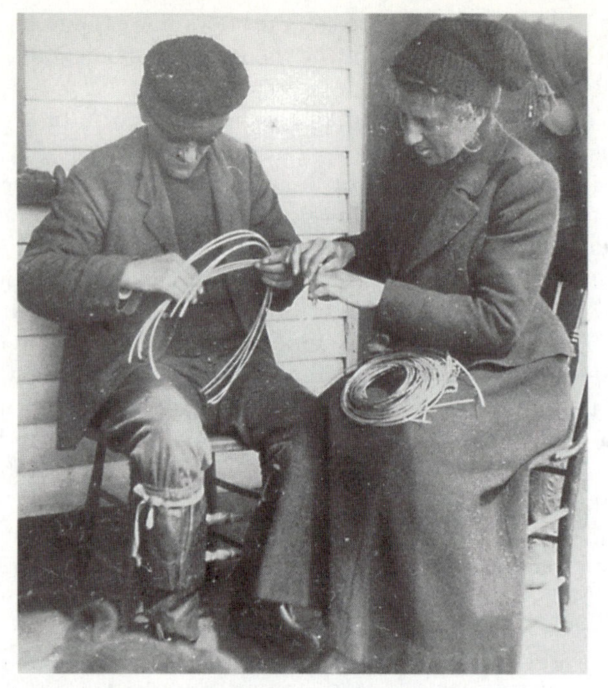

Jessie Luther teaches basket weaving to a blind patient, September 1908 (GHS)

potter's wheel. The majority, however, are built from coil foundation. He is just beginning to use decoration – line drawing on the clay with designs of animals, fish, birds and sea life. Today, he is modelling a bear in bas-relief on a damp bowl. It looks like a guinea pig.

Another industry only partially developed is metal work. As copper mines are a part of Newfoundland's natural resources, the use of copper seemed a legitimate industry. With this in mind, I brought my metal working equipment with me, including a small supply of sheet copper, and Mark, who seems to be quite a Jack of all trades, is working at this craft and has several small bowls, trays and jar covers to his credit.

While I am on the subject of individual workers, there are two blind men who have been taught to make reed baskets using the long-delayed material. One went home to his settlement on the Strait of Belle Isle in the summer, taking with him enough material to continue his work for a while after selling all the baskets he made while at St. Anthony. The other man is still here. He and several patients make baskets on the hospital porch when weather permits and seem to enjoy what is really the beginning of an occupational therapy department. Mr. Jones (Dr. Grenfell's secretary), who has charge of the co-operative store, wants some sort of industrial work connected with its activities. It is an excellent idea, really in line with the home industries that have from the first been hoped for, but it needs direction.

Miss Dwight and I visited Miss Keese's school today and were greatly entertained. We went provided with written excuses from Dr. Wakefield and Mr. Cushing for being late. The building has been repaired. The shelf along the wall is replaced by real desks. The chimney is still an iron stove-pipe through the roof with an outside ladder against it to meet the fire hazard, but we know local innovations come slowly. No one expects too much. There is great need for a non-denominational school available to children on both sides of the harbor, and I think that is one of Dr. Grenfell's objectives for the coming year.

13

Preparation for Exodus

12 September – 8 November 1908

September 12th Only a few weeks before I leave for home! This is the first time I have acknowledged discouragement, even to myself, but Mr. Holley has not touched the kiln for two days. He has evidently lost heart again. I may have to finish it myself. It is the dome. Many people have made suggestions, several insisting that a curved wooden frame should be used for support while the brick arch is being built. We experimented but found it only interfered with our improvised technique and proceeded without it. We may be all wrong, and when the last bricks are laid at the top we shall hold our breaths lest they fall in. The last straw is the discovery that the iron encircling [bands] are too unwieldy for the small shop forge, and they will have to be sent to St. John's to be bent properly when the doctor returns on the *Strathcona.* Another delay. Optimism was once one of my strong points. Am I losing it?

September 13th This afternoon, Mr. Cushing conducted an open-air service for patients on the hospital piazza facing the harbor. These short Sunday services, held when weather permits, seem to be appreciated also by the village people, for several joined the patients and sat on boards laid across boxes for improvised seats. Some of the men and boys perched on piles of lumber on the outskirts, where they could slip away quietly if the service did not interest them. Several fishermen came from the schooners anchored in the harbor over Sunday, the day when no one sails except in emergency. The simple service, with its peaceful background in the afternoon sunlight, must have been welcome after weeks of schooner life on the Labrador. Sheep and goats wandered about in a casual way, and the calves in a wire enclosure nearby all stood together at one end, looking toward the hospital with every appearance of listening attentively.

Last night at 11:30, Miss Storr and Miss Kennedy returned unexpectedly from their Labrador trip after eight weeks of wandering on any available craft – for the past eight days a fishing schooner – a trying experience, for the food gave out. One can subsist on hard tack and tea but not enjoy [it] or grow fat on it. They looked rather thin, but the inconveniences and possible hazards evidently added zest to the experience, at least in retrospect. Their tales of the coast made me more eager than ever to visit it. Perhaps I may – sometime.

September 14th Once more our family has diminished. The *Prospero* has just carried away Mr. Spencer, our agricultural expert, who has turned barren land into a garden, and Miss Storr, on her way home to England for a vacation. I shall miss her. She is the only remaining staff member connected with those interesting days of our first summer at St. Anthony two years ago. She expects to return next year for another indefinite stay.

Mr. Holley returned to the kiln today. I was about to attack the difficult job of closing the dome without his help. He still has periodic intervals of despair and more than once has pulled his hat over his eyes, declaring, "If that course ain't right this time, Miss, I'se just goin' t' tro' up th' job. Ef I'd a done dis t'ing afore, Miss, I'd a known how it's fur t' be." Then I stroke his rampant feathers and tell him I know no more about it than he, and he calms down. He's very funny when he bristles, and I have to let other things go until he can be talked into a more optimistic frame of mind than I possess myself.

September 15th This has been a "grand" day! Like many of our good times, it came as a surprise, beginning this morning when Ruth Keese looked out of the window and asked, "What schooner is that?" There at anchor in the sunny harbor lay the Mission schooner *Lorna Doone* with the Bryants and their guests returning from Labrador. The schooner was unmistakable with the kayak hoisted over her stern and the two launches on deck. They soon appeared for a call and tour of inspection to see what had been going on since their last visit. Two of them became so interested in the kiln that they stayed a long time, even making helpful suggestions when Mr. Holley got into a hole from working out his own ideas. Four of the men, John and Owen Bryant, Mr. [K.D.] Forbes and Dr. Peabody,[166] stayed for lunch and then went to the shop, where they bought most of the woven articles on hand.

Mr. Lindsay haunted the schooner all day. I heard him ask Mr. Bryant if there was any of that beer still on board. It may be some time before he is again reminded of Ireland. At four o'clock, he returned with Mr. Bigelow and Dr. Peabody, who took us with the whole party for a short run in Dr. Grenfell's new launch. Afterward, Ruth and I went with them to the wharf to see them leave for the schooner, laden with garden produce, but before nine they were back again for cocoa and stayed until eleven. This was farewell, for they sail at daybreak. We count this a red letter day. In our simple life, a visit like this assumes the proportions of an event, so who can blame us if we threw duties to the winds and made the most of it? We could not have done otherwise anyway without running away.

September 16th The reaction came at dawn, when we heard the creaking of pulleys as the *Lorna Doone* raised her white wings and flew out of the harbor before a stiff north wind, much colder than yesterday. Mr. Lindsay went back to camp today, reducing our family to three. I asked Miss Kennedy for dinner to share our new potatoes, which are just ripening. We have not as yet had heavy frost, but light frost has touched some of the vegetables, and we are in nightly fear of losing everything just as it is ready to be eaten.

Last night, the hospital had a familiar experience. Prowling dogs broke into the pantry, ate a ham, a leg of mutton (the only fresh meat for a month), demolished a lot of eggs, drank all the milk and made a mess of a lot of blueberries. They had broken the screen, gnawed the wooden slats and jumped through the window. They also broke into Rube's house, ate half a barrel of pork, and spoiled nearly half a barrel of flour. This morning an ultimatum went forth: dogs will be shot if seen at large during the night. Soon after dark, we heard a shot, followed by a dog's cry. Rube had had a visitor: it was one of Dean's dogs.[167] The men have been warned many times since early summer to keep them confined, but Rube has hated to kill them, as they are so much needed by the people in wintertime. Anything like this, however, seemed to require action. It reminded one of Indian warfare when Dr. Wakefield in his long rubber coat went out with a rifle and Mr. Hause joined him in the darkness, but we heard no more shots.

September 17th No more dog tragedies since last night, but I am subconsciously listening for gunshots. Animals have figured largely in my northern experience, the incidents being as varied as the kind of animal involved. The latest was this afternoon, when I started for Dean's [J. and F. Moore's] store and saw a great commotion in his yard, a flurry of white goats and a woman chasing them frantically around the enclosure. They were our angoras. Even the kid was there, and as I approached they ran up an almost perpendicular bank to get into the yard. They were very funny, but amusement turned to alarm when the dogs became excited, one of them starting after the kid. I thought his end was near and ran toward him shouting, of course without the slightest effect. Fortunately, Eli Curtis saw the dog and called him off while I chased the goats to the Mission enclosure.

Eli joined me on the way home. He stutters painfully but after some effort made it apparent that the goats had eaten about 200 heads of cabbage, and he was afraid he would have to ask damages from the Doctor. With regrets and apologies, as spokesman for the Mission, I assured him the loss would be made good and lay in wait for Dr. Wakefield, who said he would investigate. We agreed we had no right to kill the people's dogs for destroying our property and allow our goats to destroy theirs. Dr. Grenfell will be much concerned, and some action must be taken at once to right such a wrong.

Another source of trouble is the "saucy" bull Mr. Cushing bought for future beef and is fattening on the surrounding country until his time comes. He is allowed to roam at will and to many of us seems a menace to the community. This afternoon, I heard him bellowing and to my horror saw him approach the hospital and the tuberculosis tent. The bull, with lowered horns, pawed the earth and seemed about to charge the tent full of sick people. Men came running when they heard my shouts and chased him with sticks and stones until he walked away, still looking belligerent. Later, we saw him driven to a corner of the fence while Rube roped his horns and tied a board over his eyes. We hope it will make him less dangerous and wish his "time" may come soon, for he gives us the jitters.

The kiln is nearly finished! I can scarcely believe it. In spite of dire predictions the bricks have held together, and the dome is nearly closed. Mr. Holley called me this afternoon for a final consultation, and the puckers have left his forehead. He is so interested. He told me today, "I'se a poor man, Miss, but I'd give ten dollars rather'n see dis t'ing a failure." I have promised him a celebration of some kind when the final brick is laid, perhaps a cake decorated with a kiln in red icing as a part of it. We can hardly wait for the trial firing. What if the cement – which should have been fire cement – does not hold when subjected to heat? A question without an answer until the iron bands can be taken to St. John's on the *Strathcona* and returned. Where is the *Strathcona*?

September 18th Mr. Cushing came for us this morning to go with him to Back Cove for Mr. Lindsay and try out the new motor boat. There were blueberries to preserve, mending, cleaning and designing to be done, but Dr. Wakefield advised going, and who am I to question doctor's orders?

The launch runs beautifully, and we went down in forty-five minutes. Back Cove was nearly deserted. The last time we were there it was crowded with all kinds of fishing craft. The deer were on the hill near the corral, and it was milking time – always interesting to watch. This was our last opportunity: they stop milking this week. For an added experience, I tried to milk a deer while the Lapp held it. It was interesting, but once was enough – the creature would not stand still.

We went into Mr. Lindsay's new luxurious tent, a real habitation with a stove and even a window with a netting. George Ford,[168] a new deer herder, has just come after several years' service with the Hudson's Bay Company at posts in the far north. He is an excellent story-teller. For a long time, we sat on piles of deerskins in a corner of the tent, absorbed by thrilling tales of northern Labrador, of wolves, deer, bear, walrus, and the hazardous incidents of northern life. Tales like these intensify the lure of that region, which now suggests mystery as well as adventure. Every day my interest in what lies to the north grows stronger.

Back at St. Anthony, we met Mr. Holley on the wharf. He was beaming, and no wonder, for the kiln is finished! He has laid the last brick and will have his promised cake with a red kiln on top. Now we wait for the iron bands.

September 19th The season's first combination hail and snow storm fell this morning, lasting only a few minutes but forecasting what is to come. A cold wind has blown heavily all day, bringing reminders of winter clothing. When wind and mercury fall tonight, everything will freeze. We have picked all the peas, given away many heads of lettuce, and will cover the rest, hoping to keep them a little longer.

Photography, which has always been one of our interests, has lately become a family craze. Everyone is taking photographs of the country, the village, its inhabitants – individual or *en masse* – and of each other as members of a happy group soon to be separated. The gunroom has been a printing room ever since it ceased to be a dormitory during the summer rush, and we are now printing from each other's films as well as our own. Mr. Lindsay posed for me in his long grey coat. He even brought out his walking stick, handling it with a subtle change of manner noticeable whenever he wears real clothes. Some day before I go he plans to bring to light other items of his wardrobe found to be superfluous during his St. Anthony residence and appear in full evening dress. I have almost forgotten how it feels to wear an evening gown.

September 23rd No heavy frost after all! We went out after breakfast and removed comforters, blankets, etc., from the flowers and vegetables. Apparently, "Martin Luther" (the husky puppy I rescued from starvation and abuse some time ago) slept on the flower bed. There was a deep depression, and the flowers under it looked sad. There is still no news of the *Strathcona.* As the days pass, we have worn the Point smooth looking for it.

September 24th The *Strathcona* has come! Ruth and I heard the unmistakable whistle while we were dressing and after breakfast went to the hospital, where the men had gone after landing. They are all looking well. Dr. Little is even fat, and both he and Dr. Grenfell have long curly locks that cry for the local barber. We did not see them again until dinner, when they returned from the hospital for an evening like "the good old days" of last winter, sitting about, comparing notes, telling experiences.

It has been a day of great activity, for the *Strathcona* has to leave at once for St. John's, taking Dr. Little to transact business and collect supplies for the coming winter, which he says includes clothes, for he is destitute. Miss Dwight will also go to reach home in time for school. They have just gone aboard. They sail at daybreak. The *Strathcona* is rented to the government at $75.00 per day for use in the election,[169] and after her return from St. John's she will call at all the little hamlets for ballot boxes. The Doctor hopes to build an addition to the hospital from the proceeds.

September 25th This is a beautiful day, and I have squandered the morning roaming over the premises with Dr. Grenfell and his retainers, listening to his orders, sometimes with consternation when they indicated so much clearance of land. I have given advice or approval – when asked – and volunteered an occasional mild protest. All the tuckamore is being cleared away as far as the new barn and Dr. Stewart's house. At present, it looks very bare and stumpy, but grass may grow there by and by – some day it may be a field. Dr. Stewart's house is well under way. It will provide work for local men during the winter and be ready for him in the spring.

The Doctor is delighted with what has been accomplished since he went away: Mr. Spencer's garden, the fine work done by the college boys in clearance of land, new fences, and best of all the reservoir on the hill. There is even a trolley or foot car, run on a narrow track from the wharf to the stable, known as "the daily freight," but in the morning when it starts with the men it is "the 6:15 express." The sound of traffic on rails is queer in this remote place without the accompanying bustle of city life. It has stimulated the Doctor to plan new projects which he thinks may be possible as well as desirable and which I (also an optimist) take great pleasure in discussing.

Our tour of inspection ended at the new barn, where we climbed the ladder to the great loft, now nearly filled with sweet fresh hay. Jack came too and convulsed us with his antics. The hay seemed to excite him, for he acted like a crazy dog, leaping about, burrowing into the mow, then suddenly appearing in unexpected places. Perhaps he felt the need of violent exercise after being confined to the *Strathcona*'s rather restricted limits for so many weeks.

Paul Matteson has decided to stay through the winter. He will have general charge of the industrial work and direct the boys' classes. It was thought that Charlie might be brought to the status of assistant with the carving class, but it is now decided that he will go home to his village provided with carving tools and hard wood for making small articles during the winter. He is much better physically and does good work which can be brought to St. Anthony for sale in the spring.

September 27th I am going to Battle Harbour on the *Prospero,* which may whistle at any moment. It will be a short trip, only about an hour on shore, but this is my only chance to see the Labrador, as the *Prospero* will not cross the straits again this year. At the last moment, Dr. Grenfell decided to go too, as he wants to see his friend Dr. Armstrong[170] from England, who has just arrived at Battle Harbour to take charge of the hospital.

The steamer has just whistled.

September 30th The *Prospero* landed at Battle Harbour in early afternoon. Dr. Armstrong was at the wharf, and five minutes later it became apparent that the steamer's short stop of an hour would be quite inadequate for necessary conferences after the long interval since he and Dr. Grenfell met. A search was made to find some craft that might be starting across the straits later in the day, and Dr. Armstrong located a fisherman with a small trawler who said he would take us, so the *Prospero* departed, and with carefree minds we went to the hospital to meet Mrs. Armstrong and enjoy ourselves.

There is much to say of the Armstrongs. Dr. Armstrong is one of the handsomest men I ever met: tall, slender, fair – a fine type of Englishman with a charming, gracious manner. Mrs. Armstrong is also gracious, friendly and charming. I was told later that she is a friend of Queen Mary, but her simple, friendly attitude towards everyone suggested no consciousness of the social gulf which might be supposed to exist between an associate of the British royal family and the poor "liv'yer" of the Labrador coast.

We inspected the hospital; then Mrs. Armstrong and I left the doctors to their conference and wandered over the tiny island, one of a small group off the southern tip of Labrador. It is bare and rocky –

Battle Harbour, at the northern entrance to the tickle, September 1908 (JLP)

not a tree. Grass grows in patches and hollows, partly covering the rocks. A few little shrubs cling to the ground for shelter, but only those who have stood beside the waves on a shore like this and looked across the sea to the mysterious north can understand its fascination and lure.

A narrow "tickle" (deep water passage) separates Battle Harbour from the somewhat larger Caribou Island. Part of its rocky shore is so deep and precipitous that fishing schooners, which often come in for shelter or to "make their fish" on board after a catch,[171] can tie up to the rocks and the men step ashore. The tickle is not only narrow but tortuous, preventing large boats from passing through the southern end, where even small boats have to make a quick turn to avoid rocks. The northern end is wider, allowing coastal steamers to drop anchor there, and passengers are sent ashore in small boats.

The settlement of Battle Harbour is very small. The hospital, neat, white and rectangular, stands beside the fish trader's [Baine, Johnston & Company's] storehouse. Between it and the shore are the trader's large fish flakes. The little wharf lies beyond, and the doctor's small, comfortable house is nearby. Houses of the fishermen border the one narrow path that follows the rocky shore to the northern end of the island. Near its end, a footpath leads downward to the tiny graveyard in a quiet hollow sheltered by high rocks with the sea beyond, a lovely peaceful spot. It is here that little Prince Pomiuk, son of an Eskimo chief, is buried, the boy of whom Dr. Grenfell and others have written.[172] He was with a group of Eskimos who were brought to Chicago during the World's Fair in 1893 to act as an

Wooden grave marker at Battle Harbour cemetery, September 1908: "In Memory of John Hill Who Died December 30 1890 Aged 34. Weep not Dear Parents. For your lost Tis my Eteranel Gain. May [Make] Crist you[r] all, take up The crost that We Shuld meat again." (JLP)

attraction in the "midway" by showing their skill in snapping the long dog whip and exhibiting themselves in their fur garments during the long hot summer. When the fair was over, those responsible for their coming simply cast them adrift to make their way back to their native home in any way they could.[173]

In what way or by what means little Prince Pomiuk finally reached the north Labrador coast we do not know, but Dr. Grenfell found him there after hearing of him from an American clergyman, a Mr. Carpenter,[174] who had seen the charming, merry, bright-eyed little Eskimo at the World's Fair and had tried to trace him. Dr. Grenfell found the boy naked, diseased, half-starved, among a small group of Eskimos on the Labrador shore and took him to Battle Harbour hospital, where he partially recovered, but his injury had been too great and he died, leaving a memory of courage, endurance and a happy, loving spirit with those who knew him during his two short years among them. He was baptized before he died and given the name of Gabriel. In the little cemetery, a wooden cross marks his grave, and in the hospital is a cot donated by friends to his memory.

When Mrs. Armstrong and I returned to the hospital, we found the doctors still deep in discussions after visiting patients. Several hours had passed, and the boat supposed to take us across the straits had not materialized. Supper time came, and we lingered comfortably at the table, enjoying the congenial society of our host and hostess, but as time passed we began to think uneasily of their night's rest as well as our transportation to St. Anthony. When Dr. Armstrong went down to the wharf to make inquiries, the trawler man had vanished. Someone had seen him go out, but no one knew where or when (or if) he would return. They said he had spoken of taking us across. At eleven o'clock, Mrs. Armstrong insisted on taking me to her room to lie down and perhaps get a little sleep, saying she would wake me when the fisherman came. She left me after laying out a lovely négligée, and I gazed with admiration at the exquisite dressing table equipment which must have felt quite out of place on the background of a simple pine bureau.

At 1:30 a.m., Mrs. Armstrong called me. Carrying lanterns, we made our way to the little trawler at the wharf, climbed down the ladder to the deck and started out in the darkness through the crooked channel of the south entrance to the open sea. The weather had changed during the evening. Wind was blowing, and as we headed across the straits (rarely very calm) there was quite a "lop,"[175] to use the vernacular, and the little boat tossed like a cockle shell. It also smelled of stale fish. What followed was inevitable! We reached St. Anthony in time for breakfast, which did not interest me, but it would take more than an uncomfortable crossing to dim the pleasure of meeting the Armstrongs and roaming over the wave-washed rocky island with a glimpse of the real Labrador. I want to see it again and more of it.

[*October 22nd*][176] The kiln is fired! Dr. Little returned from St. John's on the *Strathcona* a week ago, bringing the bands, which fitted perfectly, and we prepared to fire. On account of the uncertainty, we risked only small pieces of pottery and some bricks for this first firing, but placed temperature cones where they could be seen through the peep-holes to gauge the heat and its effect on a few glazed pieces used as tests. Our fuel was wood. Men brought loads from the hill to add to the pile outside the building, and then we began to stoke. One can imagine the excitement when I lighted the fire under the kiln, while

Jessie Luther supervises the construction of the kiln at St Anthony with Mr Holley, September 1908 (GHS)

interested but doubtful people who had gathered from the neighborhood watched volumes of smoke pour from the chimney. They were sure the great heat would burn down the building and the kiln would collapse. Great puncheons of water were placed outside each window and a fire extinguisher brought into the room.

We continued to stoke for thirty-six hours. All day and far into the night the room was frequented by workmen, fishermen and convalescents from the hospital who came to see this queer "heater" really in operation. They sat on piles of brick or pieces of wood, smoking their pipes, occasionally going to the peep-holes to watch the now glowing interior, marvelling that the outer heat of the walls was so slight there was not a remote possibility of being obliged to use the fire extinguisher. A few small cracks appeared near the top, but the dome held and the draft was good. Mr. Holley was jubilant. "They's all bin tellin' me as its just foolishness an' vanity an' won't hold t'gether," said he. "But look at it now! Still standin' an' it ain't burned nothing."

On opening the kiln after it cooled, we found the heat had been sufficient for a biscuit firing[177] but not enough to mature the glaze on the test pieces placed inside it. With a few alterations it may be possible to reach a higher degree of heat, but a portable kerosene kiln of moderate size would be valuable for glazing small quantities of pottery, as it can be fired in one-sixth of the time and would be of service for experimental work.

[*October 30th*][178] The crisis of firing the kiln having passed, the days are filled with seemingly endless details to arrange before I leave by the next *Prospero.* Annie Ash [later wife of Will Simms], one of the fine village girls, is to have charge of the loom-room during the winter. There will be at least six weavers at St. Anthony, all doing good work. Another efficient weaver, Susie Denny, has just returned to her own village of Englee supplied with a loom, spinning wheel, and material to start a branch industry among the local girls.

Economic and other problems inevitably follow the development of any industry, no matter how small, and some of those problems are evident in charting the course for industrial work during the coming year. At present, it has no specified fund and no dependable income. Specially designated donations from interested persons have helped us meet running expenses. Dr. Grenfell has contributed the proceeds from several lectures, and during the summer a substantial sum was received from industrial sales, but there are no visitors during the winter, and nothing can be sent away for sale until June, with possible delay in payment to be considered.

Meanwhile, running expenses continue month by month. It is not expected that the industries can be fully established during the coming year as some of them are in their infancy. Even the weaving, which seems to be well advanced, needs further impetus, but our optimistic objective is to make the work eventually self-supporting. That, however, will not be possible until summer sales and exports of products are sufficient to pay running expenses for the year to come, including cost of material, which must be imported in advance. The running expenses to date are moderate. Only two of the workers receive definite wages. The others are paid on a piece-work basis, the best arrangement under their working and living conditions.

I think no one doubts the beneficial effect of this effort on the community. Girls who appeared careless and aimless when they joined the class became neater and more responsive. Self-respect and purpose followed the realization of their ability to accomplish something worthwhile. The boys are more courteous, co-operative and industrious. It seems apparent that such normal and interesting occupations and the need for continued effort in their accomplishment cannot fail to improve the community's social standards and have an influence on everyday life. As for myself, the year has been a worthwhile experience. Dr. Grenfell insists that unless I come again for another year the industries will wither and die. I referred him to Dr. Blumer, who would have to decide the matter. Of course, I would love to come back, and the possibility will temper my sorrow when I wave farewell to St. Anthony friends.

S.S. Prospero, en route to Exploits, off Newfoundland Coast. November 2nd, 1908 It is cold with snow on the coastal hills, but the sea is quite calm. I am alone in my stateroom, resting after the last minute rush and lively send-off from St. Anthony, feeling sure I shall be back again next year. The Doctor left on the *Strathcona* four days ago with everyone on the wharf to see him off.

It always takes some time for the *Strathcona* to start, the Doctor frequently stepping off to do some last thing, then back on board again to converse across the rail. They started, and as they moved away the Doctor waved with a parting reminder to me: "I'll meet you on the train at Notre Dame Junction." When some distance from shore, he came hurriedly to the stern and shouted again, "Oh, Miss Luther, I forgot my fur-lined coat and derby hat. I'll need them in the States. Do bring them along with your things." I nodded and waved back.

Many hands brought my many pieces of hand luggage on board, and now on my way I look around my stateroom and take stock – one small suitcase, one handbag, one large canvas travelling roll (full to overflowing), one umbrella, one pair of snowshoes, my fur-lined coat, the Doctor's fur-lined coat and his derby hat! My trunk and large suitcase are in the hold. I think of the six transfers before reaching Boston, but of course at Notre Dame Junction the Doctor will take his hat and coat and perhaps give me a hand with one of my bags, for porters are sometimes lacking.

This is really a restful trip.

On train, Reid Newfoundland Railway, en route Notre Dame Junction to Port aux Basques. November 5th, 1908 I made good connections. Deck hands and trainmen helped with luggage – all pieces accounted for – which they left beside me at Notre Dame Junction. The train came in, but where was the Doctor? The porter tackled the large bags while I struggled with the small one, umbrella, snowshoes, two fur-lined coats and the hard derby hat, which was awkward to carry with other impedimenta. I tried to plan what to do with it. I might put it on my head over my hat. I might hold it under my arm, but it requires a hand. It is non-collapsible. Perhaps I could balance it on top of the coats on my arm while holding the umbrella and snowshoes in my hand. I will try that when the time comes to move again.

On train, New Brunswick, en route to Boston. November 7th, 1908 I am safely across the Cabot Strait and have made the last change, with nothing missing. The Hat, however, had two narrow escapes. I tried balancing it on the armful of coats, but it fell off as I left the train and was rescued by a passenger. At North Sydney, a gust of wind blew it to the street, where it fell between the wheels of a passing truck but escaped injury and was again rescued by a passer-by. When we made our last change, at Truro, I pressed it firmly on top of my distinctly feminine hat and walked down the platform, ignoring curious glances from fellow passengers. I am now hoping there will be enough red caps at the North Station to get me comfortably into a cab.

Providence, R.I. November 8th, 1908 Arrived at North Station, Boston, this morning. Red caps met the train and took everything but coats, umbrella and the Hat. At the head of the platform stood the Doctor, laughing gaily and waving his arm. He dashed forward and took his hat and coat. "So sorry not to come along with you. You see, the business in St. John's was finished sooner than I thought, so I took the next train, two days ahead of you, and didn't wait. Thank you so much for bringing my hat and coat."

When Dr. Grenfell comes to Providence this winter to lecture, he will stay with us. Also, with Dr. Blumer's permission for an additional leave of absence from Butler Hospital, I have promised to return to St. Anthony next year.

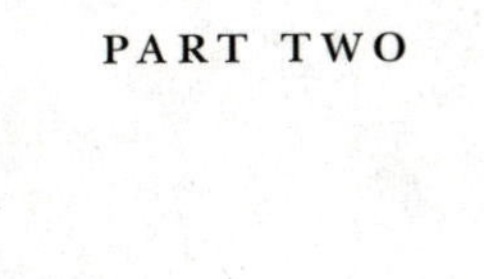

PART TWO

Introductory

During the winter following my first full year at St. Anthony, Dr. Grenfell lectured in the States and came to Providence for an engagement in February, returning in May as our house guest before sailing for England. He was to be honored by Harvard University in June, and the purpose of his trip was to bring his mother,[179] who had never before visited America, to be present when he received his degree. He wanted me to meet his mother, as well as to see him take his degree, and told me he had secured two seats for us at Sanders Theatre. With his usual enthusiasm, he was full of plans and had arranged the details of our meeting. Dr. Edward C. Moore[180] (then Dean of Harvard) and Mrs. Moore had offered to place their home in Cambridge at the disposal of Mrs. Grenfell and the Doctor during her visit, and I was to meet her there. It was a lovely plan to which I – naturally – responded eagerly. And then he sailed away.

A word of reminder accompanied by train schedules came from the Doctor on his return in June, and I went to Boston to meet him and his mother, as arranged. I fell in love with Mrs. Grenfell at once: a charming, gracious lady, and the Doctor's affectionate devotion to her was beautiful to see. We went to the theater early to find our places, and the Doctor left us there. Our seats were in the gallery, directly in front of the stage, and while we watched with interest the gathering audience of people who were utter strangers to us, I discovered in Mrs. Grenfell much of the Doctor's sense of humor and appreciation of people. We tried to fit the names of notables we thought might be there, and friends we had heard the Doctor mention, to certain individuals, impressive or otherwise, as they entered the hall. It was very amusing, but I never knew if our intuitions were reliable.

At last, President Lowell[181] and others took their places on the stage; the program began and we waited expectantly for those who were to receive degrees. When they appeared, filing in sedately to stand before [the] President, we were aware of something that had not occurred to us before – a certain incongruity in the Doctor's costume. Clothes, in their formal sense, were usually of secondary importance in the Doctor's daily routine, and it evidently never occurred to him that golden brown tweeds and very bright tan shoes might be questionable on such an occasion. Anyway, there they were! And, topped by his bright scarlet Oxford hood, he was a colorful object among the group of sombrely clad, black-shod men.

When called from the line to receive his degree, he walked forward and stood impassively while President Lowell read his citation. Then, with the parchment in his hand, he turned about face and returned to his place. Mrs. Grenfell and I, in subdued excitement, were thrilled. The great moment for her was over, and her face shone with pride. Suddenly, she turned to me. "Why," said she, "he didn't even make a bow!" I never saw the dear lady again, but the memory of our brief and cordial meeting is a very happy one.

There was no time during my brief meeting with Dr. Grenfell to refer to industrial plans, one of them being the development of brick-making in connection with pottery that we discussed after my first summer at St. Anthony. Pottery was begun as already chronicled, but experience with the wood-burning kiln built a year ago had proved that time, as well as the amount of wood required, would make its use prohibitive except for preliminary biscuit firing, and a portable kerosene kiln, which could now be transported by the Mission schooner, was necessary for the development of glazes. During the winter, I had lectured about the Mission and written an appeal for special funds to meet this need, a plea reinforced by Dr. Grenfell when he lectured in Providence, with the happy result that the kiln was shipped to St. Anthony during the summer, but plans for the brick-yard were still nebulous.

In August, Dr. Grenfell wrote me announcing his engagement to Miss Anna MacClanahan[182] of Lake Forest, Illinois. They had met on the *Mauretania* in June, when he was returning with his mother from England and she from European travel with friends. The Doctor wrote enthusiastically of his fiancée's interest in the industrial work, which he had told her about, and her desire to help. They were to be married in November and planned to arrive at St. Anthony in late December for the winter. Their house was to be begun at once.

Everyone rejoiced with the Doctor in his happiness, but it was difficult to think of him as a domestic man (although we knew he enjoyed and appreciated what comforts and amenities we were able to offer him at the Guest House), for his absorption in his work and its precedence in all his thoughts and actions was so great, irrespective of the hazards involved in implementing such work, that any thought of his connection with the possible restraints of married life had seemed remote.

At Dr. Grenfell's request, Butler Hospital had again granted me leave of absence from direction of the Occupational Department, and in October I returned to St. Anthony for another year of work and adventure.

1

Again to St Anthony

3 October 1909 – 1 January 1910

October 3rd As I travel again over the now well known journey to Newfoundland, I find that its details have become routine. There is no longer the excitement and original thrill of discovery, but its charm is not lessened and it is never monotonous. Perhaps there is even an added interest through familiarity, like the comfortable meeting of old friends.

October 7th, St. John's We arrived here at noon to find the Crosbie Hotel crowded with commercial travellers. Only wretched accommodations were available, and I left them as soon as possible to call on Mr. Peters, the Mission's business manager, to discuss ways and means as well as latest news of St. Anthony. I found Mr. Peters much perturbed. There is evidently lack of harmony on the coast, friction having developed since early summer when the newly appointed manager of local affairs [Mr. Waldron][183] arrived at St. Anthony. Making due allowance for exaggeration, it is evident there are troubled waters, and I fear Dr. Grenfell's appellation the Abode of Peace, written by him last year on the photograph of the Guest House, is no longer appropriate at St. Anthony.

As the *Prospero* is in dry dock for inspection, our sailing is delayed for a day or two. This gave me an opportunity to explore an unfamiliar part of St. John's, and this morning I walked along the top of a cliff above the harbor to the quaint fishing village of Quidi Vidi, reached by following a path among little houses and fish-flakes down to the water below. The houses are tiny, many of them built on wooden posts, and cling like snails to rocky projections on the steep hillside. A few sheep and goats wandered about, and dogs emerged from under the huts and stages, where fish were drying in the sun. Fish were everywhere. The flakes were full, and every little projection on the cliff-side held fish. They even lay on the ground on either side of the path – sometimes in it – and I stepped over them, picking my way while dogs and children ran freely among them. They were even spread on the roofs of houses. Where the path led under loaded flakes, sunlight filtering through cast interesting shadow patterns. With the lovely background of blue harbor, rocky hill and distant town, one might cover one's nostrils and accept the delusion that the walk was under a grape-covered pergola in sunny Italy.

My interesting day ended with tea in Mr. Peters's office, where I met Mr. Peck,[184] a veteran missionary from Baffin Land. He is a dear old man, very much interested in his work, simple in heart and in his faith in God. He returned last year after long exile among the Eskimos but went back this summer on the Mission schooner *Lorna Doone* to take two new missionaries to the field and leave them there for an indefinite stay.[185] It was a cozy afternoon before the open fire, but Mr. Peters's problems were uppermost in his mind. I am asking myself what problems of my own may lie ahead.

My shopping and packing are finished. We sail at 10:30 tomorrow.

October 9th On board S.S. *Prospero.* It was raining when we left St. John's, but the deck seemed more attractive than the stuffy cabin, and I stayed as long as possible, having found a comparatively dry, sheltered nook. There are no chairs, so I sat on the floor wrapped in my Jaeger blanket.

October 10th The steamer ran all night. At intervals, it unloaded freight at ports of call, and the steam winch (located over my head) clattered madly. I count the hours that must pass before reaching St. Anthony and begin to feel like the woman who was so unhappy while crossing the ocean that she stayed twenty years before gaining enough courage to return. They say we will land in the early morning.

October 11th, St. Anthony At 2:30 a.m., we felt our way into St. Anthony harbor, guided by the ship's searchlight. Through the porthole, I recognized the familiar outline of Fishing Point Head, and suddenly the searchlight picked up the church, hospital, Guest House and co-operative store. Everything seemed familiar yet strange and weird at that hour of the night. There was freight for the Church of England side of the harbor, and we stopped there to unload and await boats from the Mission side. After a long wait, two young men arrived from the Mission to meet me, and on the dock I saw with joy my old friend Mr. Ash. It was good to see a familiar face. A small schooner lay at the wharf, and we walked across her to a little ladder over the side with a drop to the deck below. One of the young men suggested that I might find it difficult, but Mr. Ash saved my reputation for Newfoundland experience by his laughing remark, "Don't worry about Miss Luther. She has climbed before."

St. Anthony looked like a metropolis. The hospital was lighted for patients, and I stopped for a word with the nurses. The Guest House was also brilliant. I was staggered by the sight of real steps leading to the wharf, never having hoped to land in any way but a scramble up slippery logs. They really suggested yachts. At the Guest House, I sensed only modified familiarity: some things were the same, but I could see the hand of the alien in certain arrangements, and the aftermath left by hordes of students and visitors during the summer was plainly discernible. The piazza is still the dining-room. The first thing I noticed in the living-room was a real piano, and I remembered how we longed for one two years ago. I went softly upstairs to Mr. Lindsay's old room, which is to be mine since (to my great regret) he has gone home to Ireland.

I met the household at 7:30 breakfast. In addition to the two young men who came to meet me there was an older man, Mr. Forbes, whose reason for being here I do not know;[186] Miss MacNair,[187] a Canadian schoolteacher; a nurse from Battle Harbour on her way home; and the manager's wife (to whom I shall refer as Mrs. Manager), who acts as housekeeper. Dr. Little is at the hospital, and the house is bleak without our dear L.M.K., who fortunately returns soon from a visit with her Massachusetts family. Dr. Grenfell, who will soon be on his way to the States for his wedding, is on the *Strathcona*, visiting his sawmill at Canada Bay. I foresee adjustments and changes, some of them difficult, but discouragement is not a part of my philosophy, and this is only the beginning of a new era.

St. Anthony has a general air of activity. This morning's early light revealed the building in progress, though unfortunately all unfinished. Both the hospital and the orphanage are being enlarged. The workshop addition is so far merely a partial frame. Half-way up the hill, on the site we selected last year as the most beautiful in St. Anthony, is the foundation of Dr. Grenfell's house. Its proportions seem quite palatial. It is literally founded on a rock, and blasting for the cellar was a lengthy process. There is to be a coal furnace, steam heat, modern plumbing, and of course electricity. A glass-enclosed piazza will protect three sides. The Doctor hopes it will be ready when he arrives at Christmas with his bride, but I have doubts.

The cheering experience of the day has been my visit to the loom-room, which looks like a real workshop. A few of the girls were spinning, some making drawn work,[188] others weaving. Annie Ash,

Robert Peary, en route from the North Pole, poses with nurses at the Battle Harbour hospital, October 1909 (GHS)

who was in charge during the past year, is finishing the homespun Dr. Little is to take with him when he goes home for the winter. Annie has done a fine job, and I am proud of her and the other girls who have worked so well. There is some fine weaving, also rugs from the branch loom-room at Englee and carved boxes made by the crippled boy, Charlie Snow. We have been busy all the afternoon pricing and packing articles to be sent to St. John's for sale, and the size of the box is quite impressive.

October 12th This is one of the kaleidoscopic periods of the Mission year, with changes of assignment for the Mission staff and departures and arrivals of nurses and helpers. Those arriving from Labrador on the *Prospero* brought news of Peary's return from his arctic expedition. The *Strathcona* was at Indian Harbour when he stopped there to send a wireless message, and some members of the Mission staff went on board the *Roosevelt* and talked with Peary and his crew. He also stopped later at Battle Harbour, and a nurse who is now here on her way home was present when Peary first saw a newspaper account of Dr. Cook's[189] own arrival at the Pole, of the honors given him, and his enthusiastic reception at New York and Copenhagen. She said they hesitated to show it to him. I suppose the reaction and expressions of any man would be regrettable under the circumstances.[190]

November 4th. On board S.S. Strathcona The Doctor is on his way to the States to be married. With the *Strathcona* free, Dr. Little and Mr. Wight[191] decided to make use of her for a long-planned deer hunt in the woods beyond Dr. Grenfell's sawmill. A branch loom-room has been started here, so they took me along to arrange for winter work. Another passenger is Miss Kennedy, who needs a vacation. So here we are!

The mill is about 60 miles from St. Anthony on Canada Bay, just a group of little houses built among spruce trees. The people are very poor, and their homes mere huts built of logs or slabs from the mill. Every house has a stove, a few chairs (sometimes only two), a home-made table and a bed, little else. Some of the floors are covered lightly with sawdust; on others are a few hooked mats. The houses are generally neat and clean, but poverty is everywhere. The mill manager met us as we landed and with true Newfoundland hospitality took us to his house to offer the usual ritual of tea, bread and butter. This family is evidently in better circumstances than the others: there are four rooms downstairs, hooked rugs on the floor, curtains at the windows, and we even had napkins with our tea.

November 5th The hunting party got away at 4 a.m. At a more seemly hour, Miss Kennedy and I rowed across the bay and visited one or two houses. Then curiosity took us along a shore path that proved to be sometimes a corduroy road, sometimes a boggy pool to be waded through. It led to the chute built by Dr. Grenfell to conduct water from a dam on a stream over a mile away, which provides water power for his mill. At one place the chute, supported by two logs, crosses a swift little river. We looked for some means of crossing and, seeing none, braved the slippery logs. It required balance, but we made it and went on to inspect a tract of land where men were clearing the tuckamore for the Doctor's latest project – community farming, which he hopes may be a reality by next summer. He plans at first to plant oats, then develop gardens. It will take time and effort but is a fine objective and typical of the Doctor's vision and optimism.

November 6th The loom-room is in a small building, formerly an engine house. There are three looms and another nearly ready, built by local carpenters. We brought a reed and harnesses for the new one as well as wool for spinning and other supplies. Susie Denny, who is in charge, has done well. Her weaving is excellent. She has good ideas and much perseverance. Moreover, her love of the work stimulates others. We spent the morning inspecting material and planning winter work, which will be under her direction, for I shall not see her again until spring. Susie plans to have a part of her little loom building partitioned for a bedroom and kitchen where she and another weaving girl will live and keep house. They are looking forward to the winter.

While we were at dinner, we heard Dr. Little's voice down the companion-way: "Hello, below there!" He came back without any deer to avoid keeping us waiting but left Mr. Wight and one guide in the woods to try to save the reputation of the expedition.

Evening, Englee. Just before we sailed from Canada Bay, a paralyzed woman was brought on board as a passenger, and she was put ashore here with great difficulty. They had to lift her into the lifeboat before it was lowered from the davits and she was terrified, for the steamer rocks violently, even while at anchor in the harbor.

After supper, Dr. Little suggested going ashore to make visits. It was rather weird starting out in the darkness, but he knew the way and we made two calls on nice friendly people in their neat houses. It is remarkable what the people are capable of accomplishing with limited knowledge and materials. One man had made practically all his furniture – a sofa, tables, even a side-board – and had built his house by degrees. He still lacked some clapboards and shingles but hoped to be able to get them another year. Dr. Little offered to loan him the money for them and gave him an order on the mill for lumber. We left a delighted and grateful man when we returned to the ship.

November 7th Before daylight we were off, the *Strathcona* tossing like a cockle-shell in the heavy seas on our way to Conche, a Roman Catholic settlement where we stopped to call on our old friend Father Thibault, so often our guest at St. Anthony. I shall always remember him as I saw him arriving with his driver, in a komatik drawn by eight beautiful black dogs, his black fur cap nearly covering his cheery face. We found him jovial as usual and eager to discuss recent happenings, but our time was short and we are again on our way.

St. Anthony. Here at last! This has been a perilous trip. The storm increased, and Dr. Little hesitated before finally deciding that with luck we would be able to reach St. Anthony before dark. Dinner was ready just as we started, and Dr. Little remarked that in about ten minutes things would be lively, so we ate fast during those ten minutes. Then Miss Kennedy and I retired to a corner of the chart room on deck and stayed there.

We did "roll wonderful," as the natives say. Dr. Little came in several times to consult the chart. Suddenly, I heard the wheel-house bell ring but no anchor went down, so I knew something had happened. I looked out: a dense fog enveloped the ship. Presently, we went on again, but one of the men came in to say we had lost our bearings and were steering merely by soundings and compass. I realized we might be out all night in the sea or drifting. The crew were very solicitous. One after the other came in to report progress. Sam's reassurance took the form of dubious retrospect: "If it had come on a heavy rain, we wouldn't a' bin able t' see t' git in. We'd a' had t' be out all night, an' an awfu' nasty night it would a' bin, Miss. We'd a' bin in th' trough o' th' sea an' a' rolled most awful." He came in afterward saying, "We're off th' St. Anthony rocks, Miss, an' almost in th' harbor. Only about t'ree more whoops (rolls) an' we'll be in."

Passengers and crew are of one opinion: we are fortunate to have arrived at all. Also, it is now apparent why "The Holy Roller" is the Mission's pet name for the *Strathcona.* This is her last trip of the season. She is to lie here all winter. The Guest House seems luxurious, and so clean!

November 8th The *Prospero* has brought an addition to our winter household, a Mr. Evans[192] from the States, whose apparent mission is to join Mr. Wight and Mr. Forbes in filling chinks and making themselves generally useful. The discovery that he has a fine baritone voice stirred us to form a quartet with Dr. Little's tenor, Mrs. Manager's soprano and my feeble contralto. We also hear that Mr. Manager can manage a bass, and there is the piano. Who will mind the weather if we can have musical evenings? The grandiose plans for turning the piazza into a winter dining-room have been abandoned. A few bitterly cold days have driven its sponsors indoors. The small living-music-dining-room is crowded but possible, and we no longer hurry to eat our food before it cools or wear heavy coats at meals.

This has been a busy day, unpacking material that came by the *Prospero,* packing clothing to be sent in payment for work – some to the loom girls at the mill. A box on the wharf is ready for the blind man at Flowers Cove in payment for basket-making. It is remarkable that he is able to do so much. Charlie, the young man crippled by rheumatism, is coming on this boat. He will stay two weeks to plan his winter work and return home with designs and material, as he did last year. In another village is a crippled boy who has learned to make baskets of narrow, thin birch strips woven like palm baskets, also plaited braid for making hats. There are birch trees nearby, and he needs only a knife for equipment. A girl from Exploits who has been here to learn weaving has gone home with a loom to start another branch. It is encouraging to see such expansion of the native industries and the increasing interest and co-operation of the people toward fulfilment of our objective on the coast.

November 20th This morning, one of the hospital maids reported that a Mission schooner, the *E.E. White*, long overdue from Boston, was about to appear. She could see what she called her "wake" on the water and pointed to a wide current streak that led to the wharf, insisting that it was a sign a boat was coming in, and sure enough! At three o'clock the schooner *E.E. White* rounded Fishing Point Head, entered the harbor and anchored off shore. Mr. Wight was on board with five deer to save the Mission's reputation for sportsmanship. He said a herd appeared two hours after Dr. Little left him.

The Mission's boats rarely come in singly, so we have looked hopefully for the long-lost *Lorna Doone* from Halifax and, with unbroken tradition, she sailed into the harbor at 6:30. Dr. Little has just brought in Captain Fradsham,[193] who told of the worst voyage in his seven years' experience at sea. They were unable to cook in the galley, seas came into the cabins, and with the low temperature the *Lorna Doone* was a mass of ice when she landed. Mr. Forbes's remark that he expected to go to St. John's by the next steamer for a pleasure trip brought a stare from the Captain, who said he would be shy of pleasure trips along the coast at this time of year. He is a jolly soul and spins lovely yarns. We gathered around him last evening to hear his tales of life and experiences among the Eskimos in Baffin Land. His account of a "royal feast," when about forty people assembled in the king's tent (called a *tepo* [*tupiq*]) and with great ceremony ate five raw seals, was something to hear! They gave him the heart to eat. He made a wry face as he told of it but said it was easier to eat when frozen. Etiquette required one to eat what was offered to avoid giving offense.

Changes in personnel continue. Mission affairs are gradually taking shape for the winter, like setting the stage for a new act. Dr. Stewart has returned from Labrador. Mr. Palmer, a member of the Mission family two years ago, is here as hospital assistant. The last steamer brought L.M.K. from her visit at home, and the general atmosphere is brightened by her cheery presence. This year, her mission will not be schoolteaching but non-professional assistance at the hospital, a post that has no rigid limits of influence and activity – interesting and sometimes strenuous.

To offset these arrivals, Dr. Little will be leaving for a winter's vacation. He has been packing at intervals, and thoughts of approaching contact with the outside world have brought a subtle change in his attitude. He is conscious of civilization's outstretched hand. This psychological reaction was evident on Sunday morning, when he appeared at breakfast clad in city clothes, even new black boots that seemed to belong to Beacon Street [Boston]. Everyone will miss him, not only as a personal friend and advisor but because of the people's dependence on his medical and surgical skill. Even his manner of working inspires confidence. Occasionally, he has allowed me in the operating room to give minor assistance or as observer, and one day I watched him perform an abdominal operation so wonderful and interesting that I forgot a human being was involved, his explanations during the performance dissipating my possible nervous reaction as they focused my attention on the cause and hoped-for cure of the patient's trouble.

On Sunday afternoons, a short service is held in the men's ward at the hospital, a service that I enjoy more than any other, perhaps because of the patients' sincere appreciation. This afternoon I sat on little Pat's bed while he curled up against me, his childish voice trying to follow the tune when we sang. Dr. Little sat opposite between the cots, his mere presence comforting to those around him. It was a beautiful afternoon, and from my place I could see the hill, golden with the setting sun as evening came on. The church and little path leading to it were illumined by the light. It seemed symbolic of the path to hope.

November 22nd The *Prospero* is still delayed. Men have been unloading freight from the *Lorna Doone.* The little trolley has been in action all day, drawn by Harry the horse until he became exhausted. Then Dr. Little and Mr. Wight took a hand. They have been hauling with an improvised harness across their chests, stripped to their underwear and bare-headed. Stores have gone to the hospital, orphanage and the Doctor's new house. The men look like wrecks tonight.

November 23rd The *Prospero,* long delayed, came in this morning bringing Mr. Manager from his St. John's conference. He reports that Dr. and Mrs. Grenfell expect to arrive by the last boat. Their house is progressing but slowly, for other building under construction is much needed.

Mr. Palmer, located in the bunk-house, has struck since water froze in his pitcher. He says if it is to freeze anywhere, he prefers to have it somewhere else and has taken his bed to the third floor of the hospital's unfinished addition, where it stands in a pile of shavings. The carpenters work over-hours until bedtime. Then Mr. Palmer takes over and sweeps shavings from the spot where his bed stands in what he boasts is the largest apartment of the Mission.

The newly arrived kiln lies on its side in the engine room, partly covered by building material, awaiting completion of the new workshop and removal of machinery from the corner where it is to stand. The lathe for polishing Labradorite and the potter's wheel cannot be set up until the room is free from the curiosity and interest of men and boys who pass through it. Tools cannot be left about until there is a safe place to store them. It suggests "The House that Jack Built"[194] or making a picture puzzle with some of the pieces missing.

November 24th Dr. Little has gone. Everyone was sure the steamer would return today, and when he appeared at breakfast with a specially polite air and clad in dignified clothes we felt the end had surely come. He has even worn his ulster, travelling cap and elegant dog-skin gloves to visit the hospital. I went with him to the steamer to say good-by. When I returned to the Guest House Mr. Evans was sitting by the fire. He said nothing for a moment, then quietly remarked that "Dr. Little is a man if there ever was one" and "A mainstay of the Mission is leaving us tonight." It is a glorious moonlight night, almost as light as day. I watched the steamer as it sailed out toward the sea in the wide path of the moon and disappeared beyond the Head.

November 25th Thanksgiving Day! We have had two thanksgiving dinners and should be doubly thankful. There was mid-day at the hospital, evening dinner at the Guest House. Venison was the *pièce de résistance* for both with potatoes and canned peas, but soup and dessert varied. Aside from dinners, the day was the same as any other, for this being British Newfoundland the natives find no special cause for thankful celebration, and we conform with local custom.

To be sure, this was the evening of our weekly staff meeting, which might be called a party, but on this occasion it proved to be disappointing. Even the light feast that usually followed it was a failure. Mrs. Manager had prepared a salad and left it in the back porch, where Reuben, coming in to get food for "Skin and Bones" (the Manager's pet dog) and finding no maids, had foraged and found what looked to him like a bowl of refuse. Skin and Bones had a fine supper and the club drank cocoa. Reuben has avoided the house and sent Jimmy Green for the milk pail. He is told Mrs. Manager is waiting for him.

December 5th There has been an epidemic of influenza on the coast but less serious than two years ago. We think the weather is partly responsible. It has been warm and damp. The early snow and ice are gone and mud is deep again.

For nearly a week it has been like late October at home. If the "mild" continues Mrs. Grenfell will find she has not come to such a terrible place after all. She and Dr. Grenfell are expected in about two weeks and are probably already on their way to St. John's. The outside world is evidently interested in their arrival and particularly their prospective living conditions. A newspaper clipping from New York has an illustrated article with a photograph of a tiny shack. It is supposed to be the Doctor's new house but is really intended for a shelter at the reindeer camp. The shack is here, piled on a trap-boat, ready to be towed to camp while there is still open water. It would be impossible for it to accommodate even half of the Doctor's household furnishings that nearly fill the Mission warehouse. The article also states that everything is to be transported to Labrador by dog-sled, and we smile as we look at the iceless harbor, the hills still bare of snow, and of course their destination is not Labrador anyway!

This report exemplifies the popular belief that heroism – even martyrdom – is inevitably associated with this northern work.[195] It is difficult to convince those who hold such a view that the interest and stimulus of life more than counterbalance inconvenience and lack of many luxuries that the dweller in a metropolis has come to consider essential for existence. There are few things that annoy and yet amuse Dr. Grenfell more than to be considered a martyr, but his protests are only half believed and really tend to enlarge his halo. We who know this life agree with him that we receive from it more than we give, but apparently only experience can bring understanding.

December 12th The last steamer brought additions to our livestock: two sturdy oxen that have already been useful in drawing heavy loads since so much freight has come for the new buildings. Mr. Manager has been eager to experiment with them and is like a child with a new toy. He is no farmer, but the vegetable garden seemed to offer possibilities, and on December 7th he tried to plough and succeeded in scratching the partly frozen ground. He is planning to use the oxen for clearing paths when the deep snow comes. It is evident he is as yet unfamiliar with winter conditions at St. Anthony.

December 20th The *Prospero* is so much overdue that the Grenfells will not be here for Christmas. This at least allows more time to prepare their house, which in spite of all efforts may not be ready for them. They may have to stay for a while with Dr. and Mrs. Stewart. Building is progressing, but there is still much to be done on all the houses. On unpacking the materials, the men found rolls of thick colored burlap for wall covering. There were three colors: red, green and light tan. Part of it was ordered for the orphanage, but nothing was found to indicate for which house a color was intended, so Miss Storr chose green for her living room. It looks very nice and she is delighted.

The snow is deep, but only the upper end of the harbor is frozen over. It is late for open water. Mr. Manager has been out with one of the oxen, trying to clear drifts from the paths. He has had an awful time, for the wind makes new drifts as soon as he passes by. He will have to resign himself to walking on drifts instead of trying to clear them away. He has also taken out the dogs for the first time and thinks he will be able to travel forty miles a day. Reuben has his doubts! He has much to learn before becoming adjusted to this northern life.

It began to snow again this afternoon and now is blowing a gale, the snow so thick the hospital is blotted out. It is what the natives call "dirty weather." At first I could not tell what that meant, for storms I thought bad from southern New England standards are considered only moderate. Apparently, no weather is "dirty" without a wind at least sixty miles an hour and snow thick enough to obliterate every building.

December 31st Last night's gale was awful. When the air cleared this morning, the *Strathcona*, the *Daryl* and two other boats were all apparently aground near shore. The upper harbor ice is gone, and that is what caused the havoc, for the wind had blown it down the harbor, taking the boats with it. Drifts are high and fantastic, but there seems to be little more snow than before. It is probably in the woods.

The work-room is still unfinished, but we met there tonight to hold our first wood-carving class for the boys. The fact that some of them are round shouldered suggested calisthenics after each lesson. It may give them better carriage and posture. A social club is also started for the loom-room girls. They will meet bi-weekly in the loom-room to play games – parcheesi, backgammon, checkers – and the girls may invite their friends. The party ends with crackers and cocoa made on the little kerosene stove. The club, which is planned on the principle of those at Hull House and other social centers, is an effort to revive the activities of two years ago that seemed to be successful. I shall be with them each evening and Mr. Wight and Mr. Evans, who are interested, may help.

Tonight, we waded through drifts to Miss MacNair's school concert at the Orangemen's Lodge, but it was worth it. The children were well trained and deserved all the applause given them by the enthusiastic audience. They evidently were delighted with their own performance and applauded vigorously when the audience did, beating a tattoo with their little skin boots. The "concert" was almost entirely recitations, the only music being the gramophone, and we left Mr. Manager to operate it while we went home prepared to sit up till midnight, very sleepy but bent upon ushering in the New Year. When the clock struck twelve, Mr. Forbes had his little signal cannon ready and fired two shots. Mr. Wight lighted a green flare that illumined weirdly the snowy hills, and Mr. Evans fired his rifle five times. He said five cartridges were enough to waste on any old year or new one!

January 1st, 1910 The New Year! and the beginning of a new diary. I contemplate its clean white pages with hope, mingled with misgivings that so often accompany the unknown. It has begun auspiciously, with a friendly family atmosphere that we hope is a happy augury for the future, and the grand event of the day, the staff dinner at the hospital, was a merry meal. We gathered around a gaily decorated table and feasted on real turkey, one of two brought from St. John's by Mr. Manager. There was also plum pudding. Think of all the sympathy wasted on us by "folks back home" on account of our meagre fare! There is also a nineteen-pound turkey sent by Dr. Little that is being saved for the grand dinner planned to honor Dr. and Mrs. Grenfell when they arrive. It is the only form of entertainment we can offer them.

2

Dr Grenfell Brings His Bride

4–24 January 1910

January 4th At last we have real winter weather, one blizzard after another, and are now walking on drifts above the tops of fences. The harbor is ice as far as we can see. If zero weather continues, the steamer that brings Dr. and Mrs. Grenfell will have a hard time getting in, although it is an ice-breaker. They will probably have to land on the ice, which may be considered a hardship or a thrilling adventure, according to the point of view of a novice.

The Doctor wrote that his house must be ready on his arrival, even with papers on the walls and furniture in place. We will do our best, but it seems impossible. There is much inside finishing yet to be done. The plumbing is not in order – the water supply is some distance from the house, and when after wading through snowdrifts one gets there it is frozen. No stoves are in place, and not a bit of furniture has been brought from the wharf where it is stored. The floors are not ready for it anyway.

To our consternation, the Doctor's letter also referred to the green burlap for his living-room, and there it was on Miss Storr's orphanage wall. Of course, it had to be removed and replaced where intended, but the strips cut for the orphanage were too short for the higher ceiling of the new house and required patching. The board walls, without plaster or other finishing, have to be covered with cloth (tacked on) and paper pasted over it before the burlap can be placed. Three layers of material! It is now almost ready on the walls of its second home but is a curiosity. The men say there are fully 70 pieces, some of them not more than an inch in size, but it is patched mostly near the baseboard, where furniture will cover it, so I hope it may not be very noticeable. The Doctor's latest word is that four rooms must be ready if the rest of the house is not – bedroom, living-room, dining-room and maid's room – the kitchen stove in place and supplies brought in.

It has been difficult. At first the weather was so bad nothing could be hauled – not enough snow for a komatik and too much for a cart. When more snow fell it was a blizzard. Not until a few days ago was it possible for the men to bring loads from the storage wharf. Throughout two days they hauled from morning till night in blinding snow and a gale of wind with dog-team and the faithful ox that works very well since the wind-packed snow is so hard he can walk over drifts without falling through. Sometimes the men were nearly blinded by the windswept drift. Today the house is full of workmen, but I am appalled at how much is still to be done. It is time for action! Tomorrow, the entire staff will lend a hand.

January 7th The blizzard has blown itself out, and this morning we saw the sun for the first time in ten days. It is very cold. Alice was late in making the dining-room fire, and the room was literally freezing. Water froze in our glasses at breakfast, and knives and forks were so cold we handled them with napkins. What an item for home consumption, to verify the popular belief that we suffer privations, a belief that persists in spite of the turkey and plum pudding we ate at our Christmas dinner.

At the hospital this morning everyone was in a state of mind. Someone had just made an awful discovery. A sample of the green burlap with a tag attached, plainly marked "For Dr. Grenfell's living-room," had been found among some papers. Everyone had felt comfortably blameless since no identifying mark had been found. We don't know whose mistake it was, and we don't want to know.

We are planning a grand reception for the Grenfells. The *Strathcona* will fly all her signal flags. Every other flag in St. Anthony will wave, and wharf and orphanage will be decorated. Mr. Forbes is to bring out the little signal cannon and fire a welcoming salute as they land, and a group will meet and escort them. The village people are busy with preparations. Men are building a triumphal arch over the path from the shore where it passes the hospital. They are to decorate it with spruce boughs, and under it will hang a large banner of unbleached sheeting with the greeting "WELCOME TO OUR NOBLE DOCTOR AND HIS BONNIE BRIDE" sewed on it in letters of turkey red calico. It represents the love the village women, who made it after weeks of effort and is the only tribute they can offer.

At the house the furniture is finally unpacked and in place. Dr. Grenfell's bedroom is in order – even the beds made (with sheets borrowed from the hospital), but the walls of that room have only the first two coverings. After our late experience it seemed well not to depend on intuition for the finishing touch. The dark red burlap for the dining-room was plainly marked and is a satisfactory background for the red mahogany furniture. The living-room is lovely and the fireplace a great success. Around the opening, where tiles might be, irregular pieces of the beautiful native Labradorite (called "bluestone" locally) are built here and there into the rough grey stone structure, the unexpected gleam of brilliant blue adding beauty and interest. Unfortunately, fireplaces are still novelties to the local builder, and the importance of a hearthstone was not apparent. The Doctor's fireplace had none, an omission that was unnoticed by Mr. Manager, who in his zeal for "clearing up" built such a large fire of packing material that sparks – even embers – flew out on the wooden floor, and there was a rush for snow to protect it. We are relieved to find that the house is comfortably warm even in zero weather, although the furnace is not installed. There are airtight stoves in some of the rooms, and the fireplace heats the living-room. The house is built on a southern hillside. It has a cellar. There are double floors, well fitted windows and a glass-enclosed piazza. No wonder it is warm!

We are so glad Mrs. Grenfell will be comfortable when established, but one problem is still to be solved – household service. Lucy (one of the loom-room girls) is to be an initial member of her staff, but she knows little of cooking and Mrs. Grenfell's powers of adjustment may be severely taxed. Dr. Stewart decided that the living-room walls must be finished tonight since the Doctor may come tomorrow, and at 12:30 a.m. we fitted and pasted the last piece of green burlap while those not engaged offered suggestions from the sidelines. Then we stumbled home along the icy path.

January 8th The steamer is expected today. This morning I decorated the komatik that will bring the bride and groom ashore. The cushions are covered with turkey red and white woven strips, and long red and white streamers float from the high back. It is very gay. The men have decorated the wharf and the orphanage. The arch is covered with spruce boughs; the banner hangs under it. Flags are flying; everyone is on the alert. Now, if they will only arrive in daylight.

January 9th No steamer after all! The decorations have been taken down because of bad weather. We tried to have a meeting of the carving class this evening to offset the general feeling of deflation, but there was no fire in the work-room, and with the mercury at zero outside, the boys' hands were too cold for work, and we gave it up after going through the dumb-bell exercises. Classes will be in abeyance until this excitement is over and we return to routine.

Wilfred and Anne Grenfell following their marriage in Chicago, December 1909 (Pascoe Grenfell)

January 10th Mr. Manager announced at breakfast that a revelation during the night had told him the *Prospero* would arrive at 6 p.m., and the committee at once went into action, again putting up the flags and decorations, but it is after 6 p.m. and still no *Prospero*. Tomorrow is now the day.

January 11th A howling blizzard, gale of wind and smothering snow. The *Prospero* is probably at anchor in some harbor. She would not venture out in such a storm as this. We do not expect her today.

January 12th The Grenfells have come! At 6 p.m. smoke was seen over the Head. Someone shouted "*Prospero*!" and there was a rush of feet. Men struggled into their coats as they ran for the cannon, rockets and guns. When the steamer's lights appeared around the point, the celebration began. Mr. Forbes fired the little cannon. All along the waterfront, puffs of white smoke and the rattle of guns gave evidence of the expectant waiting men. At the wharf, green flares blazed and rockets soared into the sky. Few of the people had ever seen fireworks, and shrieks accompanied the roar of the first rocket as it rose and burst aloft.

From the hospital, the welcoming committee – Miss Kennedy, Miss Keese, Miss Storr and I – watched the decorated komatik, drawn by twenty men, start off toward the steamer, now bucking into the ice some distance from shore, and disappear in the darkness, to emerge where the searchlight revealed figures already descending the companion-way to join the group on the ice, and presently a long dark line trailed across the frozen harbor to the path on shore and moved toward the hospital. We waited until the komatik bearing two muffled figures had passed under the arch and banner, then joined the village escort on the path to the Stewarts' house. There was no demonstration. The people were very quiet with only an occasional whoop. Several bad drifts on the way must have been trying for Mrs. Grenfell, who afterward remarked that it seemed a hard road to travel. We reached the house just as they struggled from the komatik and went inside long enough for greetings and our first sight of Mrs. Grenfell. She has a lovely voice, is very pretty, very tall and beautifully dressed. The sight of a modern hat with even a little spotted veil gave us quite a thrill. The Doctor hovered about looking very happy.

Later. Mrs. Stewart has just brought Mrs. Grenfell to the Guest House. She is crying for a cook, having failed to receive a message suggesting that she bring one with her, and we have only Lucy, the inexperienced, to offer her. It is unfortunate that so many problems await solution and so much adjustment seems necessary. It will be hard for her.

January 13th Mrs. Grenfell came in this morning on the way to her first visit at the new house and borrowed an apron. Lucy had gone on before, after confiding to us that she wept all night, haunted by the dreadful thought of cooking, which she knows so little about. The young men who have been helping Mrs. Grenfell all day, unpacking and arranging, report that she has suffered her first casualty, a fall on the ice that wrenched her ankle. We realize more than ever how difficult our travelling conditions must be for newcomers, especially in winter.

January 14th It has been a fierce day – 12° below zero and a gale of wind, the drift so thick one could scarcely find one's way, but it did not interfere with the season's great event, the grand staff dinner given at the hospital in honor of Dr. and Mrs. Grenfell. Sixteen people gathered around the really lovely table in the warm glow of Mrs. Stewart's red-shaded electric lamp, and dishes of nuts, figs, raisins and fudge surrounding it made it a real party. We dressed in the best we had with us, Miss Storr in a white silk blouse and gold locket and chain. Mr. Wight might have just stepped off Fifth Avenue, and all the men looked exceedingly uncomfortable in high, stiff white collars. Dr. Grenfell carved the 19-pound turkey with skill, although he did send to some of the guests the parts they did not prefer, but someone has to eat wings and drumsticks. It was amusing to see an occasional polite effort to pass on the parts not cared for. We drank the healths (in water) of the bride and groom and Dr. Little. I cannot yet realize that Dr. Grenfell is married.

January 16th It was clear and cold today, the harbor ice aglow with color. After church, Dr. and Mrs. Grenfell came to the Guest House for a venison dinner. At the hospital, Dr. Stewart was starting on a difficult medical trip by dog-team, and we gathered to see him off. As he and the dogs climbed Fox Farm Hill, Dr. Grenfell's favorite lookout, and disappeared over the crest, Dr. Grenfell asked me to take his wife to the top to see the view. We started, but her twisted ankle and unfamiliarity with hill climbing made the venture more of an ordeal than a pleasure, and we wandered over to the new house for tea at 4:30. It was the first time we had been there since their arrival, and we noticed with interest that most of the furniture is where we placed it in our effort to hide the patched wall. This time, our intuitions did not betray us!

The house looks home-like and attractive. Mrs. Grenfell made tea at a little table, and we sat before the fire and drank it with the proud realization that we were the first guests. The living-room is really beautiful with Mrs. Grenfell's lovely wedding presents in evidence and a great white polar bearskin covering nearly half the floor. They are not yet in residence, the problem of domestics being still unsolved, but the Doctor wants the staff to meet there on Sunday afternoons for tea. Tonight, he read to us an article he is preparing on "What the Church Means to Me,"[196] and an interested discussion followed. He is greatly pleased with our social club, the gymnastic exercises for the boys after their classes and also the dressmaking class. It is comforting to know that he approves the work already started, some of it under difficulties. To add to my pleasure he has given me a beautiful "volunteer" pin like Dr. Little's, with my name and the designation "Industrial" engraved on the back. I think it is given to only the earliest volunteers.

January 22nd Last Tuesday was a half holiday for the people of the harbor in honor of Dr. Grenfell. There was a football match on the ice in the afternoon directed by him, which was followed by the children's Christmas tree and tea in the Orangemen's Hall, but this was merely preliminary to the concert and presentation to the Doctor. The hall was crowded and the air stifling. In the center was a hot stove. Also in the center chairs were placed for the Grenfells and members of the Mission staff. We nearly melted! The rest of the people sat on benches without backs, in rows on each side of the room. Every man of the harbor was there, some even from little settlements along the coast.

Mr. Sidy,[197] the minister, presided, and after speeches and some singing, a beautiful combination barometer, chronometer and thermometer was presented to Dr. Grenfell from the people of the coast. It is a most fitting, useful and valuable gift, and the rumor that it was chosen at the suggestion of the Reverend Mr. Richards, minister at Flower's Cove, a great friend of the Doctor and often our guest two years ago, did not surprise us. Then Dr. Grenfell spoke, and I never heard him more eloquent. He was deeply touched by the tribute to him, and what he said so earnestly came from the heart, as did the fine spirit and sincerity of the people who honored him.

January 24th Mr. Evans has just reported that the dog mail will make its first trip tomorrow. It brings with it a sense of uncertainty. We feel that letters go out into the void, and faith is needed to assure us that they reach their destination. Actually, very little mail is lost, but many things can happen. The mailman may be held up by a blizzard, the dogs overturn the komatik or break through the ice or lose the mailbag. At one time, a mailman made trips when possible across the Strait of Belle Isle on the ice to Labrador, but that was very hazardous. I once saw a photograph someone took of him [Ernest B. Doane],[198] a tall, husky, bearded man, and in my eyes very much a hero with a record of adventure. There are brave and hardy people on the coast.

3

Winter Activities

2 February – 17 July 1910

February 2nd One of Dr. Grenfell's present interests is deer driving. He wants to drive them himself and expects them to go fast, at least faster than those already in use, so Mr. Wight and Mr. Evans are acting as trainers. Wild deer are brought in from the herd and they try one a day. Each man harnesses a deer to a pulka and goes out on the marsh where there is plenty of room. The deer starts off briskly, but when it sees a bare spot on the marsh and scents moss it dashes for it, regardless of obstacles or the passenger on the pulka, who often flies off at a tangent when encountering icy spots along the way. The deer also dashes for any hill it sees, thinking the rest of the herd may be in that direction; and sometimes it stops short, turns around and stares the driver in the face, the pulka also stopping among the deer's legs.

There is a turn in the path which the deer seem to consider a special race track. That is usually a Waterloo for the men, who come to grief in the tuckamore and snowdrifts. Mr. Evans came home one day with a twisted shoulder and his arm in a sling. Today he is limping with a strained knee, while Mr. Wight says he sits down with an effort and refers to contusions on his anatomy. I am going out with my camera tomorrow, and the men are to tell me where to station myself in order to get the most for my money. I want to try sitting in the pulka myself and not let them have all the experience and fun, but they say "No!" In their opinion, it is more sport than anything they have done since coming here. As they are both athletic men with standards of sport, this evidently ranks rather high, and we think the deer, in spite of their training, may give the Doctor some exciting moments.

February 12th The wide white expanse of country is an invitation to break its monotony with color, and the Mission staff is responding with an orgy of gay clothing, the brighter and more bizarre the bet-

Jessie Luther in her fawn-skin coat,
February 1909 (JLP)

ter. Some of the results are remarkable. Mr. Wight has just christened his new deerskin suit, topped with a muskrat cap. It is very superior, and if on his way to a ball in immaculate evening clothes he could not take more pride in his appearance. Mr. Evans is resplendent in a yellow dressed deerskin suit trimmed with yellow fox fur. We call him "The Daffodil." L.M.K. and I have lovely hooded coats made by the Labrador Indians. They are of fawn-skin, knee length, and we wear them with the hair outside. The white inner side of the dressed pelt is gaily painted with tribal designs, red, blue and yellow, and the attached hood is the skin of the fawn's head. The ears have not been removed, and pieces of red flannel fill the empty eye holes. With the coat we wear a colorful Hudson's Bay sash with long fringe.

This morning was gorgeous, clear and cold, 4° below zero with intensely blue shadows, a day to make one restless, eager for open spaces. L.M.K. and I decided to take a walk and started across the harbor, which was a glare of ice with spots of snow. Travel was difficult and we turned inland toward the brook, but it was still heavy going, so we slid with the abandon of children down the steep bank to the frozen stream where the snow was smooth and level. It was beautiful and we walked along happily until a roar beneath our feet indicated a waterfall, and we climbed the bank hastily with a feeling of escape. It was our first hard walk of the season and more than satisfied our longing for exercise.

Yesterday was another glorious day, and my urge to paint was so strong I took my pastel box to the piazza (for it was too cold to sit outside) and tried to paint through a hole scratched in the thick frost on a window pane, but gave it up when the clear space froze over almost at once. The color was beautiful on Old Man's Neck across the harbor. Blue shadows crept across the pale violet ice and mounted the hill to its deep rose crest. The same rosy light tinged the hills across the Bight, and along the horizon glowed a warm green sunset light. I shall have to finish it from memory. It was too fleeting to catch.

February 13th A member of the Mission staff attended Dr. Grenfell's bible class this afternoon and reported that in spite of the storm about twenty men were there, some of them from across the harbor. He said the men enter freely into the discussions and remarked, "That is as it should be if it does not disturb their simple faith in the literal revelation of the Bible." That is true, for this belief – to its minutest detail – is their stronghold. They are not of a mind to grasp the problems of modern theology. It would land them in a chaos of doubt without leading them to a solid ground of essential fundamentals, sufficient to satisfy simple minds bred in the orthodox Methodist faith, which is often influenced by its emotional expression. I think the Doctor will bear this in mind, but there are so many questions the men might ask that would be difficult to answer from the practical viewpoint of the twentieth century without shaking their faith in the literal interpretation, which is a part of their religious belief.

February 14th A day of excitement! The Manager's baby has arrived. In the early morning there was banging of doors and hurrying of feet. The light of a lantern danced on the wall; then I heard Dr. Grenfell's voice in the hall. Miss [Mary] Keating (the nurse) was there when we went down for breakfast. The house was expectant, and I prepared the baby's basket, the Doctor coming in now and then to sit by the fire, then returning to his patient. After a while Clare, the cook, came to beg me to bring my work to the kitchen and sit with her, for sounds from the floor above made her nervous. It seemed an inopportune time for an attack of nerves. We heard the baby's cry at eleven o'clock. It is a fine healthy boy. He weighs eight pounds and all is well.

Mark asked me today if he could buy some homespun for a coat, and I offered to cut it out, for he intends to make it himself. His interest suggests a possible new line of work. Why not add tailoring to the Industrial Department with Mark in charge? He could be taught to cut and plan clothes, hire women to sew them and superintend what might become an extension of the dressmaking already begun. It is worth trying, like so many other opportunities for industrial and social improvements that may be wise or otherwise when faced with reality. My tendency to rise to the bait of a new venture is often amusing, but who would be static?

February 17th Another day of excitement. Dr. Stewart returned from his long medical trip. He was utterly exhausted. Alf and Eli Reid [son of George Reid], who were with him, said it was an awful trip. They were three days on the way from Canada Bay with ten men on snowshoes and with komatiks to go ahead and break the way. Only once before has this been done, for the paths are usually hard at this season.

When they reached Dr. Grenfell's bungalow on Hare Bay, Dr. Stewart thought the worst was over and sent the men and komatiks back but on starting the next day found the going was as bad, if not worse. In his anxiety to reach home, Dr. Stewart tried to shorten the distance as much as possible and ventured to cross a frozen end of the bay. He was walking a little beyond the others, his snowshoes in his hand, when he suddenly broke through to the water. With the presence of mind born of hazardous experience, he thrust the snowshoes before him on the ice and, resting his weight on them, drew himself out, pushing them ahead of him until he reached firmer footing. But this did not end his troubles. He still had before him a walk of eight miles in his frozen clothing to Lock's Cove, where he and the two other men spent the night, then continued to St. Anthony. Almost daily we are reminded that life in this country has its cruel and hazardous side. Perhaps these hazards contribute to its many fascinations. They are surely tests of resourcefulness and endurance.

February 18th Today, on leaving the hospital, I met an old man with a note he asked me to give to Dr. Grenfell. He had walked all the way from Quirpon, eighteen miles. His beard was crusted with ice and snow, and he was nearly exhausted. The note told of a case in urgent

Anne Grenfell, in the late stage of pregnancy, makes ice cream with Loula Kennedy [?], 1910 (GHS)

need of medical help, and Dr. Stewart – in spite of his recent trip – started after supper tonight with the dogs, a typical case of emergencies to be faced by this scattered population and the courageous efforts of the Mission doctors to meet them.

February 23rd The regular Sunday afternoon teas for the staff planned by Dr. Grenfell after he and Mrs. Grenfell were established in their new home have been discontinued. We met only twice. Mrs. Grenfell has not been well and rarely goes out except when the Doctor takes her with komatik and dogs.[199] To add the rigors of a northern winter to one's first experience of this kind of life must be difficult indeed. Two of the village girls are acting as maids, but neither is an experienced cook and I fancy the food is sometimes unappetizing. Miss Kennedy and I tried today to make coffee ice cream for her. We had no cream but used the "top of the milk" and beaten raw egg. Neither had we an ice cream freezer, so we manufactured one, an empty coffee tin in a wooden pail packed with snow and salt. We twisted the tin back and forth, occasionally stirring. It came out very well and Mrs. Grenfell enjoyed it.

The young people's club met last night, and Dr. Grenfell, who came to visit, showed us how to swing Indian clubs, of which he has a large collection. Then we showed him the exercises with wands we have been practicing. He is keen for us to exhibit the "drill" (as we call it) as a part of the annual sports program, which includes an evening entertainment at the Orangemen's Hall. It is to be on the 23rd and 24th of March, and I am invited to make badges!

Dr. Grenfell came in this morning with some material for overalls, asking who could be found to make them. I told him either Mrs. Green or Mrs. Geer might do it. He looked disappointed. It was evidently not the answer he had hoped for, so I asked how it would do if I made them myself. He smiled slowly. "If you only will," said he, so I took his measurements and am in for it! It was a dangerous precedent, for Mr. Palmer asked me later if I had arranged to make Dr. Grenfell's overalls, adding that he had heard I could make fine trousers. He suggested my making some for him. It is gratifying to have one's talents appreciated, but my plans do not include general tailoring.

February 25th We went to the Bight this evening for their community Christmas tree, much like such celebrations of other years and without special incident. These belated Christmas trees are really not so incongruous as they seem. It is the event of the year for these little communities, but weather and other conditions make a date uncertain. Apparently it does not matter, provided it is sometime before Easter. They even had one last year on Good Friday.

February 27th The Doctor has organized a snowshoe club to meet Saturday afternoons. All the harbor people are to have [a] half holiday if they choose to take it and are invited to join the walk that is to start at two o'clock. In planning these walks the Doctor has, I think, tried to revive the weekly wanderings so much enjoyed two years ago by our happy winter family. It is what the active Doctor likes to do, and it must be a disappointment that Mrs. Grenfell is at present not strong enough to share this form of enjoyment with him. To the village men, the invitation to struggle without objective over snowy country where many of them have worked all the week evidently lacked appeal, and the women rarely venture beyond the village.

The party dwindled to six of the Mission staff, who walked across country. At the top of the third hill the Doctor, who carried an axe, stopped to cut off the upper branches of a tree and nailed on it an official notice that no wood was to be cut between that point and high water. He made it a ceremony. Everyone drove a nail. Then he tied one of his red cotton handkerchiefs to a stick and fastened it above the sign. Meanwhile, L.M.K. and Mr. Evans, who came on skis, had gone on, tempted by the lovely hills. Miss MacNair also went on, leaving Dr. Grenfell, Dr. Stewart and me, the Doctor stopping as we went along to cut tuckamore branches and stick them in the snow to mark a komatik road. Dr. Stewart evidently felt it was slow – perhaps there were other things that seemed (to him) more necessary – and he also vanished, leaving the Doctor and me to finish. Presently, the Doctor said he supposed he should go to cheer the football team that was even then playing on the ice. He handed me the axe, and I finished the road and the walk alone.

February 28th This is the Doctor's [forty-fifth] birthday. Local facilities for celebration are limited, but I did venture to make the traditional birthday cake. While it was baking, a message came that we were all expected at the "House on the Hill" for a buffet supper at six o'clock – a happy surprise, for Mrs. Grenfell has for several weeks been far from well.

We dressed in our best and went to the birthday party. Mrs. Grenfell greeted us in the attractive living-room, looking rather thin but otherwise quite well. She wore a lovely red Liberty velvet gown with a train, elbow sleeves and wide Irish lace collar and cuffs, very becoming to her tall figure. The Doctor came down, still in his favorite old homespun trousers, but he had put on a black coat and white collar and had waxed his moustache. Mrs. Grenfell was a gracious hostess, and the table was so beautiful with delicate linen and quantities of gleaming silver that it was hard to realize we were in Newfoundland. It was quite like a formal luncheon at home. There was creamed rabbit (mentioned officially as "chicken"), a delicious salad of tinned asparagus tips and the firm parts of canned tomato, several kinds of fancy sandwiches, ripe olives and coffee. For sweets there were little cakes, wafers, chocolates, candied fruits and ice cream. Everything was served in beautiful silver dishes that reflected the soft gleam of candlelight. It was a lovely party.

The Doctor gave a lecture with stereopticon[200] views at the Orangemen's Hall that evening, and after supper we joined the crowd that came from all over the harbor. It was for the benefit of the football club – tickets five cents each. Mr. Sidy, the minister, in his zeal for driving a moral wedge on all occasions, had selected "John the Ploughman"[201] for Dr. Grenfell to read as a starter. It was illustrated with slides, each accompanied by a short sermon. Then Dr. Grenfell said he would show them some pictures of his own, and the audience jumped from the Capitol at Washington to Westminster Abbey – from Niagara Falls, the Grand Canyon of Arizona, the big trees of California and the Petrified Forest to Pompeii, Iceland and the Bataille des Fleurs at Nice. At the next lecture, he intends to show pictures of the tubercular bacilli, of which he says 12,000 clinging to each other's tails would only cover the point of a pin. The people are incredulously looking forward to it.

March 1st Two of our young people, Will [Simms] and Annie [Ash], were married a week ago. It was a very pretty wedding in church, with Elsie and Janet as bridesmaids, Archie [Ash] and Steve [Pelley] as groomsmen. I did not go to the house, but Mr. Forbes did and returned at ten o'clock, reporting that he left the bridal party and guests playing kissing games.

Skiing is daily becoming more popular. Even the orphans have caught the fever. They have made skis from barrel staves and are having a grand time tumbling around on them. After taking a dare from Mr. Evans, I started out with him this afternoon to shoot the big snow bank and proved I could do it without disaster. The trick seems to be to double up and let yourself go. Mr. Wight preaches "abandon" and I shall try to practice it. To see Mr. Evans and L.M.K. come from the summit of the hill is breathtaking. I watched them rushing down with bent knees, feet far apart, strained eyes and a curious intense expression, then collapse in a whirl of snow and windmill of arms, legs, poles and skis.

March 2nd My club is really a grand success. Twenty-five enthusiastic young people came tonight. Their performance of the "drill" is improving daily. The girls practice wand and dumb-bell exercises at the loom-room for half an hour every day, and the boys now meet almost every evening to swing Indian clubs. Each boy takes clubs home with him for practice. They now stand very straight and swing with ease and freedom of movement. Mark is becoming an expert and coaches some of the others during the dinner hour. But we do need space. The Indian clubs can be managed fairly well in the loom-room, but electric lights have to be taken down to allow for them. It is impossible for more than six to perform at once, and as for marching one is always confronted by the chimney and stove, but that does not discourage anyone from joining in the "grand march" at the end of the evening. An exhibition of the drill is to be a part of the Orangemen's Day entertainment, and Archie Ash says all the village is agog. It will be different from anything ever given before, and curiosity is aroused.

The past few days have been reminiscent of my early St. Anthony experience. I have been cooking at the orphanage. Miss Storr's cook has pneumonia, there are 19 orphans in her family, and her hands are full. In the emergency I offered my services where most needed, which seems to be over the cook-stove, so here I am a part of each day. Never before have I cooked for a hungry horde of 19, and quantities are guesswork.

The first day there was salt fish, boiled potatoes and rice pudding, which were stretchable, the second day vegetable soup and Indian pudding, but I miscalculated the amount and had to supplement with what I could find – stale bread, apple sauce, spice and molasses, mixed together and browned in the oven. The children liked it. One of them told me, "We all liked th' puddin', Miss, an' dey all kep' sayin' Miss Luther made de puddin' an' we was all proud you made it." It was not really like the early days, however, for things have changed. There is now a big kitchen, a regular hotel range with two fires and ovens, more utensils and much space, also (Oh, joy!) hot running water – very different.

March 4th A lovely cold day and snow over the fences. The much-talked-of snow conditions, when the space between orphanage and Guest House is one unbroken expanse of whiteness, without fence, shrub or rock, has come at last. Days like this bring restlessness. Four walls seem confining and open spaces call. Something in the air, the slant of sunlight, suggests spring in spite of the snow-covered country. I took my skis and climbed the hill to watch the lengthening shadows creep across the harbor to the rosy crown of Old Man's Neck and fade in opalescent light. On the way home, I met Miss Storr in search of water for twenty-one baths, her supply – in spite of the new plumbing system – having failed to function. Those of use who live in houses without bath-tubs are spared some problems!

March 6th George Ford came in from the deer camp today and was here for tea when Dr. Grenfell dropped in. It is always interesting to hear them talk, and this afternoon they swapped stories of adventure, Mr. Ford's two experiences with savage polar bears and his narrow escape from one of them being highly exciting. He certainly has tall tales to tell of life in the far north. I am still hoping to go there – at least part way.

We went to church this evening and heard Dr. Grenfell preach a beautiful sermon. When he speaks like that, revealing the spirit which animates him, the Doctor is at his best. What he says is so simple, so practical, so suited to be helpful in the everyday life of these simple people with narrow horizons. He is troubled by cases of delinquency recently reported to him, which (as the only justice of the peace north of St. John's) he is called upon to prosecute. One of them, a case of petty larceny, is to be tried next week. In its early days, St. Anthony was apparently a law-abiding community, but changes are coming with the influx of outside influence. It may be the result of civilization (so called) often accompanied by lawlessness.

March 12th Gorton was tried for larceny today, convicted and sentenced to jail for a short term. St. Anthony has no constabulary in any form; consequently, Gorton is his own jailer. He is provided with food which he prepares himself and given the key to the jail so that he may help his family, if needed, then return and lock himself in. It seems an easy punishment, and some of the harbor people think it no punishment at all, even an encouragement to crime. Others consider the Doctor "soft," but what is one to do? The jail is a tiny building. Its one room contains a cot, a stove, a table and a chair. There is no accommodation for a jailer if there were one. Neither (as far as I know) are there funds for a police department. It is a matter of adjustment to circumstances until delinquents have outgrown the jail.

March 16th Mr. Wight and Mr. Evans left this morning for a stay at camp. They started up Fox Farm Hill sitting on the long komatik, driving two deer abreast, triumphant evidence of their successful training. Behind them followed Nat Geer, one of the herders, in a pulka loaded with 150 pounds of provisions drawn by one of the larger deer. An interested retinue of small orphans trailed along behind to escort them over the hill.

The Doctor's twenty-mile trip with the deer in double harness has not materialized, but he has organized outings and social gatherings, one of them a renewal of the Saturday afternoon snowshoe club. He is also trying to renew the weekly afternoon teas at his home and last Sunday showed us his den on the third storey: a cozy, comfortable room with shelves and cubbies in every nook and cranny, suggestive of the Doctor's familiarity with space limitation on a boat.

March 17th Mr. Manager and Mr. Ford started on a long trip today: Flowers Cove, Canada Bay, etc. It rained all night and the going was awful. Even Mr. Ford, who is "game" after 25 years on the Labrador, did not think of starting, but Mr. Manager had planned to go – and go he would. We watched them start up the hill, Mr. Manager running on ahead, calling incessantly to the dogs and breaking through the soft snow with every step. The men are betting on how far they will go today and how long it will be before Mr. Manager gives out utterly.

The storm is developing into what the natives refer to as a "Sheila" that usually comes just before St. Patrick's Day.[202] "Who was Sheila?" I asked. "St. Patrick's wife, Miss," said they, without, I am sure, realizing the implication of transgression in the breaking of celibate vows.

There are several children in the hospital, and I have often been with them, trying to interest them in making cut-outs and coloring picture books with crayons. One little girl with a tuberculous spine is rigid in a plaster cast. In another case, a small boy's legs are paralyzed from tuberculosis. On entering the ward the other day, I found

Dr. Grenfell had moved two beds together so that he could put his hands on the foot-rail of each. He was practicing swinging, raising himself and turning somersaults for the benefit of little Pat, then urging him to try raising himself to test the strength of his arms.

The Doctor's weekly stereopticon talks continue, but the five-cent admission, at first for the benefit of the football team, is now devoted to other special causes. The second was for the school and the third to supply wood for the woman whose husband is serving a jail sentence. We are constantly reminded of the practical as well as ideological aspect of the Doctor's work.

Some time ago, I mentioned a mailman who had crossed the Strait of Belle Isle in winter time. His name is Doane, and he is now staying with us after crossing the wicked strait in a small canvas boat he made himself. He crossed a week ago today in one of the worst blizzards of the season. The strait was full of ice, and he frequently had to haul his boat on an ice-pan, drag it over and launch it on the other side. He brought the mail from Labrador and has signed a government contract to bring it over every month during the winter. He is a large, strong man, very quiet and unassuming, but influential on his part of the coast. It is no wonder his face indicates firmness and strength of purpose! He is also an expert taxidermist and supplies the Smithsonian Institution at Washington with stuffed specimens of birds and animals. As the house is always full when we are all here, a bed has been made for him in the gun-room until he can continue by dog-sled on his way to Canada Bay with the mail. It is the understanding that the Mission will take him there, and Mr. Manager has agreed to do this. Meanwhile, his visit is one of the interesting incidents of the winter.

March 23rd This was the first day of the much-heralded annual sports. A final drill for the wand and dumb-bell exercises the boys and girls have been practicing was held last night. Uniforms were planned, and for days I have been making badges, flags and favors, but the drill was nearly called off.

In the late afternoon, I was at the hospital and about to leave when Dr. Stewart told me one of the deacons of the church, a member of the Orangemen's club, had just told him that if the gymnastic drill was anything like dancing it was against the rules of the club and could not be allowed in the hall. I was angry and insisted that Dr. Stewart should see me rehearse the performance to judge if, by the minutest splitting of hairs, it could be so construed. It did seem too ridiculous, especially as the moral standards of some club members are questionable in matters that are really important. Dr. Stewart's verdict was favorable and the exercises met with approval, but I was told the doubtful deacon retired behind a group of men in the balcony with his face to the wall during the performance. I suppose that was to clear his conscience, for he could say he did not see it – therefore could not be held responsible if questions were asked.

The entertainment was varied, beginning with a colored kaleidoscope which elicited "Oh's" and "Ah's" from the audience, then buildings and scenes having no connection with each other, followed by famous pictures of the life of Christ, with an intermission for the wand drill and the Indian club exhibition, which the enthusiastic Doctor ended by leading after performing wild stunts with the clubs, which the boys could not follow. The Doctor used two pairs of clubs for his performance, a light pair striped with tin foil and a huge pair that looked like pillows, decorated by the Doctor himself with stars and stripes. They weighed over six pounds. We doubted that he could swing them, but he did!

The Doctor, when told of the deacon's moral qualms concerning the drill, promptly rose to the occasion and made a speech at the end of the evening. He praised the performance and explained that its purpose was to show what was being done to benefit the health of the young people. It was not a mere entertainment. He added that the exercises I had been teaching them were used in many United States schools for the pupils' physical development. The reaction was immediate, many of the older people telling me afterward how much they had enjoyed the drill that was so "good for the health"! There will be no further need for the deacon to hide.

March 24th The second day of the sports, clear, windy and bitterly cold. Target shooting was one of the major events of the day, the men coming from all along the coast to take part. Some of them brought guns of unbelievable length and ancient powder horns slung across their shoulders. The local men say some of the guns are seven feet long and kick frightfully. There were dog-teams everywhere, many of them tied to ballicatters[203] and stumps. Fine teams of dogs but "saucy," as the natives say, and we were warned not to approach them when in harness. It was interesting after the sports to watch them with their drivers start off in all directions across the ice and disappear over the hills.

March 25th A perfect day after yesterday's bitter wind. L.M.K. and I felt restless and started for a walk, stopping for friendly visits by the way. We had not called on Lizzie (one of our girls) since her baby died and found her thin and sad, not like the rosy girl of two years ago. There was a komatik being mended in the kitchen, and when we were led into the other room, Charlie and other members of the family came too and offered me the largest chair, while they stood around the wall. Lizzie showed us a photograph of the baby "laid out" and one taken with its two grandmothers when alive. It is a curious desire to photograph the dead, much in line with the preservation of coffin plates and funeral wreaths one sees so often in Cape Cod homes.

We also called on the Turner family to see Eli. We found him in a very hot room full of people. The mercury must have been 85° and the air was full of odors, for the windows are not made to open and the family poultry live in the same room. Eli is dying of tuberculosis, and we wonder how any of the family escape. Dr. Grenfell took Eli to the hospital some time ago for open-air treatment, which he disliked so much he refused to stay and departed to his family furnace, saying he would "rather die than freeze." It was too late to save him anyway.

This evening was so beautiful I took my skis and went out on the harbor, revelling in the great whiteness. There was a gorgeous full moon and the air so still it seemed breathless. Occasionally, a dog barked on shore or the laughter of children coasting near the orphanage came to me across the snow. Not a living thing was in sight, and one felt very small in the wide white expanse under the moon. On the hill behind the hospital, hoar frost glittered like diamonds on the snow and treetops. High hills rose, one beyond another, against the starry sky, and the frozen sea stretched to the horizon. I hated to leave it and wandered here and there, saturated with its beauty.

March 28th There was an exodus of men this morning, all going in different directions – Dr. Grenfell, Mr. Evans and Mr. Palmer to Cape Norman to trace reports of deer poaching; Mr. Manager and Mr. Doane at last on their way to Canada Bay with the mail, Mr. Manager finally fulfilling his contract to deliver it there. Mr. Doane was growing desperate and the Mission staff indignant because of his needless delay.

The men started early, and I went down to pour their coffee at the 6:30 breakfast. In the midst of it, Dr. Grenfell came in to hurry them up and collect things he wanted to take with him, one of them being the big komatik box we had used for storing magazines, which were promptly deposited on the floor. The Doctor drank a cup of coffee and dashed off again. He and others of the Cape Norman party expect to return in a few days to allow Dr. Stewart to start on a very important medical trip. There was therefore some excitement when it was discovered, after they had all gone, that Mr. Manager had taken one of the best dogs, reserved especially for this difficult trip. He was well on his way before this was known, and Dr. Stewart went about wildly, looking for a komatik and dogs to overtake him. Finding them all in use he started off on foot with one of the village men, only to return in the afternoon, tired and still mad, having missed him by fifteen minutes at the deer camp.

This action by Mr. Manager is only the latest of many incidents, important and unimportant, that have contributed to interfere with the Mission's individual and co-operative effort during the past months. His assumption of unwarranted authority, unwise judgment and inconsiderate attitude, combined with inefficient management of the Guest House, have made life uncomfortable for everyone. Much as we dislike to add to the Doctor's problems we feel that for the morale of the Mission, the time has come for him to take a hand, and Mr. Forbes has volunteered to lay the matter before him in detail.

March 30th Dr. Grenfell's party returned this afternoon. The expedition was planned after a man from Cape Norman (where the deer were found killed) had brought a head with antlers to St. Anthony for the Doctor to buy, and Reuben, with true detective instinct, had discovered the Mission tag in its ear. It was considered rather a joke but not by the man, who was badly frightened. He was sure the Doctor would put him in jail. He is a decent fellow and would not have transgressed intentionally. Upon investigation, however, others were found to be not so innocent. Mr. Evans and Mr. Palmer were appointed constables to collect information and make arrests, thereby adding greatly to their importance in the community. Mr. Evans said they were fed royally all the way, but there was no outward evidence of wrongdoing. No Mission tags appeared on the luscious haunches of venison offered them by their hosts.

April 1st Mr. Manager has returned and Dr. Grenfell, after hearing Mr. Forbes' report, acted promptly and firmly with tact and understanding. He has talked frankly with Mr. Manager, whose authority and responsibilities are now limited and definitely defined. The storm is past, the air cleared and we anticipate peace if not full harmony. Mrs. Manager and the baby are going home in June. Mr. Manager will follow at the end of the season.[204]

April 3rd Mrs. Stewart, L.M.K. and I have had an exciting experience. We were lost in the wilderness. We started for an eight-mile walk to the deer camp – difficult to find, for the trail is indefinite and the camp hidden among trees. It was, however, on the way back that we realized we were completely lost. The path is very deceptive, a slight track made by dogs and deer on the hard snow. Trails led in different directions, some to dead ends where firewood had been cut.

There were many little ponds where the ice was blown clear of snow, and it was difficult to follow the trail where it continued on the other bank. There was no landmark to guide us. Twice we retraced our steps on finding ourselves headed for the dreaded White Hills. We wandered about, confused and desperate, and it was after six o'clock when Nat Geer, returning to the deer camp from St. Anthony, noticed our snowshoe tracks leading in the wrong direction and, following them, overtook us and led us safely back to St. Anthony. The men estimate we walked 22 miles. Never again will we venture forth without compass or guide.

As I am visiting Mrs. Stewart during Dr. Stewart's absence, it was breakfast in bed for us this morning, and at odd moments during the day we have compared notes as to our sore spots. Our faces are burned and swollen from snow glare, and we think twice before mounting stairs. It will be some time before the benefit of all that exercise is apparent. Only a realistic experience of northern life, we say, but the possibility of spending a night of uncertain weather in the snowy wilderness was no joke. We had planned to burrow a hole in the snow for shelter.

April 4th The Doctor has another law case. This time it is for libel, involving the quiet, earnest young minister and one of the village girls. The charges, brought by malicious gossips, have been proved false, and it is up to the Doctor to punish the offenders who started the scandal, but justice may be difficult since they are members of the Orangemen's Lodge, and certain other members by defending them are obstructing action. Local feeling runs high. The Orangemen's Lodge is closed on account of the controversy, and Reuben has had a justifiable fight with one of the obstructors, during which he resorted to primitive methods and shook the culprit by the ears until his head needed bandaging. It is significant that no countercharge of assault and battery has been made. Reuben evidently represented the majority. Is all this the result of the Mission, or does it indicate a need for more strenuous effort on our part? After all, the affair, in some of its aspects, has its humorous side if one can see it that way.

April 10th I feel the workshop should be established on the same basis as the loom-room, with several boys working regularly every day. They would enter as apprentices. It seems to be the only way branches of the industrial work can be developed. The Doctor often asks to have the boys take orders for this or that – metal, woodwork or pottery – but we have no definite, permanent place to work, which the Doctor acknowledges and deplores. Space will be provided as soon as possible.

Meanwhile, we are doing the best we can with full realization that slow movement is a local characteristic. There will be no more evening classes. Without regular attendance the boys would accomplish little, and since the days have lengthened the lure of football and social visiting during the early evening have minimized some of their winter enthusiasm, which is understandable. The plan for all-day work seems, to those who have discussed it, more feasible and practical.

During my spare time lately, I have visited many homes in the village where it is evident the stork has been working overtime. There is no danger of race suicide in St. Anthony. In nearly every home there is a babe. One of the women asked me to photograph her new baby. They call it "being sketched."[205] She has six other children, all crowded into a tiny house. While I was there the father came with a load of firewood on a komatik. He had drawn it a long distance, fully two miles up and down hills and across the harbor. He was ill during part of the winter, and while he was in the hospital his three dogs died of starvation "because there was no one to look after them," he said. I fancy his poor wife had no food for them. So he has no dogs. He was cheerful and bright. No one would imagine he had done anything that was difficult or unusual. I have seen him since then, struggling along with his load, dragging it by using a harness he had made for his shoulders. He now has occasional work at the Mission which enables him to earn something, but they are very, very poor. We are doing what is possible to help them.

Mrs. Stewart and I dined with Miss Storr the other day in her new living room at the orphanage. It is very nice, quite different from the early days. We often speak of the amusing experiences of that first summer on the coast when the staff numbered four, there were no conveniences and I did the cooking, but no other period since then has been happier or more harmonious. Also, we never worked harder or under greater difficulties. This raises a psychological question that persists. Do ease and self-indulgence, when emphasized as a goal, bring greater happiness than when they are balanced by self-denial? An uneasy question often asked. We wonder and doubt.

April 15th Dr. Stewart and Mr. Wight have returned safe but weary after their difficult trip. The short route across Hare Bay was impossible when they returned, and travelling conditions on the fifteen-mile trail around the head of the bay taxed the endurance of men and dogs. It was the last overland trip of the season. Since Dr. Stewart's return I am again at the Guest House and find a change in the social atmosphere. There are still household problems, but friction is minimized. There is outward peace and an evident effort on the part of everyone to practice charity and forbearance.

Almost every day the Doctor goes to the woods to cut trees and sends men and dog-teams to bring them out. The snow is so soft at mid-day that an early start is necessary. Yesterday at 5:30 a.m. I heard Mr. Wight and Mr. Evans stirring in their rooms, and it occurred to me it might be my last chance for a dog ride, so I hurried down to join them.

We found difficult going on the path – bare spots, stumps and rocks that had been level hard snow a few days ago. It was hard for the dogs. We walked up all the hills to relieve them, and after about four miles I left the men and walked back. On the way, I came upon a man with a load of wood he was hauling for himself. There were four dogs resting on a bare spot. One of them, a young pup, was almost gone with fatigue. I spoke to the man about him. He said he knew the dog was badly off, but he did not dare leave him behind for fear of other dogs attacking him, so I offered to take him to the harbor and he went along with me, staggering with exhaustion. We stopped often for rest, and he wagged his tail feebly when I patted him. When the man joined me at the harbor, he said the dog would go on without help from there, and he walked off toward home.

The condition of the dogs is one thing that makes life on the coast uncomfortable. I cannot calmly see animals maltreated and suffering. Some dogs are nearly starved, then beaten and even killed because they have not strength to draw heavy loads. One longs for a branch of the S.P.C.A. with a dragon to enforce the laws. At present, an answer given me in response to a protest indicates the general sentiment: "We knows it's hard on them, Miss, but we gets used to it."

April 24th Another week gone and with it most of the snow. It has been warm, the season far ahead of the past few years, and the clean white blanket no longer covers unsightly rubbish. On the hills that are now mostly bare, patches of snow assume the forms of animals, as if a polar bear, a dog or chicken were lying against the rocks. The harbor is full of holes and not fit to walk on. It is hard to realize that two years ago at this time, we went on our famous trip to Bartlett's Brook.

Now that expeditions to the woods are over, the Doctor is expending his energies on hauling logs to the top of the hill behind the orphanage, where he plans to build a tea-house. He selected the spot two years ago, at the time I took my first walk with him. There is plenty of snow on the path, but it is very steep. Thirteen dogs and four men were needed to haul each load of four logs. The site of the tea-house is a natural lookout, commanding a wonderful view across wild, hilly country and far out to sea. It should be a popular rendez-vous.

From my window, I can see the ocean through an opening in the hills. Just now the wind has blown the arctic ice toward land, and it is white to the horizon. It will continue to drift on and off shore with wind and tide for some time. That is why boats cannot arrive earlier. Mrs. Stewart wants me to take the trip down the Labrador coast to Nain this summer and stop with her at Indian Harbour over a boat en route. She thinks there will be good material for lectures and articles, also sketches. The mail boat makes two or three far northern trips during the season of open water. There may be free time in August. I hope so, for the lure of the real Labrador – the Moravians, the Eskimos and the wonderful scenery – has been very strong ever since I came north.

May 10th Our comfortable sense of isolation is over. An uneasy anticipation of impending change is in the air. The general upheaval will begin with the arrival of the *Prospero*, which is expected to bring a part of the summer contingent. We hear rumors of hordes: collegians, volunteers, nurses, medical students. There will also be an exodus. Mr. Evans will leave us and is already inspecting locks on his trunk and suitcase. His feeling savors of emancipation. Mrs. Grenfell plans to go home for a visit and return in late June with visitors for the summer. Miss Kennedy, after long and devoted service, is going home for a much-needed vacation. Mrs. Manager departs with her baby in June, and Mr. Manager will travel from place to place on schooners to oversee business affairs at little settlements connected with the Mission. The Doctor will stay here but plans to go north on the *Strathcona* after Dr. Little returns in June with one of his sisters.

An interesting arrival will be Mr. Jesse Halsey,[206] a Presbyterian minister who after an intensive course in plumbing and kindred practical crafts is to serve the Mission by superintending drainage installation. He will arrive with his bride in June to stay for a year, probably located at the Guest House. Also in June, Dr. Wakefield will reappear with his bride, and still another reappearance will be Dr. Grenfell's cousin, Mr. Spencer, who is returning to direct agricultural development and enlist enthusiastic co-operation from the students if anyone can. He has the personality to bring results. The semi-annual kaleidoscopic shift of personnel is beginning. It will be interesting to see how the pieces fall into place and what the ultimate picture will be.

Meanwhile, the workshop is at last receiving attention. The downstairs partition is up, and Mark has moved the metal and other shop outfit from the loom-room, where it was much in the way. A boy is coming as apprentice, and I expect two more, a beginning of development

The Rev. Jesse Halsey, clergyman and plumber, October 1910 (JLP)

in the downstairs shop on the basis of the loom-room. The pottery house is cleared, but the lining of the portable kerosene kiln – broken when building material was piled around it – must be repaired before firing can be attempted. So many projects seem to hinge on something else. It is sometimes discouraging. The brick-yard seems to have been forgotten, but interest in it may be revived.

May 21st A lovely spring day, bright and warm. The snow is practically all gone, only a little in hollows and sheltered places. The mud has settled, the paths are firm and dry, the grass rapidly turning green. In keeping with the season's rebirth there has been an upheaval of house-cleaning. We have been homeless wanderers during the day, but the result is worth it. We fairly shine with cleanliness.

There has been another accomplishment: the lining of the kerosene kiln is repaired. Mark and I tackled it this week. It was an awful job. Practically an entire new lining had to be put in. We had new flues and floor, but some of the supporting arches were broken and new ones had to be manufactured from bits of fire-brick and cement. However, it is done now and seems satisfactory. We have moved it into the kiln house, where it will be attached to the chimney and fired this week. Soon we shall be able to justify our effort and prove that pottery, though perhaps crude, can be a worthwhile local industry. Mark and the other boys have made a number of nice pieces, both on the wheel and by hand, that are waiting to be fired and glazed. We plan to send them to Mission friends at Montreal who have been interested in our venture and may advertise our wares. The time approaches to consider outside markets.

Mission boats are beginning to move. The *Strathcona*, laid up in St. Anthony harbor all winter, has just been on her first trip of the season although far from being in commission. She has not been cleaned, her parts are not in the best running order and her crew was missing except the two engineers. But that did not discourage the Doctor. He collected a scrub crew – part Mission men, part staff – and started for Forteau across the strait to bring Dr. Armstrong, who is to assume charge at St. Anthony while the Doctor is in St. John's during the *Strathcona*'s overhauling. The crew reported a fine trip but seemed non-committal regarding comfort. The Doctor, for some unknown reason, took only ten tons of coal for a three-day trip with no other ballast. On her return, she had also on deck several tons of sand for cement, a cow, and the household furniture of a family just arrived at St. Anthony for deer herding. The seas of the strait (that resemble the English Channel) had a fine time with the top-heavy boat and its passengers. One of the victims said he might, at the last minute, have fought for his life, but he could think of nothing else to stir him.

I have already mentioned my first meeting with Dr. Armstrong at Battle Harbour just two years ago. His wife, who was with him during the winter, returned to England last summer, and he transferred his immediate field of work to the tiny village of Forteau, where he lives alone very comfortably in his own little house, with a man and wife to care for him, and sends to St. John's for fresh fruit, fresh meat and vegetables. He has done excellent work in the community, travelled hundreds of miles along the coast on professional visits. On these trips, he always takes three komatiks drawn by fine dogs: one for himself and his driver, two for provisions, sleeping bags, wraps, etc., and provisions for the people en route.

He has just come in with Mr. Wight, and we eye him with satisfaction, for besides his good looks his clothes are just what they should be in every detail. He even carries a cane, which is something to stagger one in these parts. I have not seen a cane in any man's hand since leaving home, except when it was needed as a support for the sick or aged, and Dr. Armstrong is not in that category. He might have been walking down Piccadilly. He will occupy the Grenfell house during his stay, since Mrs. Grenfell left by the first mail boat for her home visit, and the Doctor will be away part of the time.

June 5th On Wednesday, we went to another local wedding, a young man in charge of the co-operative store and a fisherman's daughter aged 17. They were married in church with six bridesmaids and six "bridesmen." Her father did not "give her" (to quote the vernacular), for he did not approve of the marriage, but afterward at the house he appeared in his sweater and no collar, just as he came from his carpentry. The man who "gave her" was a relative who had also come directly from work. He was minus collar and needed a shave, but he had a white ribbon bow in his button-hole. The bride wore a white satin gown and white hat.

At the tiny house, the village guests were in the kitchen, but those from the Mission were asked into "the room" where the wedding feast was spread, and we sat there in chairs against the wall until joined by the bride and groom and their parents. The minister was not invited to the house. Someone suggested sending out for him, but the father objected, saying he had been paid four dollars, and that was enough.

The warm sunny days of the past week have transformed the country. All the snow that still lingered in spots is suddenly replaced by brilliant green grass. We have worn thin clothes, low shoes, sometimes even discarded coats. One day the temperature rose to 72° and porch windows were open all day. When I realize that two years ago today snow lay deep on the ground, drifted nearly to the top of fences, I was out on snowshoes and had a short komatik ride, it does not seem like the same country.

Dr. Grenfell left for St. John's on the *Strathcona* a week ago, expecting to return in three weeks. His start from St. Anthony was characteristically impetuous. At 6 a.m. he was heard shouting for Mr. Manager (who was going with him) to hurry up and come aboard. He even started the *Strathcona* from the wharf, to the consternation of Mr. Manager, who was in bed when called. Blessed calm follows the mad rush of "getting off," but this is temporary. Confusion will return with the arrival of willing helpers of varied experience and qualifications, most of them uncertain of what may be expected of them as temporary "missionaries." Ships are coming and going now. Sometimes, before a storm, several anchor in the harbor, and big icebergs are beginning to float down the coast. We have had our first salmon, flat-fish and periwinkles, a welcome change from winter diet of tinned, frozen and salted.

With Mrs. Manager's departure by the last boat, I find myself again housekeeper at the Guest House, this time with greater responsibilities than before and correspondingly less time for needed work at the shop. I have also just assumed a new kind of responsibility, that of godmother for the eighth child of the Hancock family. It is a boy, to be named Edwin Nicholas.[207] One of the other children has just come to tell me the christening is this afternoon at the little Church of England chapel across the harbor. The entire christening party will go over to the other side in a row-boat to meet the clergyman and Church of England schoolteacher (who is also to be sponsor). The men of the household tell me I am expected to give the family a barrel of flour and provide for the child if left stranded. Another family in the village has named their baby for me, which I appreciate very much as an expression of friendliness, with no definite expectations, but it brings a sense of growing responsibilities.

June 6th The *Prospero* is in. She brought a new nurse for the hospital, also two young women who are eager to take part in Dr. Grenfell's work. They have never met him, only read his books and what has been written about him. One of them, a nurse, hopes to assist at the hospital where her friend Miss Fowler,[208] from New York, may be useful as housekeeper, a position sure to tax her powers of adjustment since local housekeeping conditions vary considerably from New York standards of convenience, service and supply. It will also require tact, since heretofore the head nurse has controlled all hospital affairs in consultation with the doctors, and questions of authority may arise.

It was nearly full daylight [sic] when the steamer arrived, although only 2:30 a.m. She brought several patients, and Dr. Armstrong was at the hospital all night. He evidently enjoys acting as host at the Grenfells' house and has entertained us with modified formality. He often drops in at the Guest House for afternoon tea, and yesterday we invited him for lunch. It was a rainy day, and he lingered until six o'clock, a part of the afternoon being spent in the barber's chair while I gave him a much-needed haircut. We discussed such subjects as Great Britain vs. America and matrimony during the process, without violent disagreement.

June 9th The two unassigned young women are hard at work, Miss MacCallum as assistant nurse at the hospital, where Miss Fowler, with courage and vigor but little familiarity with domestic science, has taken over the housekeeping. One can imagine the trials of undergoing one's initial experience in a remote mission hospital. I hope she has a sense of humor.

June 11th There has been a real social event, Dr. Armstrong's grand dinner party. It was last night. He invited all members of the staff, and we went dressed in the best clothes we could get together, resulting in quite a presentable company. The well arranged table, glistening with Mrs. Grenfell's china and lovely silver, would have been beautiful in any surroundings, and Dr. Armstrong was a charming, gracious host, but for a man familiar with perfect service and standards of formal entertainment among the British aristocracy it must have been a trial to have mashed potato and cucumber salad passed with salmon (the fish course) and the two little turkeys brought to the table with strings still tied around them. The service was a joke, and after a general effort to ignore the unexpected, Dr. Armstrong gave it up and joined us in becoming hilarious over the situation. As a consequence, we had a grand time, and Dr. Armstrong proved himself a good sport. It was his last fling at entertaining, for Mrs. Grenfell and her guest arrive by the next boat, and he will be at the hospital until his return to Forteau.

The last boat brought Susie Denny, the weaver from Englee. Her arrival is opportune, making it possible to leave her in charge of the loom-room during my absence on a Labrador trip. It will also serve as a probation period to test her qualifications for management during the winter. If she justifies my confidence, it will be a happy solution of responsibility.

June 19th The weather has been awful. Dr. Stewart and Mr. Wight waited until yesterday before starting for Indian Harbour in the little yawl *Pomiuk* because of high seas. Mrs. Stewart goes by this *Prospero* to connect at Battle Harbour with the northern steamer.

My own plan is to follow the same route two weeks later and visit Mrs. Stewart at Indian Harbour. While there, Dr. Stewart is to take me on the *Pomiuk* to Cartwright, one of the Hudson's Bay Company's stations where the native women are expert in decorating deerskin and sealskin articles with beads and silk embroidery. I hope to stop there long enough to plan their work for the Industrial Department and also to teach some of them to spin the carded wool that Dr. Wakefield sent there last winter, thinking they might spin their own much-needed yarn, but there was no spinning wheel, and no one knew how to use one anyway. Dr. Stewart will stop for me on his return to Indian Harbour, where I will wait for the coastal steamer that goes to Nain, the oldest Moravian station on the Labrador coast, where there is an Eskimo village and the scenery en route is very beautiful. It is a lovely plan.

Later. How about "plans of mice and men?"[209] The *Prospero*'s arrival has brought a complete change. We hear she will make no more trips to Battle Harbour. This means that anyone wishing to cross the straits to Labrador must beg passage on the *Strathcona*, depend on a schooner, or take the mail boat half way back to St. John's to connect with the Labrador boat that does not stop at St. Anthony. It means that Mrs. Stewart, with her luggage on the dock ready to sail, must return to her house and wait for the *Strathcona* to take her across, which delays the opening of Indian Harbour hospital. As a matter of personal interest, the Guest House cook who expected to return from a visit at Battle Harbour cannot get here, and the only maids left are two loom-girls who assist with housework but cannot cook. I see myself preparing meals indefinitely for a family of five, and Labrador plans are in doubt. Dr. Grenfell is expected on the *Strathcona* in a few days, so Mrs. Stewart's problems may soon be solved, but mine are so mixed they are funny. At least I am trying to think so.

There is one bright spot – the pottery. At last the kiln is functioning! It fires well; all patches and repairs hold together and the glazes are successful. Some of Mark's "pots" (as he calls them) are lovely, and the crippled boy taken on as assistant is working with interest and some skill under Mark's direction. It is quite exciting to realize we can now send to Mission friends at Montreal and St. John's a finished product to justify our optimistic promises.

June 26th Another week gone. Mrs. Grenfell has arrived with her guest, also Dr. Grenfell's cousin, Mr. Spencer, who will renew his superintendence of agriculture. Dr. Grenfell, who returned from St. John's on the *Strathcona* a few days ago, at once took Mrs. Stewart and her impedimenta to Indian Harbour, so that uncertainty is past. For me, each day is a question mark, with a background of loom-room routine and preparation of three regular meals, but I still am an optimist.

July 10th Dr. Little has returned, also Dr. Grenfell from Indian Harbour. We hear rumors that the *Prospero* may resume crossing to Battle Harbour on her next trip, and meanwhile we have an utterly changed household.

On Wednesday, a strange steamer entered the harbor. We thought it might be the *Invermore*,[210] bound for Labrador, but it was the *Harlaw*[211] from Halifax with Mission supplies and four passengers, among them our friend of two years ago, Miss Mary Dwight. No angel from heaven could have been more welcome, for aside from her popularity she came as housekeeper, and I immediately transferred my responsibilities to her capable hands. Since then, events have crowded on each other. Last Monday, the little yacht *Yale*,[212] the latest addition to the Mission fleet, arrived with six men: the four students who brought her down from Boston and two others found at Battle Harbour, waiting for a chance to cross the strait.

The family now numbers twenty (three women and seventeen men – quite the reverse of New England ratios). The number was increased yesterday when the Grenfells' stove gave out and all the family came to dinner. Miss Dwight's responsibilities are heavy, but she rises to meet them, always calm and unruffled. Another on the *Harlaw* was Miss Olive Lesley, a kindergarten teacher, attractive, bright, and much interested in her work with the children. It is a lively household and great fun. The boys are jolly and courteous. Many of them have good voices, and the little living-room bristles with musical instruments – a mandolin, a violin, two banjos and a piano – much out of tune. The boys are singing most of the time, in and out of the house.

The *Yale* is an attractive yawl-rigged [sic] yacht, 45 feet long with an auxiliary engine. The men who brought her intend to take her down the coast for Mission use. They were to start two days ago and before leaving planned to take a party for a morning sail, returning in time for dinner. The party – Dr. Grenfell, Dr. McCabe, Miss Fowler and I – were ready to start, but at the last moment Dr. Grenfell said we would not be back until 6:30, for he wanted to go to Hare Bay to look up some sheep. It was four o'clock before we reached Lock's Cove, and then the Doctor was having such a good time he did not want to leave. We wandered about, took photographs, talked with the village people, and when we finally started off the Doctor decided it was too late to reach St. Anthony, and we would go to George Reid's house at Ireland Bight for the night. At 9:30 we approached the little settlement of fifteen houses: a lovely moonlight evening with so little breeze the engine was started. It ran for ten minutes, then became so hot we waited an hour for it to cool.

George Reid is one of the men who rescued Dr. Grenfell from the ice. In spite of the late hour, we found him and his family still up and warmly hospitable, eager to do everything for our comfort. We sat with them after drinking tea, while the Doctor and Mr. Reid compared notes about the rescue – very thrilling for the rest of the party. The story was made more realistic by going outside in the moonlight to look at the shore where the Doctor floated off on the ice-pan and the men launched their boat to go out for him.

The kind women led us to a room with two beds, and the daughter, who went in with us, stopped to talk. She showed us her knitting and crochet and still lingered, calling attention to some dolls hanging on the walls that Dr. Grenfell had given the children at a Christmas tree party. They had treasured them as something to look at, not play with. Twice she apologized for keeping us up but still stayed. Finally, I thanked her very much for all her kindness and said I hoped we had not given her too much trouble. Her reply came from her heart: she would be glad to take any amount of trouble, for we were the first people to visit the harbor in over a year. I suppose it was a red letter day.

Dr. Grenfell came ashore for us at four a.m. and we had bread and tea before starting. He told us the *Yale* had dragged anchor during the night when the wind rose and was floating out to sea with all on board asleep when discovered by a fisherman who rowed out to waken them. Something usually happens on the Doctor's parties. It adds to their interest.

St. Anthony fairly swarms with vigorous young manhood, some of them professionals or with special training, some merely filling chinks. Outward appearance does not always indicate which is which. Miss Storr has just experienced this uncertainty. There are several large stones in front of the orphanage where she hopes to cultivate grass. After waiting vainly for local men to remove them, she decided to attempt the job herself and was laboriously plying a crowbar when a fair-haired youth walked into the yard. The sight of husky, unemployed youth gave her an idea. "Here, young fellow," said she. "Will you give me a hand with this?"

The young fellow complied, and Miss Storr, noting that he was working effectively, left him to finish and went into the house. After a while, the young man appeared at the door. Miss Storr met him and asked if he had finished.

"Yes," said he. "And now can I see the children?"

"And why do you want to see the children?" Miss Storr asked. She was unprepared for the answer: "I am Dr. Fowler[213] from New York and came to look at the children's adenoids."

Poor Miss Storr! Appearances are sometimes misleading.

July 17th The *Prospero* is really going to Battle Harbour on her next trip, and at last I shall see the Labrador. She is due at any time. My bags are ready, details of work arranged at the loom-room for the weeks I shall be away.

Susie Denny, who will be left in charge, seems to qualify, and Mark will direct the workshop. In addition to pottery, he has made articles of copper – boxes, trays and bowls, also jewelry, including brooches, buttons and a necklace of polished Labradorite set in copper. His latest experiment is articles made of whalebone to demonstrate the use of another native material. There is now a wide range, from peat that can be used as fuel to articles made from mineral, as well as vegetable and animal material. We have just finished packing a box of miscellaneous products destined for exhibition and sale at Montreal and hope it is the first of many.

Dr. Little says I must not miss this northern trip, for the steamer will go beyond [Cape] Mugford,[214] the most beautiful part of the coast. It is a small steamer, taking few passengers, and its function is to take mail to fishermen on their schooners as far north as they go. I don't know where I am going or when or what connections can be made. Apparently, there are no "connections." One just goes to a place and waits for the next move. As Dr. Little says, this is really "going out into the wilds," and almost anything interesting can happen.

4

Farther North

18 July – 15 August 1910

On board S.S. Prospero, en route to Battle Harbour. July 18th, 1910 The steamer came at 3:00 p.m. and I went at once to the dock, for she sometimes waits only fifteen minutes. A retinue of collegians followed with my luggage. One trunk, one canvas hold-all, one handbag, one package of deerskins in a shawl strap, two large packages of wool for distribution among the people, a fur-lined coat, oilskins and sou'wester, an umbrella and a spinning-wheel. Mr. Spencer and Mr. Janney[215] are on board as far as Griquet with one of the Lapland deer herders and his little dog "Run." Some of the deer have strayed in the direction of Cape Norman, and the men are going across country to look for them. The interior will swarm with black flies and mosquitoes, and they will be unhappy.

Battle Harbour, July 19th We arrived at 7:30 this morning and came ashore in a little boat through the tickle where we were nearly swamped two years ago. Dr. Butler,[216] who is in charge at the hospital, was at the wharf and I asked him if he could take in a wayfarer. "Yes," said he, "if you don't mind doubling up." So Mrs. Philmore, the nurse, has installed a cot for me in her room. She and Miss Whitten,[217] the housekeeper, are warmly hospitable. The other members of the staff appeared at breakfast: Mr. Van Gorder[218] and Mr. Fallon[219] – young medical students – and to my surprise our St. Anthony friend Mr. Evans, who found Philadelphia too hot after going home and returned here to cool off. He is now waiting for something to turn up and may go to Indian Harbour.

The northern mail boat *Invermore* is on her way to St. John's, which means delay, but someone told me Mr. [John T.] Croucher, a trader for Baine, Johnston and Company,[220] had news of a little steamer that is to connect with the *Invermore* at Indian Harbour and go north to the most beautiful part of the coast. He tells me she is the *Susu,*[221] a fair-sized boat, 145 feet long with comfortable accommodations and is commanded by Captain Winsor,[222] a friend of Dr. Grenfell. I shall probably be the only passenger, the trip being purely on government business to carry mail to fishermen in the far north. Mr. Croucher says the boat will surely stop.

Battle Harbour is a rocky island, one of many, and so small one could walk the length and breadth of it in an hour. All this morning, Miss Whitten and I have explored, clambered over rocks and visited the little cemetery where Prince Pomiuk was buried. The cemetery is in a hollow between high rocks, a lovely little nook open to the sea. We wandered on beyond it, over the rocks to a spot where waves wash in at high tide to leave a deep pool of clear green water popular for bathing, and near it is a rocky bath-tub filled with rain water where one can sit on convenient projections just under the surface and really scrub. Soap is kept nearby in a special crevice known only to the elect. There is no excuse for uncleanliness in Battle Harbour in spite of the bathless state of the hospital.

Mrs. Philmore and I went across the tickle this afternoon in a trap-boat, and we climbed the high hill on Caribou Island to visit the old Scotchman's shop. It is so tiny and so low that his head nearly touches the ceiling, and from the center of the room he can touch almost everything in it. We afterward sailed in the trap-boat among the islands and passed one where many of the dogs are kept during the summer. The boatman told us they are fed with fish every few days. Apparently, that is all they think a dog needs.

We continued our walk along the shore that at the north is free of vegetation. There are high rocks cleft by deep chasms, where waves rush in and foam dashes high in the air. Several large icebergs gleamed on the blue sea dotted with small fishing boats. Occasionally, one of the boats came in, its little brown sails furled, and the fisherman sculling with a long oar. On larger boats anchored in the tickle, men were cleaning their fish. The life here is different from St. Anthony, in some ways more interesting and also more primitive. The little steamer *Susu* may come in the night. It will be a relief to be settled somewhere. This waiting for something that may happen any minute is upsetting.

On board S.S. Susu, Labrador coast. July 20th The *Susu* arrived just after breakfast. She came in by way of the strait and stopped outside the tickle. From the porch we watched the tender put off, and I prepared to depart. Everyone escorted me to the wharf, carrying my belongings. In the tender was an official with the mail-bag, whom we found to be Skipper [Moses] Bartlett,[223] one of Dr. Cook's expedition to the North

Pole, a cousin of Captain Bob Bartlett,[224] who went with Peary. A man with a jolly, round, tanned face, wearing a real straw hat and civilized clothes proved to be Captain Winsor. He said he would take me to Indian Harbour with pleasure, so I stepped into the tender, the Captain supporting my spinning wheel with one hand while he steered with the other.

There was only a rope ladder over the side, and he insisted on getting out the accommodation ladder for my benefit. I told him I was a good climber, but he said he would not like me to climb up that way, so I sat in the tender and conversed with the coxswain while seven or eight men, the Captain in his straw hat included, worked above me trying to get the platform and steps in place. It had not been used for some time and was rusty. Everyone not actively engaged leaned on the rail and offered suggestions. Meanwhile, the loading of freight for Indian Harbour was delayed, but at last I walked proudly up the steps like a personage!

The *Susu* is small, only 150 tons [sic]. She is long and narrow, "very steady," the ship's doctor says. I have a nice stateroom off the saloon and am the only female aboard. There are two other passengers, young men taking the round trip. It is a perfect day, just the sort to start on a trip like this. I have a deck chair on the sunny side, and it is so warm a sweater is my only wrap. The saloon is very nice and the food plain but good. There are flowers on the table and even oranges and nuts for dessert. My seat is at the Captain's right, and everyone is courteous and friendly. The ship's doctor,[225] who is an admirer of Mark Twain[226] and Jerome K. Jerome,[227] sits opposite me, and the constable, Skipper Bartlett and two passengers make up the table. The doctor, the Captain and I do most of the talking.

We are running quite near the coast, which looks much like northern Newfoundland – small mountains and cliffs all the way and a very irregular skyline. We have seen several small bergs and growlers but nothing remarkable. The Captain says we shall see plenty of ice farther down. We have passed a number of fishing schooners going north. Captain Winsor runs near them and shouts from the bridge that there is plenty of fish at Cape Harrison and Turnavik. The men must be glad to know this, for there is none farther south.

The sea is a brilliant blue, and on the horizon a wonderful mirage (called a *loom* down here)[228] seemingly sends ships, islands and icebergs into the sky and reflects them in the sea below. Icebergs, apparently enormous at a distance, dwindle to insignificance as we near them, and ships seem to be sailing in the air. On some of the mountainsides are great patches of snow. There are strange clouds on the horizon. I can't tell whether they are clouds or mountains – perhaps a little of both with mirage thrown in. I am deliciously lazy and have left all cares behind. The swaying of the steamer, warm sun and soft hiss of parted waves make one drowsy. If some creatures I know of with jaded nerves [at the Butler Hospital] could only take this trip!

At Batteau, 10 p.m. We arrived here for the night at quarter to nine, just as the sunset light cast a golden glory over the land and the full moon rose from the sea. This is a little village of scattered houses around a harbor formed by several islands. All the coast seems to be islands, at a distance quite indistinguishable from the mainland.

At the entrance to the harbor is an "American man,"[229] so called, silhouetted against the sky. It is a pile of stones like a cairn on top of the highest hill and is supposed to have been built by early American visitors to the coast [sic]. They are so named by the natives and serve as landmarks for sailors on the uncharted coast. It is rather comforting to think of American men serving as guides to smooth waters and a safe haven. It should be symbolic. The Captain says we will reach Indian Harbour tomorrow.

July 21st, 1:00 p.m. Just finished dinner: pea soup, roast beef, boiled potatoes, boiled cabbage, coconut pudding, oranges, nuts and tea. We are creeping along the coast at seven knots and have just passed Hamilton Inlet, forty miles wide, its mouth filled with islands. There is a following wind, and our smoke is blowing straight ahead. A little snow is on mountain tops in the direction of Cartwright, but there are no icebergs and lately no schooners except far out at sea. We seem very small and insignificant out here all alone.

The ship's doctor, who proves to be a son of Captain Kean of the *Prospero,* has been very attentive. After a long talk with him this morning, I took up my writing and he asked if I wanted to continue my work. I may have looked surprised, for he said, apologetically, "You know, we don't have lady passengers often, and when we do ... !" with a wag of the head that expressed volumes. We are nearing Indian Harbour.

Indian Harbour, night. We landed on the rocky shore at four o'clock. Everything is rocks, not a tree or shrub in sight, but the place is so beautiful one does not seem to miss them. They would almost seem out of place. At the hospital, a short distance from the shore, Dr. and Mrs. Stewart; Mr. Wight, who went north with them; Mr. Fogg, an assistant; and Miss Child,[230] a young nurse, waited to greet me. I have disposed of my luggage and told all the St. Anthony news. As at Battle Harbour, accommodations for non-patients are limited, and I am located in the women's ward that happens to be vacant. Any bed, however, would be welcome if stable. I am glad to be here.

Dr Norman Stewart, 1910 (GHS)

July 22nd A beautiful day, clear and bright. Behind the hospital, there is a high rock known as Maria's Lookout because of a former maid who used to frequent the spot. This morning, Dr. Stewart and I climbed to the top to see the view: the wild inland country, the coastline with its many islands and the vast expanse of ocean stretching to the horizon, a view typical of the region.

Dr. Stewart went up the inlet in the *Pomiuk* to see patients and took some of us along, including a young Church of England clergyman [Mr Corby] who arrived during the morning and told us he was on his way down the coast, travelling as the missionaries do in any kind of craft. He was landed here by trap-boat from Smoky, and Dr. Stewart offered to take him aboard the launch. The doctor went ashore to visit John Horan's mother. They live in a tiny isolated house and depend entirely on the fish they catch and the winter trapping. The brother, who came out for the doctor, looked half starved.

We went ashore with the parson to visit the Fillmores, one of two families who live in what we would call huts or hovels. To my great surprise I found Mrs. Fillmore[231] is a Cape Cod woman. How she ever wandered down here I do not know. Her house was scrupulously clean, and so was she. Dr. Stewart said she did good skin work and I asked to see it. She brought out a cap of fox paws. It is very pretty. Her price for it was forty cents. In the other hut, I found the girl Kirkina, whose father had to cut off her frozen feet with an axe to save her from gangrene. It is one of Dr. Grenfell's stories and always makes his audience shudder.[232] She is a well grown girl now. These people live at Indian Harbour in the winter, the only family in the place, and come to this hut for the summer. Many people go up the bays in winter to trap and to the shore in summer to fish, and I don't see why this family reverses the order.

We left Mr. Corby to be taken on by trap-boat.

Sunday, July 24th Very warm and bright, with millions of mosquitoes and black flies. Dr. Stewart held service in the chapel this evening at 6:30 with a good congregation. It is a wonder where all the people spring from, for the population of Indian Harbour is very small, but there are several schooners in the harbor and the men like to come ashore. There are also houses here and there along the coast within walking or trap-boat distance.

July 25th Dr. Stewart said today he hoped to take me to Rigolet in the launch. The trip will take two days. There is skin work to be planned here, and I am eager to start it. The Fillmore family came today, and I gave them deerskin and silks for embroidery. Mrs. Stewart will bring the finished work to St. Anthony in September.

July 26th I have helped Mrs. Stewart in the clothing store today. The women come in with skin things for sale and are paid in clothing. They are in great need of it and have little use for cash. Fortunately, a supply came on the *Susu* with us from Battle Harbour, but there are very few men's things. The *Stella Maris*[233] is supposed to leave St. John's August 1st. She is to take the place of the *Susu* for the northern trip. They say she is a larger boat, better able to take passengers, and we expect her in about a week.

Rigolet, July 28th We are here, Dr. Stewart, Mr. Wight and I. We started at about 10 with fine weather until well into the inlet. Then the fog shut down like a blank grey wall. We might have been out at sea. We passed near Black Island, and Dr. Stewart laid the course to the point of another island not far from Rigolet, allowing for a slight variation of compass [from magnetic to true north].

We sailed a long way without anything being visible. Suddenly, Mr. Wight, who was at the mast on lookout, turned and said quietly, "There's the *Pelican*!"[234] And just before us, looking up out of the fog, was the steamer at anchor near an island several miles from where we thought we were. If we had laid the course without variation, we should have missed her altogether. As it was, we were steering as directly for her as if she had been our objective.

We tied up alongside and all went on board. The *Pelican*, a Hudson's Bay Company steamer, is an old gunboat and very heavy. The barrel for lookout is still on one of the masts, and she has a high poop [stern] with the wheel astern. She was a queer object with a heterogeneous cargo stacked on deck so high one could walk over the rail. Forward, the deck was paved with coal bricks several deep, and there was an assortment of boats of all kinds, from bully boats[235] to canoes. More boats were on the bridge back of the wheelhouse, and at the foot of the steps leading to it was a large coop of poultry. We found there were also sheep.

Mrs. Swaffield[236] has told me about the people, who are in sore need. They depend upon the salmon fishery, and this year it is a failure, some families having no catch at all. They are at the starvation point anyway, and what they are to do no one knows. They say it is a case for Dr. Grenfell and for government relief. All along the inlet one sees solitary houses, scarcely more than huts. The people trap a little in winter, but unless one is fortunate enough to get a silver fox there is little return. Silver foxes bring $150 to $250, sometimes even more – an amount that will easily keep a family during the winter – but not many are caught. There is one family in particular that interests Mrs. Swaffield. The father is dead, and the eldest son, the mainstay of the family, is ill with tuberculosis. He is too weak to row a boat, probably starved as well. Mrs. Swaffield wants Dr. Stewart to take him with us to Indian Harbour. There are pitiful tales of privation on every hand. The need of clothing is very great and none has been left here, so I promised to tell Mrs. Stewart, who will send a supply from Indian Harbour.

This afternoon Mr. Corby, the young clergyman, appeared again, just arrived in a trap-boat on his way south. He is to be a passenger on our launch in the morning, so he hurried around at once to make visits and arrange for a baptism at Mrs. Swaffield's house for her youngest and some others. He will have a busy afternoon.

Later. The baptizing took place after supper in the dining room. The dining table was an altar and a china cuspidor the font. The parlor organ was brought in, and I was invited to play an unfamiliar hymn, a performance that must have convinced all concerned of the risks in being accommodating. A child of a native woman was also baptized and received into the church, a young girl who acted as sponsor stumbling through the responses. Everyone went in for the service and sat in a row along the dining room wall. It was an interesting experience and an example of a travelling clergyman's work. Afterward, Mrs. Swaffield conferred with me about the skin industry in Rigolet. She will act as local agent and collect the finished work.

July 30th A perfect summer day! Early this morning, Mr. Corby offered the last service of his visit, a communion service, much appreciated in this remote district where there is no church. We were ready to start for Indian Harbour directly afterward, but Dr. Stewart seemed in no hurry and suggested visiting an Indian cemetery nearby, which when we reached it seemed to be merely a pile of stones. Some of the graves had been excavated, the stones piled in heaps, but farther on others were undisturbed. We found bones near one of them. Stone implements and utensils had been found in those that were opened, but of course taken away. The graves are hollows filled with rough stones, level with the ground. Moss and grass have grown among them, sometimes even shrubs or small trees that provide shade where the hillside slopes to the water.

When we finally started there were passengers on board: the sick boy, Reuben Otis, who was placed comfortably in the small boat on deck; Mr. Corby, again on his unscheduled way; and Dan [Groves], a local boy who has been a trapper since his tender years and covers the region between Rigolet and Grand Falls, sometimes travelling 400 miles in a winter. He knows only a life in the wilderness but now, anxious to "see the world," is to join his brother at Montreal. He has made quite a sum from trapping and seems eager to spend it. He had a chance to go by schooner but preferred to connect with the mail boat and go first class, "like other folks." He also wants to see St. John's, not realizing that he will have less money than he thinks by the time he reaches Montreal. It was interesting to hear his tales of the woods. Each trapper has a certain district which he does not own but has the privilege of using for trapping, and no one is supposed to set traps in another's district. It is the unwritten law of the wilderness and adhered to by common consent.

The parson wanted to be put off on the south side of the inlet, at a place unfamiliar to Dr. Stewart and for which the chart on board showed no soundings, so it occurred to someone to drop him into one of the trap skiffs, where men were fishing some distance from shore. It did seem funny and rather heartless to dispose of him so summarily, but I think he enjoyed it. He must find it interesting, if precarious, to be handed on along such a coast as this.

Returned to Indian Harbour, July 31st The *Yale* arrived this morning from St. Anthony with students aboard, and the dining room had to be stretched to its limit to accommodate twelve people around the table. The men brought Mr. Fallon with them as cook. It was the only capacity in which he could be shipped, and the crew say he has won laurels. He is to be left here, he hopes. Their plans are indefinite: everything is waiting for Dr. Grenfell's decision. They are taking the *Yale* up the inlet tomorrow with Dr. Stewart on board for professional visits. It is all new country for the young men, who are alert for unusual experiences. One of them has heard of a black bear to be had somewhere along the way and expects to bring it back with him. He may, but "I don't think," as the natives say.

August 1st The *Yale* left this morning and the family is reduced to five, two of the men having stayed behind. The house seems very peaceful, and Mrs. Stewart will have a much needed rest after feeding and planning for so many. I have helped at the hospital when needed and tried to sketch the interesting country, but it is difficult to settle at anything. This afternoon, we rowed across a little harbor to the Marconi[237] station to send messages. The land is the same as at Indian [Harbour]: bare rocks in terraces, with moss, a little grass and occasional small pools. The Marconi operator lives in his little house alone and looks as if he never had a square meal. He must be terribly lonely.

We saw native trap-boats coming in before a good breeze, but bringing in little fish. Beyond the point, a fair-sized iceberg made its way south. Along the shore were several fishermen's huts made of overturned boats with a door cut in the end and turf piled on top. Many of the huts have turf roofs, and turf is banked around the sides for warmth. At a distance one can scarcely distinguish them from the land.

August 3rd Alternate fog and sunshine again. My packing is finished and I am ready to start for Hopedale, but there is no sign of a boat to take me there.

August 5th Still foggy. Still no boats. It will soon be time for the *Yale* to return from the inlet. I tried to sketch the misty hill and failed, then took a long walk over the rocks to a high hill, also rocky. Even a stroll in this region is excellent training in agility and balance. I shall soon acquire the technique of a goat! Yet this wild, barren shore, its loneliness, far distances and untouched nature, has a strong fascination for those who can feel it.

August 6th Still no boats. We are all becoming veritable "Sister Annes."[238] But there has been one arrival: a sea serpent was seen yesterday afternoon according to one of the men fishing near the coast. He told a great tale. Apparently, the creature rose from the water like a great hoop, and the men were terrified. "It was its first visit to the coast," he added, and evidently believed every word he said.

August 7th, Sunday Nothing yet and still fog, but something unusual has happened: I conducted the church service this evening. This service means much to the fishermen who anchor in the harbor over Sunday and to the people of the nearby coast, and there was discussion at breakfast as to who should conduct it since Dr. Stewart was absent and Mrs. Stewart not feeling well. Everyone agreed someone should act, but the others declined, so it seemed to be up to me. I selected passages to read from the Bible, someone found a book of prayers compiled from a ritual, and I tried to find hymns that were familiar to everyone.

The chapel was full. It was a new experience and I was nervous, especially during the extemporaneous prayer that was necessary when, because of dim light, it was impossible to read the fine print of the prayer book, also when starting the hymns, which might be pitched too high or too low. For the sermon, I read the Sermon on the Mount,[239] stopping occasionally to comment on certain passages. Aside from the agonized moments when I tried to read the fine print and one hymn which was nearly a solo performance, it went fairly well, and there will be no jitters on future occasions. Mrs. Stewart says I have now experienced all the activities of the coast, but she is wrong. I have not yet conducted a prayer meeting.

August 8th Still no boats. We took a walk this afternoon to the Roberts' turf-covered hut, where I took a photograph. Mrs. Roberts said she enjoyed the service Sunday, but when I mentioned the remark at table with an eye to my own aggrandizement someone commented that "Mrs. Roberts always talks like that" and I subsided. There is nothing like communal living to quench self-esteem.

August 9th Near a fish stage where we left our boat today, we found a tiny garden in an enclosure. There were cabbage and turnip, which are used as "greens" without maturing. In fact, *greens* always mean cabbage tops.[240] The men said they were planted in late June. One sees few gardens: the soil is mossy and the rock foundation permits little depth. Nevertheless, Mrs. Roberts had a nice little garden in a hollow between two rocks near her turf hut. Perhaps she had discovered that gardens, like other things, require care and attention for success.

August 10th The monotony is broken. This has been a day of excitement. I woke early to hear the *Yale*'s throbbing engine. Presently, Dr. Stewart came ashore with two men, bringing three patients, and immediately everyone was busy, but my assistance stopped abruptly when someone called that the *Stella Maris* was in the harbor. She did not seem real after so much waiting.

I flung my few things together, put my travelling skirt and jacket over my canvas gown and started off. The *Stella Maris* tender was in when I reached the rocks, and the men seemed surprised that I was going. I began to feel queer and asked if there were any other passengers. No. Any other woman on board? No. Mr. Wight and Mr. Fogg had rowed out to see what kind of boat it was and reported there was a fair-sized cabin. They said nothing more. I got into the boat and suddenly realized I did not want to go. The mate, who had been imbibing, sat beside me, and the longer we waited the less I wanted to go, but my pride still held me to what had been planned.

We started off, but with every stroke of the oar my heart sank deeper, and when we reached the little tub of a boat, dirty and rough, my decision was made. Mr. Jarret, one of the men on the tender, was returning to shore and I told him I was going back with him. Some men leaning on the rail as we reached the steamer's side heard me announce a change of plans, and during my transfer to another boat the steward remarked "The lady seems faint-hearted," which was quite true. Even the coxswain smiled slowly and acknowledged that he did not think I would be comfortable. Then, crestfallen and discouraged, I returned to those who had just bidden me good-by, a miserable anti-climax.

Boats usually come in pairs, so it was no surprise when the *Strathcona* appeared this afternoon. We stood on the rocks to watch her steam in, and the men went out in a body to call on the Doctor. We expected them back for tea, but it was late before they all came ashore for supper at the hospital and crowded, sixteen strong, into the little dining room. As usual, when Dr. Grenfell with his *Strathcona* group meets another group, a kaleidoscopic change takes place. The men are starting off in different combinations and directions, some on the *Pomiuk*, some on the *Strathcona*. Mr. Palmer will take the *Invermore* north on Mission affairs; only Mr. Fallon and Mr. Rowland[241] are to stay here. The Doctor will go on the *Strathcona* up the inlet for wood and a stop at Rigolet to transact business with Mr. Swaffield.[242] As the *Invermore* is due on her way north, I shall go on her to Hopedale and after a two-week stop with the Moravian Mission will return on her next trip.

Leave takings have been general, and those who are not going, as well as those who are, have gone out to the *Strathcona* to make a night of it. There will be little sleep for anyone.

August 11th This has been a strenuous day. The *Strathcona* went out at 2 a.m. I heard her start, and later the *Pomiuk* was also on her way. It was a busy day at the hospital, for the Doctor brought two new patients.

There are now seven in the wards. Some of them are operative cases, and between odd jobs I have been relieving the nurse by tending them as they emerged from anesthesia.

On board S.S. Invermore, August 15th Off at last! The steamer came before breakfast and anchored outside the harbor. We were ready, but no one heard the whistle and the jolly-boat was at the rocks and the mail on board before anyone knew of her arrival. Two men from the boat took my trunk, and Mr. Palmer and I followed with hand luggage, Dr. Stewart rowing out with sacks of clothing for distribution along the coast.

The steamer is a good-sized craft, but there are so many passengers the only accommodation is in the ladies' cabin, which I share with a rather flighty stewardess and a half-breed Eskimo girl on her way to Nova Scotia as house servant for a man connected with some lumber company[243] – with, I fancy, pulp mills in view, a sad enterprise for Labrador and Newfoundland, where the people depend upon the woods not only for fuel and shelter but as a cover for animals they trap in winter for food and fur, which is their only means of subsistence. The little half-breed girl comes from Rigolet and considers me a friend and haven of refuge because I know Mrs. Swaffield. Poor little thing! Going out among strangers! I don't wonder she feels lonely and shy. Mr. Grant, the man to whose house she is going, has had a model komatik given him with three stuffed puppies for a team. The puppies are badly done, and one of the passengers who saw them remarked on the queer shape of Eskimo dogs. They were "so like dachschunds" and they looked so gentle. Mr. Grant explained that they were not the real wolf dog always used for driving but the domestic or hunting dog. Someone asked what they hunted, bears? He couldn't say but thought it was bears – and – anything. He also told us komatik dogs could run 125 miles in a day and were not fed during the journey. Mr. Palmer and I were interested but skeptical.

5

Moravian Missions and Eskimos

16–31 August 1910

Hopedale, August 16th The weather cleared this morning and we could see the coast, still rocky and barren. At Turnavik, several of us went ashore in the jolly-boat. Captain Bob Bartlett, a member of Peary's North Pole expedition, came from there, and we saw his father's house,[244] which is also the post office. There is a large fish stage, and we found the season's catch is 200 quintals, larger than anywhere else on the coast. Innumerable icebergs gleamed in the distance, and several "growlers" (small bergs) were stranded near the shore where snow still lingered in crevices on the hills.

Everyone went ashore when we reached Hopedale. The Moravian Mission looked really imposing from the water, with its large central building; the church with its belfry, and several stores and outbuildings, all painted white with red roofs. Near it is the village of Eskimo huts. We landed by a ladder on the rocks, and two women came down the path to meet us. I asked one of them if she could direct me to Mr. Lenz[245] or Mr. Payne,[246] as I had letters of introduction to them from Dr. Grenfell and hoped they might be able to take me to board for two weeks. The woman was Mrs. Lenz herself, who introduced me to her husband, and I felt I was being weighed in the balance until he said they could manage it, although they were in process of changing rooms. I found later this was a rearrangement of housekeeping to give a newly arrived couple a *ménage* of their own in another part of the building.

Mrs. Lenz took me and other passengers around the premises. The Mission house has a fence with a red gate, and on opening it we found a dear little formal flower garden with three flower beds. In one, our common field daisy revealed what a thing of beauty it really is, and another – divided into four sections – held pansies, poppies and pink English daisies. The paths were of gravel, with pointed shingles bordering the small flower pots. There was also a greenhouse sheltering flowers and vegetables.

We entered the study, a large room comfortably furnished, with windows towards the harbor. The hall is plain with scrubbed floor, and the board walls are painted white. We went upstairs and down a long corridor lined with doors. It was like a story-book – those doors may lead anywhere! Mrs. Lenz took us to her living room. It is very German, with cross-stitch embroidery on the table covers and a pot of red geraniums on the wide window sill. There is a great porcelain stove in a corner. The windows open inward, and there are white hanging curtains and German texts on the walls. From the window, one looks out on the lovely harbor, sea and hills.

Mrs. Lenz brought out some Eskimo motto cards, postcards and a few deerskin articles which the passengers snapped up at once. She then led us along a walk bordered by trees, through a kitchen garden where lettuce, potatoes, rhubarb, turnips and even cauliflower were flourishing, then through a gate in the picket fence to another lovely spot with trees, shrubs and walks leading to seats beside them. There was even a little tea-house, and in one place, a seat and table were raised by rock and sand to a height overlooking the sea and hills. Mrs. Lenz said she and Mr. Lenz had made this themselves, and it was their favorite retreat.

Attached to the main building is the church. It is quite large, with painted walls, white window curtains, plain wooden seats and a sanded floor. At the back is an old pipe organ which one of the passengers tried by request. The notes were shrill, and one was conscious of the mechanism. It stands on a platform where the choir sits – when there is one. At present, I believe, there is none. In a corner stands an ordinary parlor organ which is used for the services. A raised desk for the preacher is at the other end of the room. It was all very interesting to the passengers. We went with them to the shore for farewells, and from the jolly-boat they looked back wistfully to where I stood with a background of the Eskimo village and high hills behind it, which they had not had time to explore. Some of them, I think, were envious of my opportunity, which seems to me understandable.

Supper, my first meal at the Mission, was in the living room on a table placed in front of the sofa, with smoked trout, German sausage, bread and butter, tea and marmalade, all placed on the table at once. Grace before and after the meal was said in English in deference to me. The Mission household consists of Mr. and Mrs. Lenz, their three children and a young German, Mr. Merklein,[247] who has just arrived with his bride to fill the position of storekeeper and businessman, for the Moravians are traders as well as missionaries. There is also

Mr. Payne, a young Englishman connected with the Mission, just now away on the launch, who is leaving next week for Nain, a post farther north. Frau Merklein sat next me at supper. Years have passed since I used the little German I knew, and English was almost an unknown tongue for her as for her husband. The general conversation was in German, my contribution to it being a slight response when recognizing a familiar word, combined with sign language and kindly interpretations by Mr. and Mrs. Lenz, who speak English very well.

After supper, Mrs. Lenz took me to my immaculate room. The window has white curtains. The two narrow white beds against the wall have no springs. There is a hard mattress with a feather bed over it and another feather bed as a covering. There are also blankets, in consideration for my American taste and habit. The furniture is mostly homemade. On the table is a cover of German linen. There is a wash stand and in a corner a barrel chair, while another corner is filled by a large stove with plenty of wood and kindling. The board walls are painted light grey-blue, and on the floor is a carpet of many pieces patched together. The general effect is very attractive and harmonious. Every provision was made for my comfort: filtered spring water in a bottle, good towels, and my bed was turned down.

Mrs. Lenz asked me to join her and her husband for cocoa at eight o'clock, and while it was being made I watched Mr. Lenz make bread in an American mixer. He makes all the bread for the household and it is delicious. At nine o'clock, I left them and returned to my peaceful room. It was very still, the silence broken only by occasional howling of the dogs that prowl over the place and faint sounds from the Eskimo village that I can see from my window and hope to explore in the morning.

August 17th A beautiful day spent in exploring and painting. Contrary to misgivings, I slept quite well in my little bed. I woke at 6:30 when the Mission bell rang to summon the men for their day's work. As if it were a signal, every dog in the village set up an unearthly howl. Our breakfast at 7:30 was whole wheat bread, oatmeal porridge, coffee and marmalade. There is only condensed milk, and none is used on the oatmeal, only demerara sugar. After breakfast, the two maids came in as usual for prayers. One of them, pure Eskimo, is short, swarthy, and wears her coarse black hair braided at each side of her forehead and brought under her ears to join the little bob at the back – a quaint and most unusual coiffure. The other is a half-breed.

The wild beauty of this place and its interesting people stimulated my long repressed urge to paint. This was an opportunity not to be lost, so after breakfast I started with my pastels for the garden, where an old Eskimo woman with a red kerchief on her head was working, and made a quick sketch. It had to be quick, for mosquitoes attacked in swarms, and the effort to paint with one hand and slap with the other soon grew wearisome. However, I tried again from high rocks behind the Mission when, just after sunset, a luminous violet light crept over the hills, and the moon cast long reflections on the still water. But the mosquitoes were still there. Frau Merklein appeared full of interest in picture painting, of which she had no knowledge. She was distressed to find me struggling with the pests that had found me out and, after trying to tell me something, dashed off, returning presently with a thick black veil for protection. She was so kind I hesitated to suggest that the landscape might look different when seen through it. It was growing too late to paint anyway.

The *Stella Maris* is in the harbor, coaling for her northern trip, and the constable and steward came ashore this evening. Even they say the boat is not fit for ladies to travel on.

August 18th The black flies and mosquitoes are worse than ever today. The only way to escape them is to stay in the house, which is screened, or sit on the hilltop, where the wind nearly blows one off the rocks.

I started out after breakfast and, finding some mushrooms, went on looking for more, mounting higher and higher to the top of a hill I had not climbed before. From it, I saw the little Mission launch round the point and when I returned found Mr. Payne, two half-breed Eskimos and a settler already seated at the table. I was attracted at once to the young Englishman, with his pleasant smile and clean, bright blue eyes. It was his boat that ran near the *Invermore* on the way down and excited the passengers when they saw the letters S.F.G. on the boat's flag and no one knew what they meant. I find they stand for Society for Forwarding the Gospel.[248] With his two men, he has just been to Makkovik, the southernmost post of the Moravian Mission. There are four more posts north of Hopedale: Nain, Okak, Hebron and Nachvak, the latter near Cape Chidley.

Mrs. Lenz thinks they may be transferred to Nachvak next year and does not enjoy the prospect, as the steamer *Harmony*'s[249] three calls during the summer will be their only contact with civilization. They have now been at Hopedale ten years without a furlough, which is due next year. The Moravian Mission requires that all children shall be sent to Germany for education when they are eight years old – a difficult ordeal to face, for it always means years of separation, sometimes even a final parting. But as Mr. Lenz says, they know this in the beginning and are prepared for it – a philosophical attitude that I fear often conceals a heavy heart.

The recent arrival of Frau Merklein as a bride led me to ask about Moravian marriages in general. I find that they are arranged through the Mission home office. The established missionary at each post is assisted by a young storekeeper or business manager, who is unmarried when he arrives since family cares might interfere with meeting the responsibilities of his new work. But after a reasonable period which might be called his novitiate, perhaps two years, he is expected to inform the home office that he is ready for a wife, and one is selected from a list that is kept on file. The prospective bride is sent (sight unseen) on the next voyage of the *Harmony*, and the marriage takes place after her arrival.

I asked Mrs. Lenz if everyone had been satisfied with this arrangement. "Well, usually," she said. "At least, they accept it without complaint. But there was one case when the young man, after sending his application to the home office, remembered a certain red-haired woman he disliked intensely, who was without doubt on the list. He was terribly troubled and prayed earnestly every day that someone else – almost anyone – might be sent, but his prayers were in vain. When the *Harmony* arrived and he went fearfully to the rocks to meet the boat, there was the red-haired woman!

"Oh! Mrs. Lenz," said I. "What did he do?"

"Well," said she, "his feeling was so strong he felt it would be a sin to marry her, and when the *Harmony* sailed away she took the woman back to her own country." The young man afterward tried again with more success and is now stationed at one of the northern posts, but the future of the poor red-haired woman is not recorded.

Mrs. Lenz spoke of her own experience, which has been a very happy one. Her father was Dr. Jannasch,[250] one of the most prominent and best-beloved missionaries on the Labrador coast. Like all the children, she was sent to Germany in her eighth year but at the end of her formal education returned to Hopedale to join her father and there found the young storekeeper, Herr Lenz, who had just arrived as assistant. He had no need to send to the home office for a wife! "It was a love match," said Mrs. Lenz with a charming smile, "and we have been very happy."

August 19th I have spent practically the whole day on the high hilltop, even climbing to the flag-pole on its summit. The view from the top is wonderful. Hills and islands appear which are unseen from a lower level, and across the blue sea I counted fifteen icebergs, some of them huge ones. On the hilltop are mossy terraces and quite large pools of clear water like bath-tubs that looked inviting in the warm sunshine. The wind blew a gale, and not a mosquito had a chance to linger.

After supper tonight, Mr. Lenz held a short service in the church. It was conducted in Eskimo with an Eskimo hymn-book from which I sang – or at least followed the tune and guessed at the pronunciation. The congregation were Eskimos and half-breeds. The women sit on one side of the centre aisle, the men on the other, and the Mission people on a bench along the wall that, on the women's side, is painted white. The organist, who is an Eskimo, plays very well. Mrs. Lenz tells me no one taught him. There are no teachers. But they are all musical, and many have organs in their homes. They just "pick it up" themselves. Frau Merklein appeared at church wearing a little triangular lace cap [*schwesternhaube*] with blue ribbons, very becoming to her German style.[251] Harriet, the half-breed servant, wore one tied with pink. I wore my hat, and the Eskimo women in general had no head covering. The caps are very pretty and different from anything I have seen before.

The German language is on my mind. For a person interested in discussion and social intercourse generally, the restrictions of an unfamiliar tongue are disturbing. What little German I ever knew is emerging from memory cells, but I never passed far beyond the early period of *Ich habe mein buch gesehen* [I have seen my book]. Herr Lenz has loaned me books and a dictionary, and I hope to enlarge my vocabulary, but Eskimo is hopeless. I do not even attempt it, although by singing with the natives one learns to pronounce combinations of letters that are sometimes whole sentences, but I have no idea of their meaning. During the day, the conversation is a mixture of English, German and Eskimo, while at prayers we use one language one morning and another the next. Mr. Lenz says the Eskimos like to be called "natives" (or their word for it [*Inuit*]) and do not wish to be confounded with the "settlers" or half-breeds.

Sunday, August 21st This has been a beautiful day in many ways: perfect weather and an atmosphere of peace. I wonder why, even here, with no traffic or sounds of industry to be hushed, there seems to be a Sunday silence. Breakfast was half an hour later than usual, the table spread with a clean cloth. Herr and Frau Lenz and Mr. Payne wore their Sunday clothes, and the children were in fresh white dresses. Even the food was different. We had no porridge, but there was sage sausage and golden syrup in place of orange marmalade.

The morning service was at ten o'clock with a large congregation, and I noticed all the women had on little caps like that worn yesterday by Frau Merklein. Frau Merklein explained the blue and pink ribbons: blue is for married women, pink for unmarried, and I believe white is for widows, although none were in evidence. I shall make a cap to wear next Sunday. The service, short and mostly musical, was conducted in Eskimo and therefore unintelligible, but words were not needed to express the reverential spirit of the people. The congregation rose and sat unexpectedly, often in the midst of singing a hymn, and at the end, Herr Lenz left without the customary benediction. Much of the music resembles a Gregorian chant, measured and dignified, some of it almost weird with strange intervals. All the people take part in the singing. They have good voices, although some of them are shrill, but they have an accurate ear and sing in tune. There was also special food for dinner: a delicious soup, partridge (canned by Frau Lenz), potato, creamed cauliflower (from the garden) and stewed dried raspberries for dessert.

In spite of this bountiful meal, the Merkleins had invited us for coffee and a house-warming party at two o'clock. We found the table spread as for a formal meal with lovely white, green and gold china and a silver coffee service. The decorations were daisies, and at each place was a dainty embroidered doily tied with pink ribbon, with a sprig of spruce thrust in it. Frau Merklein, in a black dress and frilled white apron, bustled about with a coffee pot in one hand and hot water jug in the other, pouring coffee at each place after we were seated. In the center of the table were two large *kuchen* [cakes], one with apple and almonds, the other with rhubarb. It is remarkable how one could eat such a meal so soon after a twelve o'clock dinner with the prospect of supper at five-thirty and a nightcap at nine, but we seemed to manage five meals that day without difficulty. After the table was cleared, the photograph albums were brought out, and we met all the relatives and friends, then to the church, where Herr Lenz read to the congregation after a short prayer and, at the end of the hymn that followed, everyone shook hands with his neighbor. Herr Lenz says they used to kiss as a salutation, but that is now superseded by a simple handshake.

Herr Lenz had invited me to join them again at 6:45 for communion, and I was glad to do so. Without understanding a word of the service, I never participated in one that impressed me more. Perhaps Herr Lenz's own sincere spirit and the congregation's simple reverence would contribute to the impressiveness of any service.

The women dress especially for this service and wear large white woolen shawls as well as the little caps. Frau Lenz wore one folded with a point in the back, in the manner of many years ago. Some of the Eskimo women wore elaborately embroidered dickeys [*silapaaq*] with a long tail, and one old woman, who acted as doorkeeper, had on a green and white silk skirt over a hoop, an elaborate white apron with a bib and a white shawl. On her feet were sealskin boots with white feet. She sat on the front seat with Frau Lenz, Frau Merklein and me, opening the door for Herr Lenz when she heard him on the stairs, opening it again for him with a great air of importance when he went out. Herr Lenz wore a white gown, full and belted, when he entered with the tray of wafers covered with a napkin. The communicants rose or sat during certain parts of the musical service and stood at their seats when the wafers were passed by the minister. These were placed in their mouths when the music indicated a certain phase of the service. Then the wine was passed, and afterward, each one shook hands with his neighbor, the minister left the room and the service was over.

At intervals during the day, we have heard organ music in the Eskimo village nearby. Even this evening they were playing and singing. Herr Lenz says they also perform on brass instruments and in winter have a band. They play out of doors, and when frozen moisture clogs their instruments they take them indoors to thaw on the hot stove, then go out and play again. It must be difficult to manage stops in zero temperature.

There is a full moon tonight, and at bedtime we went out to see it on the water without realizing we should find a glorious aurora streaming above us. The colors danced and swirled like living things, green, pink, crimson, yellow and violet in great waves, constantly changing and darting from one part of the heavens to another. From a crown overhead, long fingers of light fell like a fringed curtain, then suddenly changed, and a swirl of wonderful color intensified in another place. I never saw so much movement. Even those who are familiar with such displays during the winter say it was remarkable. It is still going on but is dimmer – only a wide band of light across the sky.

This has been a full day, one to remember.

August 22nd A real summer day, even by New England standards, the mercury eighty in the shade. To stay indoors is unthinkable, and I have wandered as usual, sketching and stopping on the way to gather mushrooms, which are plentiful. Some of them are familiar, and I take them with confidence, but I look askance at others and long for my mushroom book. However, as Dr. Grenfell has assured me, there is not an inedible mushroom on the Labrador coast – that he has tasted all of them. I have ventured to add them to the family menu, and no one has complained of unpleasant results. The German ladies have been shown how to prepare them and use them with toast, another unknown article of food. Frau Lenz did not know how to use the toaster that had been given her.

Mr. Payne is busily preparing for his departure for England on the *Harmony*, which is due in a few days. Most of his waking hours have been spent in packing his varied belongings, and this afternoon I have helped, receiving as perquisites three dog-skins and a sealskin that will make fine rugs in my studio at home. Some of the boxes contain interesting collections of animal skulls and paws, bird skins and curios only found in this part of the world. Tonight after service he brought out his gramophone for a final concert. He will be glad to have it with him during the long dark winter.

One of his treasured possessions is animate – a pet Eskimo dog he raised from puppyhood. She is so well trained that he brings her with him to the dining room. Her mild name is "Nellie," and she wears a collar which like her name does not seem suited to that kind of a dog. She is a beautiful great creature, very affectionate and intelligent. Mr. Payne has taught her a number of tricks. I asked him if he expects to take her with him when he goes to England on his furlough next year [i.e., week]. He was non-committal but agreed that pet cats might be in jeopardy.

August 24th The Germans love picnics, and today we went for one on an island where the Eskimos fish in summer. Herr Lenz, in the motor boat, towed a thirty-foot trap-boat carrying the women-folk and children. When we approached the cove where two Eskimo families have their summer huts, they began to run about like ants outside a hill. A flag went up (a red handkerchief tied on an oar), and they ran to the rocks to meet us with cries of *achenai*, their Eskimo greeting, then helped us over the rocks to the huts.

Inside the first one were a stove, two bunks and a table. The rest of the room seemed to be a litter of clothing and rubbish. There was one window, a square opening near the roof where some boards were left off, and stretched across it were dried seal's intestines that were thin enough to admit light. On top of the rocks was the smallest abode I ever saw, the home of an Eskimo named Matthias. His wife is a tiny slip of a girl who looked sickly. Apparently, she had on only one garment, a patched calico that clung to her body. She was carrying her baby, which looked exactly like a Japanese doll and was also tiny although nearly a year old. Their childish delight in seeing us was evident as they jabbered in their abominable language, and I responded by signs. It was the best I could do.

We brought our coffee pot and ate our picnic lunch on the rocks at a sufficient distance from huts and dogs to be fairly sanitary. There were a few tufts of dog's hair where we were sitting, but Mr. Payne said dogs were usually under the table, so it was quite in keeping with local conditions.

These people live at Hopedale during the winter and use the huts only in summer during the fishing season. It is the same custom that prevails among the liv'yers in the villages farther south. The huts are only a few old boards fastened together, with a stovepipe in the roof; but the occupants were the happiest, cheeriest lot of people I have seen for a long time. They finally shook hands all around and assisted us into the boat, again with cries of *achenai*, which seems to mean either "How do you do" or "Good-by."

This morning when I was in the Eskimo village, an old woman beckoned me inside her house to show me a "dickey" (native word for parka) such as I had wanted as a curio. It was made of thick white wool, cut with a long tail behind and a large hood gathered like a bag at the back for carrying the baby. The trimming was red braid, and the hood was bordered with fur and a wide band of gay embroidery on black cloth. I asked "Sell?" and she countered with "How much?" "I ask Herr Lenz," said I, and she nodded. Like the huts on the island, the house was filthy and in great disorder. A fire was burning in a pan near the entrance where she had obviously been cooking. The room was piled with rubbish; the two bunks, covered with rags and clothing, were evidently never made up. Frau Lenz says it is the worst house in the village. I asked her to negotiate for the dickey, which I am advised to have thoroughly cleansed. It is probably inhabited.

August 25th Today I asked Herr Lenz who are called to work when the Mission bell rings at 6:30 a.m., at 8:30 and 9:00, then again at 12:00 and 6:00 p.m. He tells me that at present there is only old Marie, who tends the garden, opens the seal fat belonging to the store and does various odd jobs. At Nain and Okak there are men who are carpenters and work in various ways. Here, the bell seems to be principally useful in telling the time.

The people live by fishing, trapping and sealing, as they do all along the coast. The fat is seal blubber, and casks of it are opened daily for the sun to clarify. After a while, the sediment falls to the bottom, leaving the oil clear and white. During the winter, the blubber is cut in pieces and pounded with wooden mallets by the women before being put in vats, where it is left three or four weeks until the fibre settles and the oil separates from it without further treatment. The oil is sent to England for soap manufacture, but the natives here use it for frying and sometimes eat the blubber, which those farther north depend on for their principal food. The Moravians act as traders for the people, and all sorts of things – furs, curios, etc., in addition to seal oil and fish – are brought to the Mission in exchange for supplies. The *Harmony*, that brings Mission supplies from England, takes the fish as return cargo, and the oil is shipped on the final trip in the fall.

Mr. Payne is in the final agonies of packing and has just closed two large boxes, one of tinned food and photographic supplies, the other final odds and ends from his best tweed suit to cartridges and deer sinew for sewing skin boots. He says he begins to feel like a free man and even took time for me to operate on his hair to demonstrate my status as St. Anthony barber. Nellie lay against the door and at times browsed around for spoils. We left the room for a moment and, on returning, found her under the bed crunching a seal's skull. The sinew and skin boots had to be packed at once for safety.

August 26th Frau Merklein called me to her apartment today to show me her linens. Such a display! Bed linen with embroidered initials and wide, hand-made insertions on the pillow slips; beautiful table linen, also embroidered; and lingerie with hand-made lace and drawn work. And there were night-caps. She put one on to show me how it looked. The linen was a fine quality and the needlework exquisite. Her pride in it was evident, and her eyes shone during the exhibition. Herr Lenz told me all the women (especially those in small towns and the country) begin when very young to prepare their household linen. I asked what happened if they did not marry. "Why," said he, "they just have it." Perhaps such a thought never accompanies all that lovely handiwork.

Herr Lenz has negotiated for the dickey. It is now on the floor in a corner, rolled up tightly to await a thorough cleaning out of doors. It is a gorgeous affair and will create interest when exhibited with my other curios at lectures. Frau Merklein's maid, Augusta, is making some boots with white feet for me. Herr Lenz has told me something of her history. She is a full-blooded Eskimo and used to be rather flighty, but reformed and has so fully turned from her sinful ways that she runs whenever a strange man appears on the premises. When Mr. Payne first came she would not enter the room, and when men came ashore from schooners anchored in the harbor she could not be driven from her house with a club. Since acting as house domestic, she sleeps at Frau Merklein's apartment, and if out later than she thinks the proprieties permit, she is careful to forestall criticism by explaining exactly how it happened.

This morning, I made one of the little caps worn by the native women. It has bright pink ribbons, and when those ribbons are tied under my chin I look as black as the Eskimos and as old as the hills, but tomorrow I shall wear it to church and become indistinguishable from the rest of the female congregation.

The *Invermore* is due on her way south. I shall return on her to comparative civilization at St. Anthony, now that I have seen the Labrador.

Sunday, August 28th A day of excitement! My brain still reels from the disruption of our quiet life. We went to church this morning, and soon after the service began four men entered and sat on the bench along the opposite wall. Half-way through the service, I looked up suddenly to find the eyes of one of the men fixed on me; then he smiled. His face seemed familiar and, to my amazement, I recognized DeWitt Scoville Clark of Salem, Massachusetts, who was at St. Anthony two years ago. Then I remembered having heard of a party that had started in the early summer to explore the northern interior to Ungava Bay.

Mr. Clark joined us after service and introduced the party: Dr. [George P.] Howe[252] and Mr. William Cabot[253] of Boston and Donald MacMillan,[254] who was one of Commander Peary's North Pole expedition and went with him as far as the 84th parallel.They had travelled across country to Ungava Bay, following the George River where the ill-fated Wallace-Hubbard expedition met disaster. A trap-boat took them out of Davis Inlet, and a passing schooner sailing for harbor on Sunday because of a leak brought them to Hopedale. They came to the Mission, bag and baggage, at the invitation of Herr Lenz, who arranged to put them up while waiting for the *Invermore* and, of course, invited them to dinner.

In the midst of a gay meal and stories of adventure, told to an audience for the first time since leaving civilization, the *Stella Maris* steamed into the harbor, bringing the startling news that she was to leave in a few minutes to connect with the *Invermore,* which was not coming to Hopedale as expected, and we would have to go on her now if at all. Then the *Strathcona* suddenly arrived! I finished flinging things into my trunk, left my curios to be sent on the *Strathcona* and followed the men to the *Stella Maris,* but was hardly aboard when a boat put off from the *Strathcona,* bringing three more passengers and Dr. Grenfell with letters for the mail. It was then that we discovered the *Stella Maris* was going only a short distance south to wait for the *Invermore,* so Dr. Grenfell, who was going directly to St. Anthony, offered to take me on the *Strathcona.* In three minutes, I was off again, leaving my interesting fellow passengers rather dazed when they saw me row away with the Doctor. Mr. Cabot, who had once been a passenger on the *Strathcona* and occupied one of the swinging hospital beds, called a parting word over the rail: "Those beds squeak!" The *Stella Maris* did not seem half bad. My stateroom was clean, the saloon was attractive and the Captain, with the cleanest kind of a shave, was resplendent in his uniform. I almost wish I had gone north after all.

Later. We went ashore for supper at the Mission, where I faced another anti-climax after good-bys, and in the evening Dr. Grenfell conducted service, followed by stereopticon views before a large Eskimo audience. I am staying at the Mission with far from a tranquil mind, for Dr. Grenfell has after all decided to go north to Nain instead of south to St. Anthony. He expects to return in a few days, but Frau Lenz predicts a week. I wish I had stayed on the *Stella Maris.* It is amusing how I coquet with her. Twice I have been on board ready to embark and have turned away. The first time was a happy decision, but now I have mixed feelings. This has been a confused and nerve-racking day. Such suddenly shifting scenes are upsetting, but I should be immune to them by this time.

August 29th The *Strathcona* left this morning. Dr. Grenfell came ashore at 6:30 to extract Frau Lenz's teeth. She was still in bed, but he was ready to start, time was precious, and he wanted to go to her room for the operation but was deterred by Herr Lenz.

There is still excitement, for this afternoon the *Harmony* came in. I recognized her in the distance by her bark rig. She has one passenger, a Mr. Stewart,[255] an Irish Protestant missionary returning to [Fort] Chimo, Ungava Bay, after a year's furlough in England. He has suffered horribly from *mal de mer* and hailed with joy Herr Lenz's invitation to stay at the Mission during the two days the vessel is in port. His work is principally among the Indians who live inland and rarely visit the coast but sometimes travel to Hebron and Okak in the winter. He says many have died from starvation on such a trip, and only six years ago a party encountered such difficulties that several died on the way. They were all starving, and the living ate the dead. He spoke also of another party who crossed the same country twenty years ago and said a woman now living in Chimo had told him, with tears in her eyes, how they had killed and eaten her mother. They became wild beasts from hunger. He says there are many wolves, and sometimes they come to play with Eskimo dogs. The only communication with Chimo is when the Hudson's Bay ship *Pelican* stops there on its annual visit. The rest of the time, travel is only by komatik to Hebron or Killinek, 400 miles away.

Many of the native women have brought articles for me to buy as curios. There are beaded wall-pockets, bags, owl's feathers, etc., some of them lovely and all interesting. Augusta beams on me. She cocks her head on one side and says "Oh," which means "I am so sorry you are going away." The inflexion is like that one might use with a dying friend.

August 30th Captain Jackson[256] of the *Harmony,* a genial, kindly man, was our guest at dinner today, and we are invited for supper on the *Harmony* tomorrow night.

This afternoon, I went to the hill for a last climb to my favorite mossy seat among the rocks. My writing pad was with me, but I could not take my eyes from the changing light on sea and land below, the islands in brilliant sunshine, or swept by deep blue shadows of passing clouds. The only signs of life were two komatik dogs that appeared over the rocks to make friends. They rubbed their heads against my knees, then sat on their haunches with heads raised to the wind, sniffing, as if enjoying the salt air. I am going to bed early to be assured of one more good night's rest, having in mind Mr. Cabot's parting inference of uncertain sleep in the *Strathcona*'s squeaking hospital beds.

August 31st No *Strathcona* yet. I looked for her all the morning when hunting mushrooms for the last time on the lower hills.

The bakeapple berry is ripe now, also blueberries and what is called blackberry. something like a partridge berry. There are a few of the delicious arctic strawberries, really more like a blackberry in form and very small. Their flavor suggests both strawberry and raspberry. Among the flowers is the arctic goldenrod, still another variety from that found at Indian Harbour. There is also fireweed, well known in northern New England, and the ground is now covered with a tiny delicate pink, bell-shaped flower having a delicious fragrance, known as Labrador tea. It grows among bushes in swampy places. There are also quantities of the white starry blossoms of the bunchberry. In some places the ground is covered with it. It will be lovely when all the white flowers have changed to bunches of scarlet berries.

The Captain came this afternoon for coffee before taking us to the *Harmony* for a tour of inspection before supper. The ship is very comfortable. In the officers' quarters is a small stove where a soft-coal fire was burning, making the cabin very cozy, and in the passenger end of the ship, where we had supper, was another fire. It was an excellent supper, served by a funny old steward who looked like an actor, and afterward we looked at photographs. The Captain has offered to take me north as a passenger on one of his trips next year. The *Harmony* will round Cape Chidley into Ungava Bay, farther north than any ship except the *Pelican.* I should love to go but fear it is only a dream.

Nellie has not gone aboard, but the kennel has a conspicuous place amidships, and Mr. Payne's 28 boxes are safely stowed. I bade a long good-by to the Captain, Mr. Stewart and Mr. Payne and added them, regretfully, to my "ships that pass in the night." The *Harmony* sailed this morning at five o'clock. There is still no *Strathcona.* I am discouraged, for it is now a matter of "killing time," an unusual and unpleasant occupation from my point of view.

At odd moments, I have talked with Herr Lenz about the Eskimos and the Mission's connection with them and was particularly interested in what he had to say about the schools. At present, there are only native school teachers, although I think Herr Lenz has taught a school at Nain, and Frau Lenz has a sewing and knitting class during the winter. The curriculum, if it may be so called, consists of reading, writing, a little arithmetic, a little geography, a few questions and answers concerning the Scriptures, and memorizing verses of hymns. The Eskimos have no literature as a background and, theoretically, it would seem advantageous to offer them a wider field for development by teaching the German language. It may be, however, that their environment and manner of life might make development difficult or inadvisable, although, judging from those of Eskimo blood who have found their way to Newfoundland, they have demonstrated their intelligence and adaptability. The school term here is from late January until June. In Nain and Okak, however, it begins in November and the school is larger: about forty children. Here there are only thirteen.

The natives have no connection with the Mission as far as their work is concerned except in a few instances. Like all others on the coast, they depend on trapping and fishing, which is only a precarious living.

The wind howls, and there is an occasional dash of rain. In such weather, the thought of the *Strathcona* has no glamor. I hope it may be calmer when we start anyway. As for the voyage, how many times will she bump on rocks or shoals? She bumped twice on the way down and was aground several hours. A trip on the *Strathcona* is never without excitement, and this is the time of year for gales.

6

Voyage on the "Holy Roller": Hopedale to St Anthony

3 September – 23 October 1910

Long Tickle, on board S.S. Strathcona. September 3rd The *Strathcona* arrived at Hopedale yesterday at about 10 a.m. and the men all came ashore, stopped long enough for a cigar, then gathered all my many belongings and we were off. I was quite ready to go. Augusta waved her final farewell from the window, and Frau Merklein came at the last moment with a plate of sugar kisses with nuts in them, a fine thing to take at the beginning of a "Holy Roller" voyage!

The wind blew heavily, but outside the harbor we went quite steadily with little motion. It was lovely on deck, and I sat on a packing box watching the blue hills fade away, and the sea, where a school of porpoises were leaping out of the water nearby, then prepared the mushrooms I had picked for supper. There are certain difficulties attending the preparation and serving of food on the *Strathcona*. The tiny galley is on deck, the dining saloon below. The actual cooking is simple when one becomes accustomed to the limited space and uncertain movement, and at least no extra steps need be taken, but serving is another matter. The deck is piled high with firewood, in some places nearly level with the rail, and passage over the logs and down the companion-way with hands carrying dishes of food requires special balance and is often precarious. However, like other things, one can become accustomed to it.

Dr. Grenfell intended to stop at Double Island, the large Eskimo summer settlement, but passed it before recognizing the opening. Here at Long Tickle, a number of Eskimos came aboard to exchange their skin articles for clothing. It was amusing to see them pawing over the contents of the clothing chest and bartering with Mr. Palmer. Clothing is what they want and need. They have less need for cash since there are no shops.

We ran into Turnavik, and all went ashore to the house of Tom Eustace, an English settler who has a half-breed wife. Their cottage is the ordinary type of settler's house, a fair-sized room with a stove at one end, a wooden bench at the side, table against the wall and a few chairs. Sometimes there are no chairs, only home-made benches. There are few hooked mats on the real Labrador. The floors are bare and the walls covered with newspapers or pages of magazines, probably given them by the Doctor, pasted over cracks between the boards. From this room a staircase led to the floor above, and there was a little bare, scantily furnished room that I did not enter. In this house lived the Eustaces and another family with two children. I found Mrs. Eustace very interesting. She surprised me by her appreciation of nature, sunsets and the country. She knew the wild flowers, though not their names, and showed me a number that she had pressed in an old magazine. Her English and enunciation were remarkably good, and she spoke distinctly, sounding all her Gs. She never went to school. Her mother and aunt had taught her to read and write, and she reads a great deal. She said she thought if she spoke the way people did in books it must be all right, and she tried to do so.

She had some good beaded skin articles ready for sale and was just sewing little tassels of wool and beads on them as a finishing touch. I bought them and offered some suggestions about the most desirable kind of work, which she accepted gratefully.

A neighbor, Mr. McBride, came to fetch Dr. Grenfell to see about his son, the youngest and only one at home. The Doctor went on ahead and I followed slowly. Mrs. McBride was outside the house and came to meet me. The poor woman was weeping bitterly, for the boy was very ill and she had been with him several nights without rest. I tried to comfort her, but what could one say or do? Dr. Grenfell said he had pneumonia and thought it best to take him to the hospital at Indian Harbour, so now he is tucked away in the hospital on board.

Dr. Grenfell told the people he would come ashore for service in the evening at 7:30, but he lingered so long with his cigar that it was 8:30 before we rowed out into the darkness. About twenty people were gathered in the Eustaces' kitchen. There was a baby to be christened, the eighteenth child of one mother, and she only forty-two. The eldest child was twenty-four. The service was very simple. Dr. Grenfell explained the meaning of baptism and performed the ceremony in his own way, perhaps more fitting under the circumstances than an orthodox service but possibly startling for a ritualistic clergyman. There was a diversion when the hanging kerosene lamp began to burn badly, and men rushed to take it outside to avoid an explosion. Meanwhile, the congregation waited quietly for the service to continue. The native people are accustomed to emergencies.

Afterward, we sat for a while and talked until the mother made a move and the gathering melted away, each one coming to shake hands with the Doctor and also with me as being "The Lady of the *Strathcona.*" It was late when we stumbled through the low fish stage and down the slippery logs to the boat, though light still lingered in the sky. It may have been the aurora. We see it often these days.

My little cabin is very comfortable, though piled full of several people's belongings: Mr. Palmer's suitcase and various curios, my bags, Mr. Wight's hat, the shotguns and rifles of some of the men. The shelf that serves as a dressing table is littered with cartridges, also Mr. Palmer's toilet articles, and there are numerous unclaimed boots and Dr. Grenfell's lantern slides. A pair of trousers (owner unknown) hang beside my one good skirt, and my wrapper is on the same hook with Mr. Palmer's khaki overalls. The bed is very hard but comfortable after one is in it, and I slept well.

All night we lay at anchor and in the early morning ran to West Turnavik, a still smaller place, only notable for being the home of Wilson Jacque,[257] one of the St. Anthony men. His father and brother came on board, and I talked with them and the sister-in-law, an attractive girl who said she found life very lonely. She had no children, which is unusual on this coast. The people are anxious for clothing and magazines and bring out all sorts of skin work for exchange – good, bad and indifferent. The women are poorly clothed, and I have parted with my garments, one by one, all along the way, regretting that my trunk has gone on with articles yet to be disposed of.

We arrived at Long Tickle in the afternoon. The families here are terribly poor. Dr. Grenfell has practically saved them from starvation. He is buying furs from them all along the coast, and some of the men have made good catches during the winter. One of them is the father of a man who works at St. Anthony. His wife wore an old black hat with a pitiful feather, which she held on because she had forgotten her hatpin. Then her hair came down because she had but one hair-pin. I produced a paper of hairpins and suggested she might tie her handkerchief over her head instead of around her neck, an idea which she hailed joyfully. Evidently, the hat was donned only on state occasions like the arrival of the mail boat or *Strathcona.* She also wore white cotton gloves, much too large for her. She wanted to talk and came to sit beside me, telling in detail of the deaths of her son and daughter. Poor woman! It was probably a relief to express her grief, even to a stranger.

There have been many calls for Dr. Butler's services. Teeth and fingers seem often to be the ailing members, though there have also been "feelin's in th' chist." The clothing in the great box is all gone – one man tonight took the scrapings, small articles that he said would do for patches. A woman came on board afterward with some excellent beaded sealskin, which I paid for with cash although she would have preferred clothing.

The McBride boy is very ill. Dr. Butler and Mr. Martin[258] take turns in watching him, and they fear he may die. The crisis is near. The doctor wishes he were safely at Indian Harbour hospital.

A squall this afternoon prevented our starting out, and we are here for Sunday.

September 4th Rain woke me this morning, pattering on the roof of the little deck cabin, and it was a dreary outlook from the window, but it cleared in the afternoon and we went ashore for prayers at 3:30. A flag raised at half-mast on a pole near one of the fish stores indicated that the meeting was to be held there. A few men came from the schooners lying at anchor in the tickle and practically all the settlers, many of whom are half-breeds. The room was a mere shed of rough, blackened boards. An old sail covered the floor, and overhead on the rafters were stored a canoe, oars and bits of fishing gear. The furnishing consisted of a plain deal table and wooden benches borrowed from the natives' kitchens. Dr. Grenfell is at his best in this kind of life, and it is a privilege to see him in this role, so like what is often written about him. In some ways he is a different Dr. Grenfell from the one most people meet in a more sophisticated environment. I looked around the poor, miserable little place with its congregation of half-starved people, their scanty clothing, undernourished bodies, and realized more fully than ever before what the Doctor's annual visits mean to them.

After the service, several of the people followed us to the *Strathcona* and came aboard to wander about and talk. One of the women saw my woolen gloves and said, wistfully, that she wished she had a pair of warm "cuffs" [mittens]. I thought of an extra pair I had with me, and while I searched for them she looked around the little cabin at the bunk made up with sheets, pillow slips and folded blanket. "This is where I sleep," I told her. "Oh," said she. "How clean! Just like the hospital!" One has only to see the beds of the average settler and Eskimo to realize the full meaning of that awed utterance. Mr. B., her husband, looks half starved. His cheek-bones are prominent, his eyes sunken and his lips drawn tightly across his teeth. His gaunt body has the slight stoop of most fishermen.

The McBride boy is passing through his crisis. The doctor is anxious but thinks he may recover after all.

September 6th We came to anchor in this little harbor at midnight. Early in the morning I heard movements in the cabin and on deck, men's voices, then the sound of oars and all was still. We had just begun breakfast when the Doctor returned from shore, having been called by one of the fishermen to baptize his child, the youngest of eight. The Doctor was in high spirits and entertained us throughout the meal with his experiences. He said the parents had no name for the child (perhaps their favorite names were exhausted) and depended on him for advice. He did not say what name was chosen but remarked that it might be an excellent idea to have a list of names on file for use in emergencies and mentioned several not in general use on the coast, such as Pneumonia, Insomnia, Hysteria, and Hernia as being appropriate for a medical missionary to suggest.

The weather is clearing, and we hope for a fair day tomorrow to continue on our way to Indian Harbour, where the sick boy, who is improving, will be left to recover and return to his home.

September 7th A glorious day. I am sitting on the roof of the cabin on a coil of rope as we steam out of the inlet where we spent the night. The hills and distant mountains behind us, the cape and islands ahead, are a beautiful soft blue. The brilliant sea is dancing in the sunlight, and with the fresh wind my Jaeger sweater and leather coat combination is very comfortable. I woke with the dawn to hear Dr. Grenfell's voice: "All right, go ahead." Then came the *Strathcona*'s heart throbs, the voices of Bill [Simms] and Abe [Mercer] as they stumbled over the wood-pile on deck. The anchor came up and we were off. From my window, I watched the shore as we passed by, the jagged skyline of the coast and islands now beginning to look familiar, then dressed and went on deck.

Breakfast was late, and it was nearly nine when we finally sat down owing to delays on the part of the Doctor who, on his way from deck to cabin, found something to distract him from his original purpose. We had barely started on our oatmeal when Bill called that the *Yale* was alongside, and we slowed down while Dr. Stewart climbed on deck with amazing agility. The doctors had a short conference; the sick boy was taken aboard; Dr. Stewart dropped over the side. The *Yale* sailed away like a great white bird, bound for Indian Harbour, and we are again on our way. Even after similar experiences, these unexpected and casual encounters on the high seas still seem exciting.

We have all been busy today: the Doctor with plans for disposition of furs to the Hudson's Bay Company and other business matters, Mr. Palmer with clerical work. Mr. Wight has responded to frequent calls from the engine room of "more wood" and is taking it from one place on deck and piling it in another. There is now a passage about a foot wide to walk on as far as the chart room. My own occupation has been domestic. Dr. Butler brought me his coat to mend, buttonholes to repair. I have hungered for darning, mending and buttons after so much of it all winter. If the men realized how much I like it, they would tear holes to be mended and pull off buttons to be sewed on again!

We are nearing Cartwright and will anchor there for the night.

Cartwright, September 8th At dawn, I heard the voices of Dr. Grenfell, Mr. Wight and Dr. Butler, calls for Mr. Palmer to get to work on a letter for Mr. Shepherd,[259] the Hudson's Bay Company factor, then the sound of oars, and silence. After a long time the men returned, but we did not start at once. Boxes were being landed on the wood-pile and stored in the hospital amidships, which is now a freight hold as well as Mr. Wight's sleeping quarters. Mr. Shepherd was aboard and had just finished his business transactions with Dr. Grenfell when I went on deck in time for hail and farewell. The Doctor was jubilant. He had disposed of some furs for $450.00 and left the Hudson's Bay Company bearing him no ill will for having established an independent fur trade on the coast. He said he knew his gift of a box of cigars at a critical stage in the transaction contributed greatly to its success. Mr. Palmer, however, seemed pessimistic concerning the Doctor's abilities in striking a sharp bargain, skepticism which may or may not be justified.[260]

We have made a straight run for Grady [Harbour]. The clouds that were threatening in the early morning have blown away, and it is beautifully clear. We are passing many islands: inside some of them, outside others. It is hard to tell what is island and what mainland. Our course, known as the "inside run," is taken to avoid rough water. There is a ground swell but fortunately for us not much of a roll, for there is still a great pile of wood on deck with scarcely any coal in the bunkers, and the top-heavy condition brings an apprehensive thought to even the most hardy and optimistic traveller on the "Holy Roller." There have been dire predictions of foundering in a heavy sea.

Grady, September 9th This is the largest fishing port on the Labrador. There is a good harbor, and we found a number of schooners at anchor, on their way home from northern fishing grounds. The shore is lined with fish houses, some of them in the dilapidated condition of many we had seen but most of them in good repair. Many have turf roofs, and some of the tilts are banked with turf along the sides with long grass growing on the roofs. It is hard to believe they are human habitations. Dr. Grenfell wanted to go ashore, so I went too. Mr. Wight rowed us and we landed on the rocks where waves were breaking. The

Doctor disappeared along the path leading to one of the houses and went in while I wandered on to a place where the sea dashed high on the cliffs. Fish were drying on the rocks, and many of the men and women were busy turning and stacking them. The drying flakes (stages) were large, and the whole place appeared more prosperous than anything I had seen.

Mr. Wight called "all aboard," and I went down to the boat and got in. Dr. Grenfell was on a large schooner at the wharf. He started toward us through the fishing stage, and then we lost him. After waiting half an hour, a woman appeared on the wharf above us and in answer to our inquiries told us the Doctor was at her house having dinner, and they must be nearly finished for she had just given them their second helping. It was past our own dinner hour, so Mr. Wight rowed me out to the steamer and returned to shore with the promise that something would be kept hot for him. Half an hour later they arrived, the Doctor full of apologies and convinced that his visit had been worthwhile from the point of view of policy [prudence].

There was a brief stop at Indian Tickle, a few calls on the schooners at anchor on their way home with fishermen and their families. Then we started out at dusk with Bill at the wheel, and presently we knew we had come to the shoals, for he sent for Dr. Grenfell. Bill is a timorous soul, but I fancy in this case he had good reason to be, for it was only a few minutes later that the engine slowed down. I went on deck. The Doctor was in front of the wheelhouse, calling directions to Bill at the wheel. It was something like this: "A little to starboard. Now a little to port. Have we been that way, Bill? There's a shoal off there. You can go either side of the island now. Bill! Stop her! We must go the other way, toward that light!"

To the right there was a roar of waves on a reef, and a row of breakers gleamed white in the darkness. We turned almost at right angles and steered toward the bright light that proved to be a schooner burning seal-oil lamps on deck while the crew baited their trawls for the morning. The sail was up, and in the reflected light it looked as though the schooner was on fire. It led us to safety, for we cleared the shoals. We are now anchored at Domino for the night. The wind is rising, and the clouds are dark. It looks like bad weather.

September 10th It is bad weather. We have kept to the inside run, but through open spaces between islands, the wind and high seas tossed our little craft unmercifully, a reminder of its lack of ballast, and we cast an apprehensive eye on the overloaded deck. In such weather, only travel by day is possible. We anchored at night and reached Battle Harbour after a two days' run, with a final buffet in the open sea before entering the tickle about noon. Many affairs awaited the Doctor's attention, and after important matters were disposed of he became interested in making designs for hooked mats that are being prepared for distribution among the local women who work for the Industrial Department. The ample surface of the dining room table made an excellent designing board, but its use so delayed our supper that it was nearly sunset before our start for St. Anthony.

The storm had passed and the sky cleared, but the sea was still boisterous, as we realized more and more when we passed Belle Isle and started across the straits, which are rarely calm. Bill was at the wheel when we started, but as our tossing became more violent the Doctor came forward. "I'll take it now, Bill. It's going to be bad, I'm afraid, but we can't go back." I stood in the doorway of the wheelhouse, watching the turbulent sea until it became too much for me, and I lay in the little deck cabin just behind the wheel. The door was open, and I could hear the men's voices.

The *Strathcona* tossed, pitched and performed its antics. Sounds of crashing china came from below, accompanied by banging of dislodged freight. On deck, the wood-pile began to move about. Logs rolled and pounded against the galley, chart room and deckhouse. The *Strathcona* rolled over to the point of foundering, then slowly righted itself. A huge wave fell on her with a roar. Bill rushed to the wheelhouse door.

"Oh, Doctor. Do you think we'll make it? Are we goin' down?"

Then the Doctor's calm, cheerful voice: "Well, Bill, I don't know. It is pretty bad. We may not make it, but we can't help it now. It would be worse to go back. Anyway, they say drowning is a pleasant death. And just think – all our troubles would be over!"

In spite of discomfort and apprehension, I smiled. It was so like the Doctor. Then I lay and waited, hoping all might still be well. One's mental reaction when confronted with an emergency is unpredictable. It is the way one faces it that brings calm or panic. This was an emergency. There was nothing one could do to avert or mitigate it. Lying on the narrow berth, my body braced to prevent being thrown out by the violent motion, I looked the Thing in the face, and suddenly I was not afraid. Then, presently, I slept.

Cessation of motion woke me. From the little window, I could see Fishing Point Head towering above us as we passed into quiet St. Anthony harbor, while overhead, from a luminous crown in the zenith, the curtain of a wonderful aurora fell to the surrounding hills. Stars shone dimly through it, and the night was glorified by vibrant colored light. Its weird radiance illumined faintly the church and little buildings on the shore as we passed, with only the sound of the engine's throb over the tranquil water to our landing. We were safe at home.

St. Anthony, October 23rd Over a month has passed since the *Strathcona* came to a safe landing after our perilous voyage, and now I am going home. These have been busy weeks, for necessary plans and preparations before leaving the industrial work for the winter, after so long a residence, seem endless. Susie Denny has returned after a vacation to continue in charge of the loom-room, and conferences regarding details have been the order of the day.

Much of my packing is done, many old clothes given away, and the resulting vacuum is being filled with curios and industrial products for sales in Boston and Providence. Dr. Grenfell, who has gone to St. John's on the schooner *Lorna Doone,* took with him samples of the industrial work: a lovely copper bowl, pottery, a basket and pieces of polished Labradorite – all of it Mark's work – and a fine assortment of woven articles, including homespun, went by the last *Prospero* for a public exhibition and sale. Dr. Grenfell plans to show all these to Sir Edward Morris,[261] who is well known in St. John's, to demonstrate what is being accomplished here. The homespun is already well advertised. Earl Grey,[262] the Governor General of Canada, whose home is in Ottawa, bought a piece for himself, also Mr. Amery,[263] the editor of the London *Times,* who was a member of the Governor's party when he stopped here. Other pieces were bought by Sir Ralph Williams,[264] Governor of Newfoundland, and many others, among them students representing numerous colleges who have been here during the summer. If a definite market for the products can be secured in St. John's and elsewhere, there is I think no doubt of the permanent success of the work.

During any absence of several weeks beyond the reach of definite communication, one can always count on changes – sometimes even startling events to be reported. On returning to St. Anthony, I found this to be true. There had been arrivals. Mr. and Mrs. Jesse Halsey (who were expected) had arrived and already become popular as well as valuable assets to the community. Dr. Wakefield had also arrived with his attractive bride,[265] her lively personality and beautiful voice adding to everyone's pleasure as well as her interest and co-operation in local affairs. Many of the young students – among them one from my own city, John K.H. Nightingale, Jr.[266] – had gone home after performing valuable service as day laborers, coal heavers, boatmen or hospital assistants with a spirit of interest and co-operation creditable to them and to the Mission they served. But other news paled in interest and importance beside the double visit of the stork! Two weeks ago, a son was born to Dr. and Mrs. Grenfell[267] and, a few days later, a daughter to Dr. and Mrs. Stewart, bringing great happiness to both families.

Mrs. Grenfell, looking very well but rather thin, came to see us yesterday for the first time, and we have seen the baby, Wilfred Thomason Grenfell, Jr., a fine, sturdy little fellow who looks like his mother. I can imagine the Doctor's happiness. Mrs. Stewart, who is recuperating rapidly from her more difficult confinement, will soon be able to resume her active participation in Mission affairs with the added responsibility of her second child. The baby is beautiful. I am not always enthusiastic about the appearance of the newly born, their beauty – according to enthusiastic statements of relatives – often seeming overrated, but this baby girl is really a beauty. I fell in love with her at once.

The *Strathcona,* which came in on Tuesday bringing Dr. and Mrs. Wakefield back from their trip up the bay, started off again to get whale meat for the dogs' winter supply. It is expected back tomorrow and will then be laid up at St. Anthony for the winter, a month earlier than usual. I took a trip on her last year in November. We are hourly expecting the schooner from Halifax, chartered by the Mission to bring coal and general supplies for the winter. Our present supply of coal is practically nil, and the new steam heating plant in the hospital is run only on the coldest days to conserve fuel. A heating plant has also been donated for the orphanage. Dr. Grenfell's house has had steam heat from the first, so with full bunkers they will all be comfortable. The Guest House still depends on stoves for heat, and water will continue to freeze in bedroom pitchers at night, but we know one can be healthy and happy in spite of such conditions. The weather is stormy and growing colder. There has been one real snowstorm, the country as white as in winter. The ground, however, has been frozen many times, and one morning the harbor was covered with beautiful "young ice." Winter will soon be closing in.

I am leaving by the next boat. Something tells me that the past winter was my last on the coast. I shall never again see the country half submerged under its beautiful white blanket of snow, enjoy the thrill of travelling behind a dog-team or know the comfortable intimacy of a small group, isolated far from the outside world. Increasing responsibilities at home will prevent another long period of volunteer service during the winter, but there are free summers, and if all goes well another June will find me again headed north – a cheering thought of hope and anticipation to take with me as I wave good-by to friends on the wharf and sail out of the harbor towards the problems of a more complex civilization.

My real mission to Newfoundland and Labrador is accomplished, a mission to introduce weaving and other handicrafts among the people. At St. Anthony, they are already established. Some tentative efforts for extension have begun, but a definite attempt for expansion and development along the uncharted coast still lies ahead. Possibly this may be a part of my work during the coming summers, combined with direction of local industries at St. Anthony. At least, it is a stimulating plan to have in mind.

Retrospect

Written at Providence, R.I., 1916, final year of active Mission service My intuitions on leaving St. Anthony in the fall of 1910 were verified. The preceding winter was my last on the coast, but fortune has been kind. During the four succeeding summers, each June has found me again on my way north, and the nebulous plans for development of local industries have taken form and meaning. Year by year, they have been extended. Each year, activities connected with the several branches have led me farther afield, even to northern Labrador and west to the Canadian shore of the Strait of Belle Isle, where uncertain accommodations and means of transportation often resulted in interesting experience and adventure. Throughout this period certain events are associated with each year, and individuals more or less well known have crossed or lingered on the stage to play their parts.

The summer of 1912 brought Dr. Harry L. Paddon,[268] one of the most notable characters of the Mission, and Miss Mina Gilchrist,[269] a nurse from New England who afterward became his wife. Like Dr. Grenfell, Dr. Paddon came to make the Mission his life work. He arrived in Newfoundland after a stormy voyage from England and stopped briefly at St. Anthony on his way to the Labrador coast, which was to be his field of work. The districts surrounding his headquarters at Indian Harbour in the summer and at North West River in winter are only a part of the field covered by this remarkable doctor on his errands of mercy. For it extends from the Moravian Mission settlements in northern Labrador to the Battle Harbour district in the south: in winter, on trips by dog-team, with their accompaniment of peril and privation to be met with courage, resource and determination; in summer, at Indian Harbour, where with the assistance of his devoted, efficient wife and an occasional helper he ministers to men of the wandering fishing fleet as well as to the families on shore.

Another important event for all of us is associated with that year. This was romance, the marriage of Dr. Little and Ruth Keese (L.M.K.) after three years of happy association. They were married by Mr. Jesse Halsey at the assistant doctor's house, and the union of two persons so beloved by everyone was hailed with joy. The wedding was one of the simplest and most beautiful I ever witnessed, just a small gathering of friends brought together informally in the little room decorated with spruce boughs and yellow-leaved birch branches. Ruth Keese, in her familiar blue linen gown, and Dr. Little, in well known tweeds, entered the house together and after chatting with guests came forward with joined hands to stand before Mr. Halsey when he entered in his clerical gown to perform the short ceremony. In the dining room, Mrs. Grenfell had prepared a beautiful table for the wedding feast, to which all members of the staff contributed. Then, after congratulations and best wishes to the newlyweds, we left them.

Another exciting event of that summer was the arrival at St. Anthony of a beautiful three-masted schooner[270] with a party of Mrs. Grenfell's friends aboard, bound for a trip down the Labrador coast. The ship, donated to the Mission by Mr. George B. Cluett[271] of Troy, N.Y., and built with special reference to its requirements, had been chartered by Mr. W.R. Stirling[272] of Chicago for a Labrador cruise, and this was her maiden voyage. At Boston, she had taken on a cargo of Mission supplies, and her stop at St. Anthony was to deliver them as well as to pay a social visit.

Mr. Stirling's small party, which included his daughter, Miss Dorothy Stirling, Miss Harriot Houghteling[273] and other friends of Mrs. Grenfell, spent their time on shore, and St. Anthony was gay until they sailed away after a brief visit, taking Dr. Grenfell with them as skipper for the northern voyage. One can imagine the enjoyment of such a trip under the Doctor's guidance and the thrilling anticipation of possible hazards from icebergs or shoals along the uncharted coast.

For my own part, the hazards of a voyage on the *George B. Cluett* became my most notable experience of the following year, 1912. In late June, with other Mission personnel – doctors, nurses, collegians going north for a summer of volunteer service – I sailed from Boston for St. Anthony on the schooner instead of taking the conventional route by rail and steamer. The voyage was uneventful, but the last day our calm security vanished as we faced threatened disaster from fire, shoals, fog, heavy seas and finally near collision with an iceberg, a twenty-four-hour experience which, in detail, would be worth telling.

This was also a year of industrial expansion in southern Labrador, with which the name of Minnie Pike,[274] a native woman of Red Bay, will be associated. During a visit at St. Anthony, she learned to weave, then returned to her home with a loom and material to start another branch of the industry in her neighborhood. There are few persons I have known with more pronounced characteristics than Minnie Pike. Her enthusiasm in trying earnestly to solve technical problems of the weaving industry under discouraging conditions is worthy of praise, and a short visit at her home when I was helping her get started is one of my distinct memories of that summer.

The most extended venture of this period was in 1913, when varied objectives, including interests of the Industrial Department as well as my own curiosity and urge for adventure, led me to at last board the *Stella Maris* for a voyage along the uncharted Labrador coast to the far north. Twice before she had eluded me (or I had deserted her), but this time I was on board, having found improved conditions that made passage tolerable. Five other passengers (among them Miss Alice Appleton,[275] at that time teaching school at St. Anthony), the Captain and his wife filled the available accommodations. As before mentioned, the *Stella Maris* had no time schedule or definite ports of call, her objective being delivery of mail to fishermen on schooners that might be found anywhere in sheltered bays or the lee of small islands as far north as Cape Chidley. It is a coast of wild beauty, a savage coast of icebergs, shoals, stormy seas, roaring streams rushing down mist-obscured mountainsides to the sea. Isolated groups of Eskimos, a few Moravian missionaries and Hudson's Bay factors are its only inhabitants. It was an adventurous voyage, accompanied by weather referred to by the captain after landing as "stiddy gales an' seas mon*tain*-eous," a description giving only an inadequate impression of situations I am glad to have experienced now that they are over.

In contrast to this journey of comparative inaction, on board a vessel bound in one direction but with no definite schedule or destination, the summer of 1914 was like a merry-go-round, continued movement and change of location, always with a plan and destination in view. It began with a trip to Montreal for conferences to discuss industrial affairs with friends of the Mission and Dr. Mather Hare,[276] one of its important men, at that time on holiday from his post at Harrington hospital on the Canadian Labrador. After this brief visit, the real movement began: by rail to Gaspé to meet Commander Wakeham,[277] the Canadian fish commissioner, a friend of Dr. Grenfell who had offered to take me on his comfortable little government steamer across the Strait of Belle Isle to Harrington hospital and wait for me there during my inspection of the newly installed weaving room, then on again to Bonne Espérance, where I waited at the hospitable home of Mr. and Mrs. George Whiteley,[278] also friends of Dr. Grenfell, to intercept the little steamer *Home* on her way to Bay of Islands for connection with the cross-country Newfoundland railway to St. John's. There were business affairs to transact during my two days' wait for the *Prospero*'s sailing; then I again started north on the four-day voyage to St. Anthony, arriving just before a British battleship [HMS *Essex*] carrying the Duke of Connaught[279] appeared off the coast and, amidst great excitement, anchored outside the harbor.

The duke, followed by his retinue and the ship's band, came ashore for a brief visit and inspection of the hospital, orphanage, industrial rooms and other Mission buildings, ending with tea at Dr. Grenfell's home. Meanwhile, the band played outside the hospital for the patients' benefit, and the village people gathered at a respectful distance or hung on the fence to gaze with awe at the uniforms and bright instruments as they listened to a real band for the first time in their lives.

There was a brief stop at St. Anthony. Then my movement began again, this time on the *Strathcona* to Red Bay on the Strait of Belle Isle, where Dr. Grieve, who was temporarily in charge at Harrington hospital, met us with the Mission yawl for a brief conference with Dr. Grenfell and took me on board to return with him to Harrington, stopping on the way at St. Augustine to visit a Hudson's Bay post and Indian encampment, an interesting experience and my first contact with Indian life and customs.

It was at Harrington that news reached us by wireless of the outbreak of the war. There were few details, but all boats were ordered to remain in port. There were rumors of submarines off shore, and the frightened natives were looking for Germans on every rock. Uncertainties of future movement were apparent: no steamer stopped regularly at Harrington, only small motorboats, and the problem of return to St. Anthony and the United States under any foreseeable circumstances was disturbing. After a tense interval, boats were again allowed to leave port, and a Church of England missionary travelling in his tiny launch took me to Battle Harbour with stops for the night at native homes by the way. Tenseness persisted, however, after the *Prospero*'s arrival, when we ventured (without lights) across the Strait of Belle Isle to St. Anthony. The *Prospero*'s lights were still covered and her officers and crew alert for submarines when three weeks later I sailed on her out of the harbor after my last year of activity on the coast.

Each of these four summers would in detail be a story in itself, not only because of incidents but personalities closely interwoven with them. In addition to those who already made the background, there were others. Mr. Jesse Halsey, only briefly mentioned at the end of my second year, was a Presbyterian minister who came for a year of volunteer service not only to give spiritual guidance but to install plumbing, a shining example of practical Christianity who dressed in overalls for emergencies and donned his clerical gown when occasion required. A notable visitor was Dr. Joseph Andrews[280] from Santa Barbara, California, who came every summer to meet one of the people's greatest needs – an oculist, a fine man, devoted to the Mission and beloved by everyone, without whose help many on the coast would be without sight.

Other names, some well known, others only of local or transient interest, are memorable. There was Professor Fred C. Sears,[281] representing the agricultural department of Amherst College, a friend of Dr. Grenfell whose interest was garden development. Cecil S. Ashdown[282] and Albert T. Gould,[283] two of the finest men I have known, began their devoted and invaluable services to the Mission during this period. Another name bringing happy memories is Richard Clipston Sturgis,[284] a son of Boston's well known architect, whom I found at St. Anthony on returning from one of my northern trips. His interest and valuable assistance in the weaving room, his co-operation in other branches of the Mission work, his sense of humor, imagination and charming personality made the end of that summer a very happy one to remember.

At St. Anthony, the personnel included devoted and efficient helpers. Miss Katie Spalding[285] came from England to relieve Miss Storr at the orphanage and remained to continue her excellent work and important place in the Mission family. During one summer, the Misses Mary and Sarah Schwall of New Bedford, Massachusetts, came to act respectively as Guest House housekeeper and director of the loom-room during my absence. Mr. Alfred Blackburn[286] arrived from England for secretarial and office work. Ted McNeill, after courses at Pratt Institute, returned to St. Anthony as construction engineer and is doing fine work with increasing responsibility.

Many young volunteers, self-designated "Wops" (workers without pay),[287] arrived from far and near: England, Canada, the United States. My own state was represented by John K.H. Nightingale, Jr., from Providence, Rhode Island. Names with familiar association were frequently recognized. One of the workers on the land was Leverett Saltonstall[288] of Boston, and after grime was removed from a husky young man unloading coal from the Mission schooner one discovered a son of the famous Rockefeller family.[289] Among the group were other sons of well known fathers, individuals with familiar names and students representing many schools and colleges, all doing their bit with cheerful co-operation while gaining unfamiliar but valuable experience.

Among roving visitors bent on pleasure, curiosity, or possible enterprise was Clarence Birdseye,[290] with a plan for extensive refrigeration of fish and, possibly, venison as a business venture, his odd name making as much impression as his business among those who, at the time, were not financially interested. And as for Commander Donald MacMillan, one might count on meeting him unexpectedly on any part of the coast, which I frequently did.

This is the second year of the war. It has taken its toll from those we know at St. Anthony. Archie Ash, one of the first to respond for overseas service, has died in France. Other boys are still in the conflict. Our friend Dr. Armstrong, who was with us a few years ago, has also died in France. The war is far from over, and we dread the future and what it may bring to us all.

Once more during this period, I made a short trip on the *Strathcona*, always memorable for revealing the Doctor's personality and character more clearly, I think, than any other environment. On the deck of the *Strathcona*, on shore with the simple people, in their homes where they gather around him to gossip, to tell their troubles and joys, to ask for advice, legal opinion or solution of their problems, the Doctor is always ready to respond. Often during a call, the gramophone is brought ashore for a concert in a little crowded room, and the visit finally ends with a short prayer meeting. Then we are off again.

I wish all those who have known the Doctor merely by reputation, seen him only on a speaker's platform or at public functions, could share the rare privilege of meeting him under circumstances which, it seems to me, more fully reveal his lovable and dynamic personality, a personality that combines impulsiveness, inconsistency, omissions and sometimes errors of judgment, the minor weaknesses of man, with the attributes of vision, optimism, determination, great courage, interwoven throughout with love of mankind and an abiding faith in God – a character truly great in its sincerity and simplicity. This is "the Doctor" as I have known him.

Some people once asked me what were the essentials for life with the Grenfell Mission. I told them a sense of humor, a good digestion and love of one's fellow men. After several years of pioneer work and four summers of wandering along the coast of Labrador, I would add patience, powers of adjustment, and resignation. With these, one can face the inevitable uncertainties and hazards with a smile. Without them, one would be lost indeed.

Afterword, 1952

In writing this book, I have relived those years so full of vital interest and effort. I see again the faces of my associates in the work, my friends among the people of the coast – I can even remember some of their voices. In reviewing the letters and daily records written at the time, I can see in the perspective of years and the light of later experience certain psychological aspects and reactions of which I was unconscious when events and contacts with the people and environment were actually taking place. As so often happens, understanding comes with perspective.

It was a great experience, this opportunity to share in the service rendered by a great man to those in need of help and encouragement and one that has meant much in my life. The years have passed and with them many changes. On the top of Fox Farm Hill at St. Anthony, overlooking the Mission and the far reaches of wild country and open sea, a large boulder marks the resting place of Sir Wilfred ("the Doctor," as we loved to call him) and Lady Grenfell, and near them lies Dr. John Mason Little, a man beloved by everyone, who gave over ten years of his unselfish, strenuous life and his great ability to the people's service and so inspired their confidence that they almost believed he could restore life to the dead.

Dr. Harry L. Paddon, whose field was the true Labrador and who, like Sir Wilfred, gave his whole life's service to the Mission, has passed, leaving a record of skillful, sympathetic, unselfish years of accomplishment whose importance and influence cannot be estimated. Mrs. Paddon and their son, Dr. William Anthony Paddon,[291] are carrying on, the son following in his father's footsteps. At St. Anthony, Dr. Charles S. Curtis,[292] the logical successor of Sir Wilfred as superintendent, after over twenty-five years of devoted service to the Mission is ably continuing its medical supervision, guidance and development.

Jessie Luther aboard the *Kyle*, en route to St Anthony for one of her summer tours (JLP)

From time to time, a reminiscent word has come to me across land or sea, and I have met face to face some old associates nearer home. But whether they are far or near, I feel that among all who have shared the experience of life in the Mission, and especially those who labored together in the early difficult years, a bond exists that will endure. To those now living who shared this experience and to the families of those who are now a memory, I send greetings.

Jessie Luther
April 1952

Reindeer herd grazes in front of the St Anthony hospital, 1908 (GHS)

Notes

ABBREVIATIONS

AB Bachelor of Arts
ADSF *Among the Deep Sea Fishers*
AMRC American Medical Reserve Corps
B CHIR Bachelor of Surgery
CBE Commander of the Order of the British Empire
CHB Bachelor of Surgery
CM Master of Surgery
CMS Church Missionary Society
HBC Hudson's Bay Company
IGA International Grenfell Association
LFPS Licentiate of the Faculty of Physicians and Surgeons
LOA Loyal Orange Association
LRCP Licentiate of the Royal College of Physicians
LRCS Licentiate of the Royal College of Surgeons
MA Master of Arts
MB Bachelor of Medicine
MD Doctor of Medicine
MRCS Member of the Royal College of Surgeons
OBE Officer of the Order of the British Empire
RAMC Royal Army Medical Corps
RISD Rhode Island School of Design
RN Royal Navy
RNMDSF Royal National Mission to Deep Sea Fishermen
Toilers *Toilers of the Deep*
USAMC United States Army Medical Corps

INTRODUCTION

1 Cartier, *Relations*, 98. See also Messet, "Les rochelais à Terre-Neuve," 137–45.
2 Harrisse, *Découverte de Terre-Neuve*, 319.
3 See Morandière, *Histoire de la pêche*, 1:390–7.
4 John J. Mannion, "Settlers and Traders in Western Newfoundland," in his *Peopling of Newfoundland*, 248–9.
5 "Report Upon the Newfoundland Fisheries, 1872, by Captain A.H. Hoskins, RN, HMS *Eclipse*, and Commander Charles G.F. Knowles, RN, HMS *Lapwing*," *Journal of the House of Assembly* (1873), 723.
6 McGrath, "France in Newfoundland," 47–8.
7 Gobineau, *Voyage à Terre-Neuve*, 230–1. "Under the terms of the treaties, our crews cannot winter on the shore. However, they are obliged when they depart in the autumn to leave behind them, with their houses, a great number of boats that it would be very cumbersome and very costly to take back to France on every trip; then, finally, there are the furniture, the utensils, the nets, the piles of salt. So that all this booty shall come to no harm and not be pillaged during their absence, since time immemorial they have been in the habit of entrusting it to the care of custodians who are English subjects, and who can give rise to no disputes by wintering there" (Gobineau, *A Gentleman in the Outports*, 144).
8 [Feild], *Journal of a Voyage of Visitation*, 96.
9 Chimmo, *Journal of Voyage to N.E. Coast*, 66.
10 Crowdy, "1873 Labrador," 26–7.
11 Thoulet, *Voyage à Terre-Neuve*, 90. "Sailors and officers, French and English, Catholics and Protestants, a Villaret de Joyeuse between a French quartermaster and an English boy seaman, sleep side by side as equals."
12 Kirwin, "Fisheries Synonyms," 10.
13 Hare, "The Climate of Newfoundland," 47.
14 Laut, "Cruising on the French Shore," 385.
15 Grant, *Quaternary Geology*, 15.
16 Murray and Howley, *Geological Surveys of Newfoundland*, 500.
17 Grenfell, "Jottings from Newfoundland," *Toilers* 15 (1900): 107.
18 Grenfell, "Newfoundland and Labrador Jottings," *Toilers* 19 (1904): 191.
19 Grenfell, "At Work in Labrador," *Toilers* 20 (1905): 187–8.
20 Society of Arts and Crafts Archives, Jessie Luther file.
21 Williams, *Mary C. Wheeler*, 193.
22 *Catalogue and Circular of the College for Women*, 16.
23 Ibid., 29.
24 Luther, "Hand-weaving," 39.
25 Boris, *Art and Labor*, xv.
26 Society of Arts and Crafts Archives, Jessie Luther file.
27 Addams, *Twenty Years at Hull-House*, 64.

28 Wynne, "Exhibition of Chicago Arts and Crafts Society, 126.
29 See *Hull-House Bulletin* 5 (1902), passim.
30 "The Hull-House Labor Museum," 60.
31 Luther, "The Labor Museum at Hull House," 2.
32 Washburne, "A Labor Museum," 573.
33 Annett and Lehtinen, *History of Jaffrey*, 1:523.
34 Lears, *No Place of Grace*, 50.
35 Hall, "The Systematic Use of Work as a Remedy in Neurasthenia and Allied Conditions," 30.
36 Northend, "Where Women Work and Rest," 343.
37 She was ultimately given a "B" designation in neuropsychiatry by the American Occupational Therapy Association on the basis of successful experience. See *National Directory of Qualified Occupational Therapists*, 9.
38 Blumer, "The Commitment, Detention, Care and Treatment of the Insane in America," in Blumer and Richardson, *Commitment*, 164–5.
39 Luther, "In Retrospect," 123.
40 Luther, "Hooked Mats," 106.
41 Luther, "In Retrospect," 124.
42 Luther, "Industrial Work," 27.
43 Jessie Luther Papers, Addams to Luther, 6 April 1914.
44 The deletions reveal the level of her distress and disappointment.
45 Jessie Luther Papers, Luther to Wilfred Grenfell [1915].
46 Luther, "Dr Little at Work," 128.
47 Jessie Luther Papers, Wilfred Grenfell to Luther, 20 October 1915.
48 W.R. Stirling Papers, Anne Grenfell to Stirling, 6 August 1916.
49 Ibid., 1 September 1916.
50 Mills, *Discourses of Difference*, 74–7.
51 Schriber, *Writing Home*, 6.
52 Ibid., 75.
53 Public Record Office, London, Army Form B111, William George Lindsay.
54 Thuen, *Quest for Equity*, 24.

JESSIE LUTHER AT THE GRENFELL MISSION

1 The "Ark," named for its immense proportions, was not a sanitarium as such. The rooms were let on an informal basis.
2 Grenfell recorded in *Toilers*, 20 (1905): 157, his visit to Salem on 15 March 1905: "The next day I spoke at Salem by the sea at a gathering in the house of Mrs. Francis Seamans. A large number of invited guests attended, and considerable interest has been manifested as the outcome of this gathering.

The next morning I visited Marblehead, the seat of a large fishery in former days, and still a beautiful harbour, and the resort of the Yacht

Squadron of New York in the summer. Here I was initiated into the art of weaving with hand-looms, and have arranged for a lady who is to come to help me in establishing the craft in Labrador, to come here and take a course with the looms, six of which are being given me to start a work among the women on the coast."

3 The two young women, Clara Koonz and Isabel Harris, were featured on the front page of the *New York Times*, 15 April 1905, where it was reported, "Miss Harris is said to be independently wealthy. She has for years been giving her time to Settlement work exclusively, and is now connected with the Protestant Episcopal Church of the Land and Sea, Henry and Market Streets. Miss Koonz is connected with the Tenement House Commission, and is regarded as an expert in the industrial system."

4 Edmund Burke Delabarre (1863–1945), professor of psychology at Brown University, had written of Grenfell's work, "It is undoubtedly the most important and promising feature connected with life on the Atlantic side of Newfoundland and Labrador, being an economic and moral force of the greatest significance" ("Report of the Brown-Harvard Expedition to Nachvak, Labrador, 1900," *Geographical Society of Philadelphia Bulletin* 3 [1902]: 157–8).

5 No Grenfell Mission office existed as such. Emma E. White (d. 1944), assistant librarian at the Congregational Library from 1888 until her retirement in 1934, had formed a committee with the Rev. C.C. Carpenter to support Grenfell and urged him to come to Boston to raise money. Miss White, as "secretary," conducted Grenfell's business there after hours. It remained his "office" until 1915, when Grenfell opened a separate one at 20 Beacon Street. See Brown, "History of the Boston Office," 1–6.

6 The *Ladies' Home Journal* was a tastemaking journal influenced by arts and crafts style. It is not clear whether the writer was interested in Miss Luther's work or in Labrador crafts.

7 SS *Blue Hill*, 98 tons, 135 feet, built in 1887 at East Boston, Massachusetts.

8 Margaret Warner Morley, *Down North and Up Along* (New York: Dodd, Mead 1903).

9 Alexander Graham Bell (1847–1922), inventor of the telephone.

10 SS *Portia*, 205 feet, 978 gross tons, a coastal vessel of C.T. Bowering & Co., built at Port Glasgow, Scotland, in 1904, with accommodation for 150 passengers. The first master was Captain Abram Kean, who remained in command for the next fifteen years. Her sister ship, SS *Prospero*, was identical. See entry for 25 May 1908.

11 SS *Bruce*, 1,100 gross tons, built in 1897 for the Reid Newfoundland Company's "alphabet fleet." The first vessel to provide a link between the railheads at North Sydney, Nova Scotia, and Port aux Basques, Newfoundland, she was lost in the ice near Louisbourg in 1911.

12 Wreck House, seven miles north of Cape Ray, so named because of the havoc caused by high winds funnelled down from the Table Mountains.

13 SS *Clyde*, 155 feet, 439 gross tons, built at Glasgow for the Reid Newfoundland Railway in 1900.

14 Jabez or Ches Manuel, one of the brothers operating one of the two principal merchant firms in Exploits. The Manuels had previously housed Norman Duncan and maintained a friendship with him.

15 Norman Duncan (1871–1916), Canadian journalist, had arrived in Newfoundland in 1900 in search of Grenfell but did not meet him until three years later in New York. He wrote numerous articles and books on Newfoundland, a controversial novel based on Grenfell's life, *Dr Luke of the Labrador* (New York: Revell 1904), and the more adulatory *Dr Grenfell's Parish* (New York: Revell 1905).

16 Abram Kean (1855–1945), mariner, politician, Orangeman, and temperance advocate, had been acquainted with Grenfell from the beginning. He was later celebrated as the first sealer to take a million seals. See Abram Kean, *Old and Young Ahead* (London: Heath Cranton 1935).

17 The Rev. A. Pittman, Church of England rector of Tilt Cove, on the western side of Notre Dame Bay, site of the first commercial copper mine in Newfoundland. The mine began producing in 1864, but by 1906 it had passed its peak years. See Wendy Martin, *Once Upon a Mine* (Montreal: Canadian Institute of Mining and Metallurgy 1983), 12–15.

18 The French were not permitted to settle on the French Shore, as Miss Luther believed, but were granted landing and drying privileges.

19 Queen's Theological College, sometimes known as St John's College.

20 The Rev. Samuel Dwight Chown (1853–1933) was in fact secretary of the Methodist Church's Department of Evangelism and Social Service. He was later general superintendent of the church, which he led into union with the United Church of Canada in 1925.

21 The Rev. John Thomas Richards (1875–1958), a former fisherman, witnessed Grenfell's arrival in Labrador in 1892. Encouraged by Grenfell to return to school, he subsequently entered the Church of England ministry and began work in the Flower's Cove mission in 1904. Ordained a priest in 1905, he remained at Flower's Cove for forty years.

22 William Lamson Soule (1871–1965) of Waterville, Maine, graduated from Colby College in 1890, then received the MD from Boston University (1896) and Harvard University (1908). He subsequently qualified as an anaesthetist and was clinical instructor in anaesthesia at Cornell Medical School, 1914–27. He retired in 1951.

23 SS *Strathcona*, the first power-driven hospital vessel of the Royal National Mission to Deep Sea Fishermen, was built by Philip & Sons, Dartmouth, and launched in 1899. A vessel 97 feet long and 18 feet at the beam, she

displaced 130 tons and was fitted up with swinging cots and x-ray apparatus, perhaps the first so designed for use at sea.

24 Miss G. was probably interested in the native crafts of Labrador. George Francis Durgin wrote in 1903, "The women do beautiful embroideries in silk on caribou skin and make Ranger seal-skin bags worked with beads, also tobacco pouches, bags of loon skins and moccasins, which the Mission send abroad for sale and offers to the few tourists who venture into these Wilds." See Durgin, *Letters from Labrador* (Concord, N.H.: Rumford Printing 1908), 51.

25 Eleanor Storr (1873–1962) had previously worked with the Bermondsey Medical Mission, London, with her friend Miss Bayley (see n26). Dr Little wrote his mother on 14 June 1908, "Miss Storr has had much experience, has a midwife's certificate and apothecary's and is ready for anything." She remained as superintendent at her own expense until 1914, when she volunteered to go as district nurse at Paul's Island, Quebec, under the direction of Dr Mather Hare. Following the outbreak of war, she did orthopaedic ambulance work in Europe with Sir Robert Jones before returning to St Anthony. A skilled craftswoman, she adapted the Bradford and Pyrford orthopaedic frames to accommodate children. When Grenfell opened an office in London in 1926, in direct competition with the Royal National Mission to Deep Sea Fishermen, Miss Storr enlisted her uncle, Canon V.F. Storr, theological writer and canon of Westminster, to secure a one-room office in Victoria Street. She also brought in a cousin as honorary treasurer and another cousin as honorary solicitor.

26 Ethel Kittie Bayley (1871–1955) practised at the Bermondsey Medical Mission in East London, founded in 1904 by Dr Selina Fitzherbert Fox to work mostly with women and children. The sick were visited in their homes, or they attended clinics in the outpatients' hall. A large number of surgical dressings were performed daily, minor operations were performed once a week, and district maternity work was carried out. All clinics were preceded by a short service of worship, and on Sunday evenings a service was held by the staff, many of whom went abroad to serve in foreign missions. Sister Bayley opened the service with Dr Fox, whom she had known at Wimbledon High School, and stayed with the mission until 1920. Like Miss Storr, she was well prepared for the circumstances in St Anthony.

27 *Warps:* threads to be stretched lengthwise in the loom.

28 The Methodist church was built in 1899–1901, principally by John Moore.

29 This practice is called *raising* or *rising* the tune, the terms being interchangeable in everyday speech.

30 The "after meeting" or "after service" was an occasion for prayer, testimony, and hymn singing following the Sunday evening service. While the practice was not unique to Newfoundland Methodists, the name was.

31 By contrast, Dr Little, writing to his mother on 3 February 1908, had this to say: "The girls and young women in this country are very finely built and generally have not an ounce of fat on them. They are muscular, their bellies hard as boards like a man's – not that their muscles show, but their flesh is very firm. Of course, they all work, and their food, bread and tea with boiled dinner [of] pork or salt beef or fish twice a week does not predispose them to fat."

32 St Mary of Nazareth Hospital, founded in Chicago by the Sisters of the Holy Family of Nazareth in 1894.

33 The sawmill was funded by Lady Roddick of Montreal, near what is now Roddickton.

34 *Bakeapple:* cloudberry (*Rubus chamaemorus*).

35 *Heddles:* small cords on the loom through which the warp is passed.

36 Edward Rainsford Mumford, LRCP, LRCS (Edinburgh), LFPS (Glasgow), had been a student of the Edinburgh Medical Missionary Society and worked with the Livingston Medical Mission, Cowgate. He had served with the RNMDSF for two years before coming to St Anthony. In 1906 he went to India with the Salvation Army and did eighteen years' service, then entered private practice there, still doing medical missionary work.

37 Although this kind of hospitality was legitimate, the RNMDSF Hospital Committee had previously made a rule about visitors. "The Hospitals exist for the relief of the sick," it declared, "and visitors are to be entertained only under exceptional circumstances. Tourists and others visiting the coast for pleasure or business should not be received as boarders at the Hospitals unless they are stranded, and in actual need of accommodation." See RNMDSF, Hospital Committee Minutes, 8 July 1904.

38 Florence Bailey (d. 1952), the first nurse to take charge of a nursing station for the Grenfell Mission, came to Labrador from the Mildmay Mission Hospital in East London, where a range of social services were provided. A skilled midwife, she left Battle Harbour for Forteau and handled all the obstetrical cases on the Labrador coast.

39 At this time, the RNMDSF council feared that the Labrador mission was expanding too rapidly and was likely to become permanent. It recorded that "the danger we could not help feeling existed from help being identified with the name of Dr. Grenfell rather than that of the Society." The council also noted that Grenfell's rule against visitors was being observed by everybody except him: distinguished visitors and teachers seemed to be welcome, and Yale and Harvard students were turning up without any clear purpose. See RNMDSF, Council Minutes, 29 September 1905.

40 John Bryant (1880–1935) of Cohasset, Massachusetts, had been educated at Milton Academy and Harvard University, where he graduated in 1903. Now a medical student, he would receive his medical degree in 1907 and return to St Anthony as a physician in 1908. He would be best man at

Grenfell's wedding (see *New York Herald*, 19 November 1909). During the First World War, Bryant served in the office of the surgeon-general with the rank of major, his specialty being gastroenterology, especially problems of convalescence, and he was president of the American Gastro-Enterological Association, 1933–35. See *Boston Traveler*, 19 September 1935.

41 This entry appears in the manuscript outside the chronological sequence.

42 The hymns "Jesus loves me, this I know," by Anna B. Warner, and "Hold the fort, for I am coming!" by P.P. Bliss.

43 *Quintal:* measure of salted or dried cod equalling 112 pounds.

44 Elizabeth Patey, wife of Edmund Patey (1860–1929), whose property was adjacent to the Grenfell Mission.

45 Isabella Ash, wife of Albert Ash (1851–1934), later skipper of Grenfell Mission vessels.

46 Reuben Simms (1864–1937).

47 SS *Virginia Lake*, built in 1882, maintained a mail service between St John's and St Anthony. From 1901, it had been used in the seal fishery, but during the 1909 season it caught fire and sank.

48 SS *Home*, 155 feet, 439 gross tons, built at Glasgow for the Reid Newfoundland Railway in 1900.

49 *Turkey red:* a brilliant colour produced on cotton with a long process using alizarin or madder.

50 *Machine:* term applied to any unnamed object or phenomenon.

51 A hymn by Anna L. Coghill.

52 *Mug-up:* a casual cup of tea and snack.

53 The mail boat used to dock at Boyd's wharf.

54 Mark Penny (1887–1968). Miss Luther added in *ADSF* (April 1909): 31–2, "Some really beautiful pieces of pottery have been made by Mark Penny, a young Newfoundlander, whose physical condition would not allow him to fish, and who showed ability from the beginning. The pieces he has made are decorated with designs of local interest – seal, bear, reindeer, fish, etc., and are well shaped and finished, though not glazed – that must come later." Penny was later sent by Grenfell to study at Bishop Feild College, St John's, for three years and received his diploma in 1916. That same year, he was granted a teaching certificate and began a career in Labrador communities. In 1919, he married Elsie (Bradbury) Croucher, a widow of twenty-two, in Cartwright. As a lay reader of the Anglican Church, he was also granted a licence to marry.

55 William Lansing Hodgman (1854–1936), president of the Title Guarantee Company of Rhode Island and linked to numerous other corporate and philanthropic institutions. He was also a trustee of the Butler Hospital and a member of the corporation of the Rhode Island School of Design.

56 William Henry Grueby (1867–1925), distinguished New England potter, who began making architectural *faïence* in Revere, Massachusetts. After encountering the French *flambé* technique at the Chicago world's fair in 1893, he established Grueby Faïence in South Boston, where he produced architectural *faïence*, terra cotta tiles, and Chinese ceramics. In 1897, he discovered the matt glazes for which he became best known and began to produce art pottery in the Arts and Crafts idiom. His art pottery carried the Grueby Pottery mark, the Grueby Faïence mark appearing on architectural *faïence*.

57 *Uncle*: like *skipper*, a term of respect accorded to older men in Newfoundland.

58 Charles Wilson Brown (1874–1940?), associate professor of geology at Brown University. While studying for the master's degree at Brown, he had been assistant to Professor Alphaeus Packard, professor of zoology and geology and author of *The Labrador Coast* (New York: N.D.C. Hodges 1891) and had arrived at Brown in 1905 to replace him.

59 The Church of England cathedral of St John the Baptist, rebuilt after the fire of 1892.

60 Perhaps the cruiser HMS *Calypso*, permanently based at St John's for the training of the Royal Naval Reserve.

61 The Roman Catholic cathedral of St John the Baptist, consecrated in 1855, where the funeral of Daniel Murphy, formerly of the Royal Naval Reserve, took place. The gun carriage was hauled by naval reservists. The band, of which Mr Murphy had also been a member, belonged to the Catholic Cadet Corps. See *Evening Telegram*, 4 November 1907.

62 William H. Peters (1875–1957) of Peters & Son, manufacturers' agents, local secretary of the RNMDSF. The Mission maintained an office at 347 Duckworth St, St John's.

63 Sir William MacGregor (1846–1919), governor of Newfoundland 1904–9. A physician by training, MacGregor had taken a special interest in Grenfell's measures to combat tuberculosis and had visited St Anthony in 1905. At the time, he was encouraging Grenfell in his scheme to import a herd of reindeer. See "Report of an Official Visit to the Coast of Labrador by the Governor of Newfoundland, During the Month of August, 1905," *Journal of the House of Assembly* (1907), 300–69.

64 Herbert Anthony Sawyer (1847–1918), retired colonel with thirty-three years' service in the Indian Army, had been commandant of the Bengal Infantry. He was involved in the Burmese Expedition, 1887–89, and had fought in the defence of the Malakand and the relief of Chaldarra Fort, 1897. Sawyer married Lucy Hutchinson, a grandchild of Grenfell's uncle, George Hutchinson, his mother's brother. Sawyer's daughter Justine later married Martyn Spencer (see n138), another grandchild of George Hutchinson, who arrived at St Anthony from New Zealand in 1908.

Dr Little wrote his mother on 13 October 1907, "There is an English colonel retired here, a cousin of Dr Grenfell's, who has come out to help and possibly start a social settlement and mill. He is very charming and a great addition to the company."

65 Hymns by the American revivalists Dwight Lyman Moody (1837–99) and Ira David Sankey (1840–1908).

66 Like the prodigal son of the parable in Luke 15:11–32.

67 Norman Burgess Stewart (1878–1951) of Maybole, Ayrshire, educated at the Royal High School, Edinburgh, and the University of Glasgow, where he received the MB and CHB in 1903. After becoming house surgeon at the Glasgow Royal Infirmary, he joined the RNMDSF as a medical officer and with his wife Janie, a nurse, divided his time between St Anthony in the winter and Indian Harbour, Labrador, in the summer throughout 1906–10. Two daughters were born at St Anthony. Stewart joined the English Presbyterian Church Mission at Wukingfu, Swatow, in 1911 and during the First World War served with the British Expeditionary Force in Macedonia and Bulgaria as a major in the RAMC. He returned from China in 1925 and settled at Edinburgh.

68 *Guest House:* Miss E.E. White claimed that this was an adaptation of a Moravian term, *Gästehaus*, "the home of those who are in one way or another helping in the work and are able to find here a place to call home ..." See E.E. White, "A First Visit to the Mission in Newfoundland and Labrador," *Toilers* 24 (1909): 40.

69 Ruth Esther Keese (1884–1977) of Chelsea, Massachusetts, educated at Cushing Academy, Ashburnham, and the Vermont State College. Miss Keese had hearing difficulties resulting from a bout of typhoid fever. After marrying Dr Little, she returned to the United States and, following the death of her husband in 1926, took her children back to Ashburnham, where she built a replica of a colonial house in which she lived until her death.

70 Loula Esdale Kennedy (1874–1958) of Middletown, Virginia, typifies the kind of public-minded woman recruited by Grenfell. After attending the Girls' Latin School in Baltimore, she received the AB from Goucher (then Woman's) College in 1896 and graduated from the Johns Hopkins School of Nursing in 1903. Miss Kennedy was public health nurse at the Henry Street Settlement in New York, 1907; nurse-in-charge at St Anthony hospital, 1907–10; assistant supervisor of nurses at Johns Hopkins, 1911–12; educational secretary of the Tuberculosis League of Pittsburgh, 1912–13; public health nurse, Clarksburg, West Virginia, 1913–18; instructor at the Army School of Nursing, Fort Mead, Maryland, 1918–19; acting director of Simmons College School of Public Health Nursing, 1919–21; head of nursing education at Johns Hopkins Hospital, 1921–44, and student counsellor 1944–46.

71 Cuthbert Crumett Lee (1891–1971) of Hyde Park, Massachusetts, educated at the William Penn Charter School, Philadelphia. See his verse "Off Fortune Head," ADSF (July 1914): 44; and his memoir *With Dr Grenfell in Labrador* (New York: Neale Publishing 1914). After graduation in 1912, Lee entered the advertising department of the John Lane Co., New York, and served in France as an officer of the U.S. Army, 1918–19. He subsequently studied trust administration at New York University, became a trustee, and wrote a manual of trust administration widely used by American banks and trust companies. A keen portraitist, he also published two studies of American portrait painters.

72 The Rev. C.H. Brown, Methodist minister.

73 Reindeer introduced by Dr Sheldon Jackson, a Presbyterian missionary. See Gilbert H. Grosvenor, "Reindeer in Alaska," *National Geographic* 14 (1903): 127–48; and Jacobs, ed., *A Schoolteacher in Old Alaska.*

74 The reindeer did eventually breed with the caribou, importing thereby the deadly infection cerebrospinal elaphostrongylosis, caused by the nematode parasite *Elaphostrongylus rangiferi.* See M.W. Lankester and D. Fong, "Protostrongylid nematodes in caribou *Rangifer tarandus caribou* and moose *Alces alces* of Newfoundland," *Rangifer* 10 (1998): 73–83.

75 The Anglo-Newfoundland Development Company was formed by Alfred Harmsworth, Viscount Northcliffe (1865–1922), and his brother Harold Harmsworth, Viscount Rothermere (1868–1940), who decided to build a paper mill at Grand Falls in 1904 but did not open it until 1909.

76 For a full account of the preparations and the contract, see Wood, "Reindeer, Dr Grenfell, and Labrador," 176–8.

77 John Mason Little (1875–1926), son of John Mason and Helen Beal Little of Swampscott, Massachusetts, was educated at the Noble and Greenough School and graduated from the Harvard Medical School in 1901.

78 A device made by the Aeolian Company of New York, a firm of player piano manufacturers who also made automatic organs. The self-playing mechanism was almost always mounted in a separate "attachment" about the size of a sideboard.

79 Bioluminescent dinoflagellates. A wide range of marine organisms are capable of producing light in surface waters.

80 Daylight hours are considerably longer at this time of year. Miss Luther is probably referring to hours of direct sunlight.

81 Not an unusual home remedy for relieving the symptoms. See Crellin, *Home Medicine,* 169.

82 G.A.A. Jones, Grenfell's secretary. The Rev. Edward Moore wrote in his journal, 26 July 1905, that Grenfell had taken on Jones for Jones's sake, "rather than for the help which he himself so much needs."

83 Hugh Henry Wilding Cole (1883–1960) of Farnham, Surrey, educated at Rugby, joined the Anglo-Newfoundland Development Company in 1905 and was subsequently logging superintendent.

84 A "common" or "amalgamated" school made possible by the Education Act of 1903, which provided for the sharing of resources where different denominations agreed.

85 *Canfield:* a type of solitaire arranged as a competition for two players or more.

86 Alf Slade, Grenfell's dog team driver.

87 In the poem "Excelsior," Henry Wadsworth Longfellow writes of a youth "who bore, 'mid snow and ice, / A banner with the strange device, / Excelsior!"

88 The blackened face was one of the disguises adopted by Newfoundland mummers. See J.D.A. Widdowson and Herbert Halpert, "The Disguises of Newfoundland Mummers," in Halpert and Story, eds., *Christmas Mumming in Newfoundland,* 148–9.

89 This chapter is missing from the typescript. It was borrowed by the International Grenfell Association for publication and appeared as "Landing of the Reindeer," *ADSF* (January 1954): 126–8; (April 1954): 29–31, the text followed here.

90 Drusella Simms (1868–1938).

91 Morris Sundine, Swedish mechanic, one of a second generation of Swedish families brought by Lewis Miller to Lewisporte and Millertown for logging operations.

92 *Grand:* not "big" or "expansive" but a weak exclamation of general approval. See this usage in entries for 6 January, 3 February, and 15 September 1908.

93 On 31 May 1907, in an audience with Edward VII, Grenfell had received the insignia of a companion of the Order of St Michael and St George.

94 The MD (*honoris causa*) was conferred on Grenfell by Oxford University on 28 May 1907, the first honorary MD so conferred.

95 Wood, "Reindeer and Labrador," 41–3. For an account of the reception of the Lapp family at Newcastle en route, see Jarvis, "Laplanders for Labrador," 255–6.

96 The letter was published as "Grenfell's Reindeer Land," *Boston Evening Transcript,* 25 January 1908, before it was reprinted in *ADSF.*

97 Miss Luther may not have this correct. *Wonderful* in Newfoundland usage is sometimes an intensifier requiring an adjective to complete the sense. Sometimes it may mean "amazing, extraordinary, extreme." It is one of the Newfoundland words easiest to misinterpret as arousing wonder or curiosity but not enjoyment or approval. This more ancient sense, in educated usage on both sides of Atlantic, has since become diluted to mean "really good" or "thoroughly approvable." See, for example, entry for 7 November 1908.

98 The firm of James Henry Biles (1841–1915), an Englishman who had jumped ship and later became one of the first St Anthony fish merchants. Biles married Ellen Janes (1846–1913) of Carbonear and fathered twelve children. A picture of the Biles premises appears in Biles, *My Town, My Province,* 14.

99 Maude Biles (1889–1908), daughter of James Henry and Ellen Biles.

100 The Loyal Orange Association, force of conservatism. See Philip Hiscock, "The Loyal Orange Association in Newfoundland," in Burford, *The Ties That Bind,* 129–34.

101 Alfred, Lord Tennyson, "The Brook" (1855): "For men may come and men may go, / But I go on for ever."

102 *Nunny-bags:* knapsacks of sealskin, burlap, or canvas used for carrying food and personal equipment.

103 Either Joseph or Frederick Moore of J. & F. Moore, merchants.

104 Edgar "Ted" McNeill (1884–1970), formerly of Island Harbour, Labrador, had received some formal schooling at the Moravian boarding school, Makkovik, and learned basic carpentry and drafting from the Rev. Hermann Jannasch. He was later trained at the Pratt Institute and subsequently became designer and superintendent of the Grenfell Mission buildings. In 1955 he was elected mayor of St Anthony.

105 Psalm 148:10 and Isaiah 54:13.

106 Cecilia Williams had first arrived in Labrador with Grenfell in 1893, when she was nurse at Indian Harbour.

107 Grenfell, in fact, was tone deaf.

108 *Boston Evening Transcript,* 25 January 1908.

109 Wilfred Grenfell, "A Man's Faith," *Congregationalist and Christian World* 93 (1908): 726–8, 771–3, 820–2.

110 Frank Stockton, *The Casting Away of Mrs. Lecks and Mrs. Aleshine* (New York: Century 1886).

111 See Cole, Diary of reindeer trek, Memorial University of Newfoundland, Folklore Archive, MS 77–61.

112 Booker T. Washington (1856–1915), black American leader.

113 Mary Lizzie Macomber (1861–1916), New England painter of decorative symbolic panels.

114 Mary Baker Eddy (1821–1910), founder of the Christian Science Church.

115 William Jerome and Jean Schwartz, "Mister Dooley" (Chicago and New York: Shapiro, Bernstein 1902). It begins:

> There is a man that's known to all, a man of great renown,
> A man whose name is on the lips of everyone in town.
> You read about him every day, you've heard his name no doubt,
> And if he even sneezes they will get an Extra out.

116 *Pitch down:* reduce; *find:* suffer from.

117 The Rev. Georges-Abner Thibault (1874–1934), Acadian priest born at Salmon River, Nova Scotia. He was ordained at Harbour Grace, Newfoundland, in 1906 and appointed pastor at Bonavista but was sent to Conche in 1907. He served there for over sixteen years, then spent the balance of his pastorate in Conception Bay.
118 The Rev. C.H. Brown, Methodist minister.
119 "Dr Grenfell's Log" was the running title of a series of reports contributed by Grenfell to *ADSF*.
120 This would soon become the site of Grenfell's house.
121 Orange Lodges typically held annual entertainments to which the whole community was invited. The day's activities might consist of a church service, a parade, and a public supper. These events were properly held on Orangemen's Day, 12 July, but in smaller communities where men were away at the fishing grounds during the summer, the event might be held nearer Christmas or on New Year's Day. The boys, in this case, were probably members of the Orange Young Britons. See Halpert and Story, *Christmas Mumming in Newfoundland*, 53.
122 See entry for 20 December 1909.
123 The first line of the chorus to the popular sentimental ballad "In the Baggage Coach Ahead," by Gussie L. Davis (New York: Howley, Haviland 1896).
124 Like the detective agency founded by Allan Pinkerton in the 1850s.
125 W.T. Grenfell, "The Ocean Mammals," in Grenfell et al., *Labrador: The Country and the People*, 352–73.
126 Psalm 23:4 and Psalm 90:10.
127 *Civil weather:* with no wind.
128 Grenfell readily gave away his own clothing to those who needed it.
129 George Reid, fifty-eight, and his son Bill; the brothers William and George Andrews and their nephew, Levi Dawe.
130 From a hymn by Charlotte Elliott, which begins:
My God, my Father, while I stray,
Far from my home, on life's rough way,
O teach me from my heart to say,
Thy will be done.
131 Dr Little wrote his mother on 14 May 1908, "His hands and feet were frozen and he looked twenty years older. I had to give him 30 grams of bromide and ½ gram of morphine that night to make him sleep. He is all right now again. This is one of the little incidents of traveling and treating the sick up here. The opinion of all here is that any other man would have perished."
132 *Steady:* a small pond or a stretch of still water in a river.
133 Wilfred T. Grenfell et al., *Labrador: The Country and the People*, to which Grenfell contributed ten chapters.

134 Laura Elizabeth (Howe) Richards, *The Golden Windows: A Book of Fables for Young and Old* (Boston: Little, Brown 1903) and *The Silver Crown: Another Book of Fables* (Boston: Little, Brown 1906).

135 *Cossacks:* winter footwear created by cutting the tops off a pair of long rubber boots.

136 Probably for "A Voyage Perilous: Thrilling Experience of Labrador's Mariner Missionary," *New York Tribune Sunday Magazine,* 2 August 1908, 3–4, 15; or *Adrift on an Icepan* (Boston and New York: Houghton Mifflin 1909), to which Miss Luther contributed the last part.

137 A reference to the brief drinking song at the beginning of act 3 in W.S. Gilbert's satiric farce *Engaged* (1877), which begins, "Says the old Obadiah to the young Obadiah."

138 Charles Martyn Spencer (1887–1977), son of Grenfell's aunt, Evelyn Hutchison, and the Rev. Frederick Hamilton Spencer, a New Zealand missionary. Spencer visited England with his family in 1905–6 and met Justine Sawyer, later his wife. He then entered Macdonald College, McGill University, graduating with a degree in agriculture in 1911 before returning to New Zealand and establishing a fruit orchard at Henderson, near Aukland. After serving with the New Zealand forces at the Somme and Ypres, he married in England and resumed fruit farming until the late 1930s.

139 John Grieve received the MB and CHB (Edinburgh) in 1904. He served at Battle Harbour at various times between 1906 and 1915 and at Harrington Harbour, 1913–14. He was appointed secretary and business manager of the mission office in St John's in 1918.

140 Paul Matteson (1884–1978) of Providence, Rhode Island, had been educated at The Hill School, Pottstown, Pennsylvania, then graduated from Brown University in 1906. After spending a year at St Anthony, he returned to Brown in 1909–10 to study geology and botany. On 15 February 1910, he gave an address to the Providence Franklin Society on his year in Newfoundland and Labrador, focusing on the destitution and disease of the north. A Providence newspaper reported, "The speaker declared that after a year's sojourn in the Far North, he was at a loss to understand what ever brought human beings to these two countries, in which he could see nothing to attract and whose future seemed hopeless." In 1911–12, Matteson was at the Harvard Law School. He remained in Cambridge and spent most of his legal career in banking.

141 Frank E. Hause, twenty-two, of Interlaken, N.Y., a graduate of the Pratt Institute's Applied Electricity course in June 1908. After Grenfell had lectured at the institute in November 1907, a group of students from the second-year class tested a generator at the factory of August Mietz in New York. This apparatus, together with a switchboard and electrical supplies, was used to provide light and power to the mission buildings.

142 D.M.H. Cushing, volunteer electrical engineer.

143 The name Holley is rare at St Anthony but prevalent in Labrador, notably at Fox Harbour.

144 Arthur W. Wakefield (1876–1949) of Kendal, Westmoreland, son of William Wakefield of the Wakefield and Crewdson banking house. The Wakefields were Quakers. Arthur was educated at Sedbergh School, Yorkshire, and admitted a pensioner at Trinity College Cambridge in 1895, proceeding to BA in 1898, MA, MB, and B CHIR IN 1906, and MD in 1909. At Cambridge, he won blues in boxing and cycling but interrupted his medical training to fight in the Boer War as a trooper, then qualified as MRCS and LRCP in 1904. That same year, he was the United Hospitals heavyweight boxing champion, captained the London Hospital rugby team, and established a mountain-climbing record on Sca Fell Pike. He then studied surgery with Lister at the University of Edinburgh and at Heidelberg, speaking fluent German, and served in the North Sea with the RNMDSF before coming to Labrador, 1908–14, as a voluntary medical officer. While working in Labrador, Wakefield disappeared in the summers to climb in the Rockies. Passing through Montreal, he met his future wife, whom he married in 1910, and his eldest son was born at Forteau. He subsequently served in the RAMC, 1914–18, and was mentioned in dispatches. He joined the Everest Expedition in 1922, then settled at Keswick as a general practitioner.

145 *Labrador tea:* evergreen of the genus *ledum.*

146 Francis Bowes Sayre (1885–1972), an undergraduate at Williams College and the son of Robert W. Sayre, vice-president of Bethlehem Steel, typified the kind of genteel American Grenfell liked to recruit. Sayre had met Grenfell the previous year during his lecture tour. "No one I had ever met had impressed me so deeply," he wrote (Sayre, *Glad Adventure,* 3). At his graduation in 1909, Sayre was valedictorian when Grenfell received an honorary degree. He then spent the summer aboard the *Strathcona* as Grenfell's secretary and general assistant, and in August Grenfell put him aboard the *Roosevelt* as secretary to Peary (see n158). In November, Sayre was to act as usher at Grenfell's wedding. Clearly, Grenfell had put pressure on Sayre to join him in the administration of his work, but after a period of uncertainty he declined. In 1913, he married Jessie Wilson, daughter of the U.S. president. This time, Grenfell was best man, and Scoville Clark (see n153), another Grenfell volunteer, was one of the ushers. Sayre subsequently pursued a long career as lawyer, professor, author, and diplomat. He was, among other things, assistant secretary of state in the administration of Franklin D. Roosevelt, high commissioner to the Philippines, and chairman of the Trusteeship Council of the United Nations.

147 John Nevin Sayre (1884–1977) had graduated from Princeton in 1907. In 1911 he received a degree in divinity from the Episcopal Theological School, Cambridge, Massachusetts, and was instructor in biblical litera-

ture at Princeton, 1911–15. A priest of the Episcopal Church, he is perhaps best known as a leader in the movement for world peace through the ecumenical Fellowship of Reconciliation. He was active with Roger Baldwin in the founding of the American Civil Liberties Union.

148 Mary Lane Dwight, a schoolteacher from Roselle, N.J., had taught at Mrs. Dow's School, Briarcliff Manor, N.Y. She taught at West St Modeste, Labrador, during the summer of 1908, and an account of her teaching experience occurs in *Children of Labrador* (Edinburgh and London: Oliphant, Anderson & Ferrier [1913]), 68–77.

149 The mission installed 220 volt DC. Lights came on at dusk and went off at 10 PM.

150 *Passe-partout:* picture frame consisting of two pieces of glass fastened together.

151 George Andrews (1888–1973), originally of Ireland Bight.

152 Added as an appendix to Grenfell's own account, *Adrift on an Icepan.*

153 DeWitt Scoville Clark Jr (b. 1887) of Salem, Masschusetts, was still an undergraduate at Yale, where he was busy with sports and missionary activities. Obviously influenced by Grenfell, his expressed intention was to be a medical missionary. Clark received the MD from Harvard in 1913.

154 The Spanish-American War of 1898, in which modern steel-hulled ships destroyed the Spanish fleet in Cuba.

155 Four Yale undergraduates, after hearing a lecture by Grenfell the previous fall, had volunteered to sail *Pomiuk* to Newfoundland under the command of Captain Laurie Hayes. They consisted of John T. Rowland (see n241), Robert Carpenter (who later qualified in medicine at Tufts Medical School), Scoville Clark (see n153), and Sheldon Yates (see n156), as well as Charles E. Richardson (who later qualified as an engineer). See Rowland, *North to Adventure*, 50–73.

156 Sheldon Smith Yates (b. 1887) of Yonkers, New York. A classmate of Scoville Clark, Yates was active in sports at Yale and spoke to groups in neighbouring towns to promote Grenfell's work. He later worked in the Far East and in South America before associating himself with a series of manufacturing firms.

157 Sidney Douglass Palmer (1884–1954) of Oak Park, Illinois, undergraduate of Williams College, where he was a fraternity brother of Francis B. Sayre. Palmer first served at St Anthony in 1909 while still an undergraduate, and the following year he lectured extensively on the Grenfell Mission to religious organizations. See Douglass Palmer, "The Doctor of the Labrador," *Outlook* 91 (1909): 710–19, with photographs by Sayre. As a student of Cornell Medical College, he again volunteered throughout 1909–14, particularly with Dr Mather Hare at the mission hospital at Harrington Harbour, Quebec, and graduated in 1915. See also Douglass Palmer, "With Dr Grenfell in Labrador," *Outlook* 95 (1910): 848–9. In

1917 he was sent to France as a medical officer with the 108th Infantry Regiment and was decorated for bravery at the Somme. He then began private practice in New York as a gastroenterologist, becoming one of America's foremost authorities on tropical diseases.

158 Robert Edwin Peary (1856–1920), civil engineer of the U.S. Navy and arctic explorer. He had left New York on 6 July 1908 on his last attempt to reach the North Pole and passed through the Strait of Belle Isle, sending a boat ashore at Point Amour Lighthouse with telegrams before proceeding up the Labrador coast.

159 *Lorna Doone:* two-masted schooner, 76 feet and 73 gross tons, built at Essex, Massachusetts, in 1887.

160 *Jigging:* method of fishing whereby an unbaited, weighted hook is jerked sharply upwards where fish are swarming.

161 *Blake:* two-masted schooner, 81 feet and 91 gross tons, built at Shelburne, Nova Scotia, in 1902.

162 After McNeill and Ash enrolled in the machine construction course, McNeill left in June 1909 with a certificate, Ash without one. However, Ash returned in 1911–12 for a course in tanning and again in 1912–13 for a course in applied electricity.

163 The ceremony and sense of mutual aid which characterized activities of the LOA extended to the members' funerals and the tending of their graves.

164 Probably William Duff Reid (1866–1924), vice-president and general manager. The Reids had bought the steam yacht *Fedelma,* 108 feet and 96 gross tons, and renamed it *Fife* as part of their Alphabet Fleet.

165 One of numerous writing manuals with this or a similar title.

166 Probably Francis Weld Peabody (1881–1927), Harvard medical graduate, 1907, finishing an internship at the Massachusetts General Hospital.

167 Solomon Dean (1847–1939), originally of Carbonear, Conception Bay, captained one of the vessels of J. and F. Moore Ltd and perhaps managed the store.

168 George Ford (1857–1918), highly respected interpreter, postmaster, and clerk of the HBC post at Nachvak, Labrador, virtually from 1877 until 1908, when he retired and found employment with the Grenfell Mission. He returned to Labrador briefly as HBC clerk in 1916 but died suddenly of the Spanish flu in 1918.

169 The general election of 2 November 1908.

170 Walter Seymour Armstrong qualified as MRCS, LRCP in 1903. He worked as a volunteer, spending one winter at Battle Harbour, one at Forteau, and one at St Anthony. He was killed in action in Europe in 1916.

171 *Make:* preserve by salting and drying.

172 See William Forbush, *Pomiuk, a Prince of Labrador: A Brave Boy's Life for Brave Boys* (London: Marshall Brothers 1903), and Wilfred T. Grenfell,

"Little Prince Pomnik [sic]," in *Off the Rocks* (Philadelphia: Sunday School Times 1906), 33–48.

173 For a discussion of this practice, see Curtis M. Hinsley, "The World as Marketplace: Commodification of the Exotic at the World's Columbian Exposition, Chicago, 1893," in Karp and Lavine, eds., *Exhibiting Cultures,* 344–65.

174 The Rev. Charles Carroll Carpenter (1836–1918) of Amherst, Massachusetts, a former Congregationalist missionary in Labrador, 1858–65, and one of Grenfell's earliest supporters in the United States.

175 *Lop:* a rough sea marked by short waves.

176 This entry is dated 22 September in the manuscript.

177 *Biscuit firing:* the firing of ware at a temperature lower than that required for glaze firing, the lowest possible being at about 750° C.

178 This entry is dated 30 September in the manuscript.

179 Jane (Hutchinson) Grenfell (1832–1921) lived at Mostyn House School, Parkgate, Cheshire, where Grenfell's brother Algernon was the headmaster.

180 The Rev. Edward Caldwell Moore (1857–1943), formerly Congregationalist pastor in Providence, Rhode Island, author, and professor of theology and history, had spent the summer of 1905 on the coast with Grenfell.

181 Abbott Lawrence Lowell (1856–1943), elected president in January 1909.

182 Anna Elizabeth Caldwell MacClanahan (1886–1938) had graduated from Bryn Mawr in 1906 with a degree in politics and economics. The engagement occurred on 23 July. See *New York Times,* 24 July 1909.

183 Mr. Waldron is mentioned in Peggy Hitchcock's letter to Jessie Luther, 3 May 1951, but not identified elsewhere.

184 The Rev. Edmund James Peck (1850–1924), grammarian of Inuktitut, translator, and editor, had served in the RN before entering the CMS and was sent to the north in 1876. Peck wrote Grenfell as early as 1900 to request the use of the *Strathcona* for transport to and from his mission post.

185 Peck had sailed in July with J.W. Bilby and Archibald Lang Fleming (later the first Anglican bishop of the Arctic) aboard the *Lorna Doone,* commanded by Captain Albert Fradsham (see n193). Fleming wrote of his colleagues: "[Peck] was a short, stocky, bespectacled little man with an abundance of the most beautiful white hair and a heavy white beard. But I did not make the acquaintance of Bilby until we arrived in St. John's just prior to our sailing. Naturally I was somewhat curious about him. He was an Englishman of average height and well built, with heavy chin and nose, good brow, thick hair greying in the temples, and a black mustache. His nine years' experience at Blacklead Island Mission (the forerunner of the present Pangnirtung Mission) with Mr. Peck and Mr. [E.W.T.] Greenshield had given him an intimate knowledge of the Eskimo and their language" (Fleming, *Archibald the Arctic,* 37). See also Julian W. Bilby, *Among Unknown Eskimo* (Philadelphia: J.B. Lippincott 1923).

186 K.D. Forbes worked in the St Anthony office and store, 1910–12.
187 Alice MacNair, kindergarten teacher at St Anthony, 1909–10. See Alice MacNair, "The St Anthony School," *ADSF* (April 1910): 21–2.
188 *Drawn work:* a method of ornamenting fabrics by drawing out certain threads and fastening the remaining ones with fancy stitches or by drawing them to one side and replacing them with others of different colours.
189 Frederick Albert Cook (1865–1940), physician and explorer, had accompanied Peary on two previous arctic expeditions. Cook claimed on 1 September 1909 that he had reached the North Pole on 21 April 1908, five days before Peary. He later wrote, "Following Mr. Peary's return, I found myself the object of a campaign to discredit me in which, I believe, as an explorer, I stand the most shamefully abused man in the history of exploration. Deliberately planned, inspired at first, and at first directed, by Mr. Peary from the wireless stations of Labrador, this campaign was consistently and persistently worked out by a powerful and affluent organization, with unlimited money at its command, which has had as its allies dishonest pseudo-scientists, financially and otherwise interested in the success of Mr. Peary's expedition" (Cook, *My Attainment of the Pole,* 8–9).
190 Grenfell records his meeting with Peary in *A Labrador Doctor,* 339–42. He wrote in the *New York Times,* 13 October 1909, "To us up here the whole controversy is a mystery – now we are persuaded one way, another day the other way. But apparently when we reach civilization we shall learn the truth from the examination of the documents and records – we can only hope now that all parties are right."
191 Warland Wight (d. 1954), employed at Indian Harbour and St Anthony, 1910–11.
192 John J.A. Evans of Philadelphia was in St Anthony 1910–14. He acted as Grenfell's secretary until 1911 and was placed in charge of the reindeer herd, 1911–12.
193 Albert Fradsham (1883–1941), a native of Cornwall who later settled at Coley's Point, Newfoundland. After serving as an able seaman in the Royal Newfoundland Naval Reserve, he acquired his master's certificate and commanded the *Lorna Doone.* Fradsham had sailed to Baffin Land the year before. Archibald Fleming described him as an "Elizabethan buccaneer" who eschewed navigational instruments. "He worked entirely on dead reckoning," wrote Fleming, "and like all the schooner men was never happy out of sight of land" (Fleming, *Archibald the Arctic,* 48). Fradsham subsequently accompanied Grenfell on numerous voyages in the *Lorna Doone.* He finally settled in Corner Brook, where he died.
194 The subject of the accumulative rhyme beginning "This is the house that Jack built."
195 See, for example, "Dr Grenfell and Bride Go to Labrador on Honeymoon," *New York Evening Telegram,* 19 November 1909: "In the section to

which the young couple will journey the sun's rays seldom are seen shining over the barren landscape, and there are no comforts even such as poor people in this part of the world are able to enjoy. It will be a remarkable transition for Mrs. Grenfell, who has always been surrounded with luxury and every comfort that wealth could buy. Mails are delivered only at intervals of months, so that the bride will be out of communication with her friends for long stretches at a time. She will be forced to subsist on the crudest kind of food. There is no vegetation in the country, and fresh food supplies are not often received."

196 Grenfell, *What the Church Means to Me.*

197 The Rev. John Sidy, Methodist minister.

198 Ernest B. Doane (1866–1945) left Nova Scotia at an early age. He subsequently studied taxidermy and made a living by supplying specimens to scientific institutions and carrying the mail. This same year, Doane made the first recorded crossing of the Strait of Belle Isle on foot with a canvas boat and sled. See below, entry for 17 March.

199 Anne Grenfell was by this time two months pregnant. See below, entry for 23 October.

200 *Stereopticon:* an effect created by a double magic lantern so as to combine two images of the same object upon the screen.

201 Charles Haddon Spurgeon, *John Ploughman's Talks* (1869) or *John Ploughman's Pictures* (1880).

202 Also known as "Sheila's blush" or "Sheila's brush."

203 *Ballicatters:* ice in fragments formed on the foreshore by the action of tides raising and lowering it over shoals and reefs.

204 Grenfell also had other problems to contend with. He later wrote, "The next discovery was that the manager of the St. Anthony store [Mr Jones], who had been my summer secretary before, and was an exceedingly pious man – whose great zeal for cottage prayer meetings, and that form of religious work, had led me to think far too highly of him – had neglected his books. He had given credit to every one who came along (though it was a cardinal statute under his rules that no credit was to be allowed except at his own personal risk). The St. John's agent claimed that he had made a loss of twelve thousand dollars in a little over a year, in which he professed to have been able to pay ten per cent to shareholders and put by three hundred dollars to reserve" (Grenfell, *A Labrador Doctor,* 345–6).

205 *Sketch:* to photograph, as in the phrase "sketch off."

206 Jesse Halsey (1883–1954) graduated from Princeton Theological Seminary in 1908 and was subsequently awarded a doctorate in divinity by Wooster College, Ohio. Ordained to the Presbyterian ministry in 1910, he spent three years with the mission as a plumber. He was later pastor of the Seventh Presbyterian Church, Cincinnati, from 1913 to 1941, and

professor of Pastoral Theology, McCormick Theological Seminary, Chicago, from 1940 to 1952.

207 Edwin Nicholas Hancock (1910–1974?), son of William and Bessie Hancock. At the age of fifteen, Hancock wrote Miss Luther to express his interest in studying art and was awarded a scholarship by the International Grenfell Association for a year of study at Boston, with the condition that he teach for two years in the Industrial Department. Recognizing the quality of his artistic work, Miss Luther later found him a scholarship at RISD, and he received his diploma in 1934 after a four-year program in fabricated design in all materials, jewellery design, jewellery making, hub and die cutting, and tool making. See *Providence Sunday Journal*, 22 May 1938. Hancock was elected a craftsman of the Boston Society of Arts and Crafts with Miss Luther's sponsorship and in 1938 mounted an exhibition featuring sculptures of northern figures. He became a naturalized U.S. citizen, serving two and a half years overseas during the Second World War, and was an inspector at the Reece Corporation, Waltham, Massachusetts, making precision-built buttonhole and special-purpose sewing machines until his retirement in 1971.

208 Emily Anderson Fowler (1880–1970), housekeeper in 1910, returned to take charge of the orphanage, 1913–14. She went to France as a nurse in 1916 and returned to administer the clothing store in St Anthony, 1923–24. In the U.S., she remained active as a member of the Grenfell Association of America.

209 Robert Burns, "To a Mouse" (1786): "The best laid schemes o' *Mice* an' *Men*, / Gang aft agley, / An' lea'e nought but grief an' pain, / For promis'd joy!"

210 SS *Invermore*, 250 feet and 975 gross tons, built at Glasgow in 1881, was acquired by the Reid Newfoundland Company in 1909 as part of the Alphabet Fleet, replacing the *Virginia Lake*. *Invermore* provided a fortnightly passenger, cargo, and mail service for the next two years.

211 SS *Harlaw*, 165 feet, 451 gross tons, built in Port Glasgow, Scotland, in 1881 and owned by William A. Black of Halifax, Nova Scotia, was lost at sea in 1911.

212 *Yale*, a 44-foot ketch, was built by Rice Brothers, East Boothbay, Maine, from the efforts of Yale undergraduates. Equipped with a ten-horsepower engine, it was to be put at the disposal of the medical services in Hamilton Inlet, Labrador. See Rowland, *North to Adventure*, 83–96.

213 Robert Fowler, nose and throat specialist and brother of Emily Anderson Fowler.

214 To venture beyond Cape Mugford, which lies roughly halfway between Okak and Hebron, was unusual for Newfoundlanders, for it was regarded as the northern limit of the floater fishery.

215 James C. Janney (d. 1956), who volunteered during the summers 1909–11, qualified in medicine and returned in 1914, when he met and married Maud Alexander, one of the Grenfell nurses.

216 Ethan Flagg Butler (1884–1964) graduated with an AB from Princeton in 1906 and an MD from Johns Hopkins in 1910, after which he spent three summers with the Grenfell Mission. He was subsequently director of American Red Cross in Serbia, 1914–15, and served with the USAMC, 1917–20. He was with the U.S. Public Health Service as chief, Polyclinic Hospital, New York City, 1920; director of chest services, Arnot-Ogden Memorial Hospital, Elmira, N.Y., 1929–36; chief thoracic surgeon, New York State District Tuberculosis Hospital, Ithaca, N.Y., 1936–43; medical director for New York State, Veterans Administration Branch, 1946; and medical director, Buffalo Veterans Hospital, 1949. He retired in 1954. See Ethan Flagg Butler, "Labrador," in *Triennial Record of the Class of Nineteen Hundred and Six* (Princeton 1909), 129–34.

217 Annie L. Whitten of Holyoke, Massachusetts, housekeeper at Battle Harbour during the summers, 1910–11, and matron at St. Anthony, 1911–12.

218 George Van Gorder of Pittsburgh volunteered for the summers of 1909, 1910, and 1917. He later qualified in medicine.

219 Louis F. Fallon (1891–1936) of Wayne, Pennsylvania, educated at the William Penn Charter School, Philadelphia, and by then a medical student of the University of Pennsylvania. Fallon spent five seasons at St Anthony and stayed one winter. He was resident in charge of the dispensary, University Settlement House, Philadelphia, 1915–16, and served in France as a surgeon with the USAMC, 1917–19. He ultimately settled at St John's, Newfoundland.

220 Baine, Johnston & Company, St John's, acquired the Slade premises at Battle Harbour in 1871 and shipped Labrador fish to international markets. Its manager, Walter Baine Grieve, was an early supporter of Grenfell. He donated the first hospital in Battle Harbour in 1893.

221 SS *Susu*, a three-masted, steam-driven merchant vessel, 135 feet and 280 gross tons, built at Port Glasgow, Scotland, 1893.

222 Probably Captain Jessie T. Winsor (1874–1933) of Wesleyville, Newfoundland.

223 Captain Moses Bartlett (1857–1937).

224 Captain Robert Abram Bartlett (1875–1946) accompanied Peary on three attempts to reach the North Pole, the second and third in command of the *Roosevelt.*

225 Samuel E. Kean (1883–1937), son of Captain Abram Kean of the *Prospero.*

226 Samuel Langhorne Clemens (1835–1910), American humorist.

227 Jerome K. Jerome (1859–1928), British humorist.

228 *Loom:* glow over a distant icefield, also known as *ice-loom* or *ice-glim.*

229 *American man:* a pile of rocks or cairn erected as an aid to navigation, possibly a corruption of "markin' man."
230 Theodora Child, nurse in Indian Harbour during the summers of 1910 and 1911, married Warland Wight (see n191), who eventually went into the real estate business in Boston.
231 Miss Luther seems unaware that Edith (Fillmore) Hancock (d. 1955) was the nurse in charge in 1910.
232 Kirkina, an Inuit whose feet were cut off by her father, was brought to the hospital at St Anthony and fitted with a pair of prosthetics. She later spent two years in New York, where she was educated in the public schools. See *ADSF* (January 1904): 22–5; *ADSF* (January 1909): 10; Grenfell, *A Labrador Doctor*, 245.
233 SS *Stella Maris*, 250-ton wooden vessel, a refitted British gunboat.
234 SS *Pelican*, 290 tons, a gunboat built at HM Shipyard, Devonport, and launched in 1877. It was bought from the Admiralty by the HBC in 1901 and converted for cargo, then used until 1916 to supply northern posts and transport furs. J.S.C. Watt wrote, "The *Pelican* crew were a remarkable crowd – mostly all young jolly fellows, if I except the boatswain Mr. McPhail, who was a regular old-timer that might have been shipmate with Long John Silver. The ship had a proper band with a conductor who wielded a baton – 'The *Pelican*'s Foo Foo Band' painted on the big drum" (Watt, "Labrador Year," 23).
235 *Bully boats:* two-masted, decked boats used for fishing or transporting fish.
236 Mrs Swaffield, wife of the HBC agent William E. Swaffield (see n242).
237 Part of the transatlantic wireless telegraphy network of Guglielmo Marconi (1874–1937), who had received the first signal at St John's in 1901.
238 *Sister Annes:* held captive like Sister Anne in the folktale of Bluebeard.
239 Matthew 5–7.
240 *Greens* may also refer to turnip tops.
241 John Tilghman ("Pete") Rowland (b. 1888) of Greenwich, Connecticut, had been educated at Andover. He enrolled in the U.S. Navy on 6 April 1917, the day war was declared, and again served during the Second World War in the rank of commander. He was subsequently a designer and builder of small boats at Newcastle, Maine, and a journalist on maritime affairs.
242 William E. Swaffield (1867–1952) of Westholme, Dorset, had been in Labrador with the Hudson's Bay Company since 1891. He was at Cartwright, 1901–11, and served as district manager at Rigolet and Cartwright, 1911–18. Swaffield and his wife raised eight children, one of whom was named Wilfred, born in 1897.
243 The Grand River Pulp and Lumber Co. (Dickey Lumber Co.) had begun logging operations in Hamilton Inlet in 1902.
244 The house of Captain William J. Bartlett (1851–1931).

245 Berthold August Lenz (1873–1960) served in Labrador, 1899–1932, as missionary and trader in Nain, Hopedale, Okak, Killinek, and Makkovik. In 1900 he married Ingeborg Margarete Jannasch.

246 Payne, formerly of the CMS, probably worked with the ecumenical Moravian Church through what was called *doppelmitgeliedschaft* ("double membership").

247 Wilhelm Michael Merklein (1882–c.1959), Moravian trader in Killinek and Hopedale, 1906–24.

248 The Society for the Furtherance of the Gospel, the British branch of the Moravian Mission, which held a monopoly on trade with the Inuit.

249 *Harmony*, 403 gross tons, the fifth and last Moravian supply vessel so named, was originally the barque *Lorna Doone*, built at Dundee in 1876 and subsequently fitted with auxiliary steam power. The ship was purchased in 1901 and sold to the HBC in 1926, when the Moravian Church leased its trading stations to the HBC for twenty-one years and accepted the company's offer to purchase it. It was then registered as *Bayharmony* but was sold for scrap in 1927.

250 Hermann Theodor Jannasch (1849–1931) came to Labrador in 1879 and remained until 1903.

251 For further information on this style of dress, see Hahn and Reichel, *Zinzendorf*, 255.

252 George Plummer Howe (1879–1917) of Lawrence, Massachusetts, educated at St Paul's School, Concord, New Hampshire, and a graduate of the Harvard Medical School in 1904. Howe was an experienced northern explorer. After serving two years at the Boston City Hospital as house surgeon, he joined the Anglo-American Polar Expedition led by Ernest de Koven Leffingwell and Ejnar Mikkelsen in 1906. See Howe's medical notes in Mikkelson, *Conquering the Arctic Ice*, 451–5. After practising medicine in Concord for two years, he returned to Harvard to study archaeology and organized an expedition to the Yucatan in 1911. In 1917, as a volunteer with the AMRC, he was killed in action.

253 William Brooks Cabot (1858–1949), American engineer and scion of a banking family of Brattleboro, Vermont, pursued his special interest in the north, including eight summers, 1903–10, among the Labrador Innu. In 1903 he was invited by Leonidas Hubbard to join the exploration of the George River on which Hubbard died of starvation, but he declined. On this last expedition in 1910, Cabot had brought his party across Labrador to Indian House Lake and back. See Cabot, *In Northern Labrador*, 264–84.

254 Donald Baxter Macmillan (1874–1970), American explorer, had gone with Peary to the North Pole in 1908–9. He later designed the schooner *Bowdoin*, in which he engaged in twenty-six scientific voyages of exploration in various northern locations, 1921–54.

255 The Rev. S.M. Stewart, Continental and Colonial Church Missionary Society missionary to the Inuit of Ungava. See "To Labrador on the 'Harmony,'" *Periodical Accounts Relating to Moravian Missions* 8, no. 85 (1911): 16–18.

256 John C. Jackson, a member of the Leominster congregation, took command of *Harmony* in 1902 and remained in command until 1926.

257 Wilson Jacque (1886–1952), son of Joseph Jacque, was born at Kippokuk, Labrador, and was enrolled in the Machine Construction Course at the Pratt Institute, 1909–10. After eight years of employment with the Grenfell Mission in St Anthony, he emigrated to Thorold, Ontario, where he worked for the Provincial Paper Company. See "Jacques of Labrador," *Them Days Magazine* 21 (1996): 20–5; and *Thorold Standard*, 26 May 1952.

258 Perhaps Stanley Haveland Martin (1890–1941), a rare Newfoundland volunteer, educated at the Methodist College, St John's. Martin entered the medical school of Queen's University in 1911 and graduated in 1915, during which time he spent his summers with the mission. After a year's internship at the Montreal General Hospital, he spent a career as a medical missionary in Korea.

259 Christopher C.J. Shepherd (1887–1913), HBC clerk at Cartwright, had been in Labrador since 1906. He was later clerk at Wolstenholme, Hudson's Straits, from 1911 and drowned in 1913.

260 According to the HBC records, Grenfell sold three silver foxes for $450; that is, $150 or £30 per skin. Without knowing the quality of the pelts, it is difficult to say who got the better of the deal, for buyers had to base their prices on what furs had been realizing at auction. A statistical chart prepared by the HBC, displaying the average price of silver fox, 1901–13, shows that the price reached a peak of £85 in 1910 and declined to £61 in 1911. See HBC Records, HBC Archives, A63/8, f. 108.

261 Edward Patrick Morris (1859–1935), prime minister of Newfoundland, 1909–17, created Baron Morris in 1918.

262 Albert Henry George, 4th Earl Grey of Howick (1851–1917), had visited St Anthony in the summer of 1909.

263 Leopold Amery (1873–1955), *Times* editorialist and correspondent, had toured Canada with Earl Grey between July and October 1909. After touring the Moravian stations, he wrote, "I won great popularity as the possessor of quite a stock of old safety razor blades and was given a pair of Eskimo fishing waders, of thin scraped sealskin, feather light, and so finely sown [sic] as to be absolutely watertight. I first put my new treasures in my cabin, but a few hours later thought they had better stand in the passage outside ... They were lovely things, but for sheer, quiet *crescendo* of odoriforousness I have not come across their equal" (Amery, *Days of Fresh Air*, 294). Amery was elected to the House of Commons as a Conservative, 1911–45, and served in the cabinets of Lloyd George, Bonar Law, Baldwin, and Churchill.

264 Ralph Champneys Williams (1848–1927), governor of Newfoundland, 1909–13. His tribute to the Grenfell Mission appears in *How I Became a Governor* (London: John Murray 1913), 434–8.

265 Marjorie Wakefield (1886–1976) held a degree in French from McGill University and took a year's training in dentistry before arriving on the coast.

266 John K.H. Nightingale Jr (1890–1969), son of John K.H. Nightingale, an executive of the Rhode Island Hospital Trust Company, and Maria (Moale) Nightingale, Warwick Neck, Rhode Island. Nightingale *fils* graduated from St George's School, Middletown, in 1909 before going to St Anthony. Leaving Williams College in 1913, he entered his father's firm and later served out his career as a stockbroker in Providence. Nightingale was better known, however, as a strong advocate of "home rule," or improved local government, and in 1951 the General Assembly passed a home rule amendment to the Rhode Island Constitution.

267 Wilfred Thomason Grenfell Jr (1910–95).

268 Henry Locke Paddon (1881–1939) declared his intention of becoming a missionary when he heard Grenfell speak at Repton. After graduating with a BA from Oxford in 1906, he studied medicine at St Thomas's Hospital, London, qualifying MRCS, LRCP in 1911, then sailed in the North Sea with the RNMDSF before arriving at Indian Harbour, Labrador, in 1912. That summer, he met Mina Gilchrist (see n269), and the two were married the following year. In 1915 they occupied the first cottage hospital in North West River, which became their home and that of their four sons. Paddon died prematurely in 1939 and was succeeded by his son, Dr. W. Anthony Paddon (see n291).

269 Mina (Gilchrist) Paddon (1880–1967) of Fredericton, New Brunswick, a graduate of the Massachusetts General Hospital School of Nursing. She was made an OBE in the New Year's honours list of 1949 in recognition of twenty-five years of nursing, including her superintendency of the North West River station during the Second World War.

270 *George B. Cluett*, three-masted schooner, 135 feet and 180 tons, with a 75 horsepower engine, built by A.C. Brown and Sons, Tottenville, Staten Island, New York.

271 George B. Cluett (1838–1912) of Cluett, Peabody & Co., supporter of various religious and humanitarian efforts.

272 William R. Stirling (1852–1918), partner in the Chicago firm Peabody, Houghteling & Co. and formerly vice-president of the Illinois Steel Co. Stirling was secretary of the Chicago Grenfell Association.

273 Harriot Houghteling (d. 1951) returned to St Anthony several years later as a volunteer and married Dr Charles S. Curtis (see n292).

274 Minnie Pike (1873–1943) established a weaving centre at Red Bay, Labrador. She was later sent to Berea College, Kentucky, to study weaving.

275 Alice Appleton of Cambridge, Massachusetts, a teacher in Forteau during the summer of 1911 and at St Anthony, 1911–13. She later married A.C. Blackburn (see n286).

276 Mather Hare (b. 1863) of Halifax, Nova Scotia, Grenfell's lieutenant at Harrington Harbour, Quebec, from 1905 until his resignation as a result of his wife's ill health and his own heart troubles in 1915, when he departed for Florida to take up farming. A graduate of Dalhousie University and New York University, where he qualified as a physician in 1889, he had seen service as a medical officer during the North-West Rebellion in 1885 and had spent seven years in charge of the Canadian Methodist Mission hospital, West China. He was staff interpreter to the 2nd Brigade, China Expeditionary Force, 1900.

277 William Wakeham (1844–1915), MD, CM, inspector of fisheries for the Gulf of St Lawrence and Labrador, graduated from McGill University in medicine in 1866 and practised in the Gaspé until appointed inspector in 1879 with the powers of a police commissioner. In this capacity, he was authorized to enforce maritime laws, monitor international fishing conventions, and issue fishing permits.

278 George Carpenter Whiteley, OBE (1874–1961), fish merchant, son of William Henry Whiteley, one of Grenfell's earliest supporters.

279 Arthur, duke of Connaught and Strathearn (1850–1942), governor-general of Canada, 1911–16.

280 Joseph A. Andrews (d. 1928), ophthalmic surgeon, who was forced by ill health to give up his New York practice and retire to Santa Barbara. At the time of his death, he had spent sixteen summers in Labrador at his own expense.

281 Fred Coleman Sears (1866–1949), head of the Department of Pomology at the Massachusetts Agricultural College, advised Grenfell on agricultural matters during the summers 1928–39.

282 Cecil Spanton Ashdown (1875–1948), British-born public accountant sent by Price, Waterhouse to run their branch office in New York, became interested in the Grenfell Mission. During the First World War, he served as a civilian volunteer with the US Navy – one of the so-called Dollar-a-Year Men. He joined the Remington Typewriter Co. as comptroller and vice-president in 1920, but ill health compelled him to retire to Bermuda in 1928. A founding member of the IGA, he became executive director in 1936 when Grenfell resigned as superintendent.

283 Albert T. Gould (1885–1947) of Thomaston, Maine, was educated at Phillips Academy, Andover. He graduated *summa cum laude* from Bowdoin College in 1908 and from Harvard Law School in 1911. While at Bowdoin, he was a student volunteer at St. Anthony and secretary aboard the *Strathcona*. Gould later became an international authority in admiralty law. He was president of the New England Grenfell Association, 1926–38.

284 Richard Clipston Sturgis Jr, volunteer at St Anthony during the summer of 1913, at the end of his sophomore year at Harvard. He died suddenly of a cerebral haemorrhage in 1913, four days after returning home.

285 Katie Spalding (d. 1954) assisted Miss Storr at the St Anthony orphanage. After the First World War, she spent three years in Russia doing relief work with the Society of Friends, then returned to Forteau and to the St Anthony orphanage as superintendent. She wrote with Anne Grenfell *Le Petit Nord: Annals of a Labrador Harbour* (London: Hodder & Stoughton [1921]), a record of her first winter. When Grenfell opened a London office in 1926, she ran it as a volunteer, and during the Second World War she kept it open during the Blitz.

286 A.C. Blackburn (d. 1971), bookkeeper at St Anthony, 1912–28.

287 *Wop:* there is disagreement about the use of the term in the Grenfell Mission. American volunteers doing hard physical labour are said to have borrowed this familiar slur against Italian immigrants, prompting Grenfell to invent the acronym.

288 Leverett Saltonstall (1892–1979), governor of Massachusetts, 1938–45, then U.S. senator until 1967.

289 Both Nelson Rockefeller (1908–79) and Laurence Rockefeller (1910–) worked aboard the Mission vessel *Maraval* with Dr Harry Paddon during the summer of 1929.

290 Clarence Birdseye (1886–1956), American pioneer in the development of quick-frozen and dehydrated food, benefited from his visits to Labrador as a trapper and fur farmer, 1912–17. Installed at Muddy Bay in 1916, he apparently learned the method of quick-freezing fish from Garland Lethbridge, patented the process, and later sold the rights to General Foods.

291 W.A. "Tony" Paddon (1914–95) was born in Labrador and sent to the Lenox School in Lenox, Massachusetts. After graduating from Trinity College, Hartford, Connecticut, he trained at the Long Island College of Medicine, whose president was Dr Frank Babbott, his father's first medical assistant in Labrador. After serving as a medical officer of the Royal Canadian Naval Volunteer Reserve during the Second World War, Tony Paddon took charge of the Grenfell Mission's medical service at North West River until his retirement in 1977. He was lieutenant-governor of Newfoundland, 1981–86.

292 Charles S. Curtis (1890–1963), a graduate of Clark University and the Harvard Medical School, arrived on the coast in 1915 and took charge of the hospital at St Anthony in 1917. In 1928 he married Harriot P. Houghteling, a friend of Anne Grenfell, and when Grenfell resigned in 1934 he was appointed superintendent of the IGA. He was elected chairman of its board of directors in 1953. For his considerable public service, he was made a CBE in 1948.

Bibliography

MANUSCRIPT SOURCES

Blumer, Dr G. Alder, Papers. Isaac Ray Library, Butler Hospital, Providence, Rhode Island

Bronson, Elsie S. "The Rhode Island School of Design: A Half Century of Record (1878–1928)." Unpublished typescript, n.d. Rhode Island School of Design, Providence, Rhode Island

Canadian Handicrafts Guild Records. Canadian Guild of Crafts Quebec, Montreal, Quebec

Census of Newfoundland and Labrador, 1921. St John's 1923

Cole, Hugh. [Diary of reindeer trek down the Northern Peninsula of Newfoundland, 4 March to 30 April 1908]. Memorial University of Newfoundland, Folklore Archive, MS 77–61

Crowdy, Dr Frederic Hamilton. "1873 Labrador" [diary of voyage to Labrador, 17 June to 9 September 1873, in the schooner *William Stairs*]. Provincial Archives of Newfoundland and Labrador, St John's, MG 970

État des pêches françaises sur les bancs et à Saint-Pierre. Archives nationales, Paris

Fonds des colonies. Archives nationales, Paris

General Register of Marriages, Burials and Baptisms, 1880–1940. Records for the Church of England Mission of Griquet/St Anthony. Provincial Archives of Newfoundland and Labrador, St John's

Grenfell, Wilfred Thomason. Papers. Sterling Library, Yale University, New Haven, Connecticut

Hudson's Bay Company Records. Hudson's Bay Company Archives, Winnipeg, Manitoba

Little, John Mason. Correspondence. Thomas M. Smith, Cambridge, Massachusetts

Luther, Jessie. Papers. Martha Gendron, Swansea, Massachusetts

Moore, Rev. Edward C. Papers. Andover-Harvard Theological Library, Harvard University, Cambridge, Massachusetts

Registrar General of Shipping and Seamen, Certificates of Vesses Registry. Maritime History Archives, Memorial University of Newfoundland

RNMDSF Minutes and Records. Royal National Mission to Deep Sea Fishermen, London, England

RNMDSF Staff Records. Grenfell Regional Health Services, St Anthony, Newfoundland

RNMDSF Office Records, St John's. Sterling Library, Yale University, New Haven

St Anthony Methodist Church Baptismal Register. 1907–24. Newfoundland and Labrador Conference Archives, United Church of Canada, St John's

St Anthony Methodist Church Register of Deaths. 1907–80. Newfoundland and Labrador Conference Archives, United Church of Canada, St John's

Slocum, Grace L., ed. "Who's Who in Rhode Island Art: A Supplement to *Two Centuries of Art in Rhode Island, 1782–1928.* Unpublished typescript, 1945. Rhode Island School of Design, Providence, Rhode Island

Society of Arts and Crafts, Boston, Archives. Boston Public Library, Boston, Massachusetts

Stirling, W.R. Papers. Sterling Library, Yale University, New Haven, Connecticut

PRINTED SOURCES

Addams, Jane. *Twenty Years at Hull-House.* New York: Macmillan 1910

Albee, Helen R. "A Profitable Philanthropy." *Review of Reviews* 22 (1900): 57–60

– *Abnákee Rugs: A Manual Describing the Abnákee Industry, the Methods Used, with Instructions for Dyeing.* 2nd edn. Cambridge: Riverside Press 1903

Allaire, Abbé J.-B.-A. *Dictionnaire biographique du clergé canadien-français.* St-Hyacinthe: Impriméerie de "La Tribune" 1908

Allen, Mary E. "Handicrafts in Old Deerfield." *Outlook* 69 (1901): 592–7

Amery, Rt. Hon. L.S., MP *Days of Fresh Air, Being Reminiscences of Outdoor Life.* London: Jarrolds 1939

Among the Deep Sea Fishers (Toronto and Boston: International Grenfell Association). Vols. 1–78. 1903–81

Annett, Albert, and Alice E.E. Lehtinen. *History of Jaffrey (Middle Monadnock) New Hampshire.* 3 vols. Jaffrey: Published by the Town 1934–71

Baggs, Arthur E. "The Story of a Potter." *Handicrafter* 1, no. 4 (1929): 8–10

Baxter, Sylvester. "The Movement for Village Industries." *Handicraft* 1 (1902): 145–65

Beardsall, Sandra. "Methodist Religious Practices in Newfoundland." PHD dissertation, University of Toronto 1996

Bénézit, E. *Dictionnaire critique et documentaire des peintres, sculpteurs, dessinateurs et graveurs.* 10 vols. Paris: Librairie Gründ 1976

Bernet, Etienne. *Bibliographie francophone de la grande pêche: Terre-Neuve – Islande – Groenland.* Fécamp: Musée des Terre-Neuvas de la pêche 1998

Bicknell, Thomas Williams. *The History of the State of Rhode Island and Providence Plantations.* 5 vols. New York: American Historical Society 1920

Biles, Wesley. *My Town, My Province, My Country and I.* St Anthony: Bebb [1998]

Bliss, William D.P., and Rudolph M. Binder, eds. *The New Encyclopedia of Social Reform.* 3rd edn. New York and London: Funk & Wagnalls 1910

Blumer, G. Alder, and A.B. Richardson, eds. *Commitment, Detention, Care and Treatment of the Insane.* Baltimore: Johns Hopkins Press 1894

Boris, Eileen. *Art and Labor: Ruskin, Morris, and the Craftsman Ideal in America.* Philadelphia: Temple University Press 1986

Brandt, Beverly Kay. "Mutually Helpful Relations: Architects, Craftsmen, and the Society of Arts and Crafts, Boston, 1897–1917." MA dissertation, Boston University 1985

Brière, Jean-François. "Pêche et politique à Terre-Neuve au XVIIIe siècle: La France véritable gagnante du traité d'Utrecht?" *Canadian Historical Review* 64 (1983): 168–87

Brown, Linwood L. "History of the Boston Office," *ADSF* (April 1973): 1–6

Bunkers, Suzanne L., and Cynthia A. Huff, eds. *Inscribing the Daily: Critical Essays on Women's Diaries.* Amherst: University of Massachusetts Press 1996

Burford, Gayle, ed. *The Ties That Bind.* St John's: Jesperson Press 1997

Byers, Mel. *The Design Encyclopedia.* New York: John Wiley 1994

Cabot, William Brooks. *In Northern Labrador.* Boston: Gorham Press 1912

Carson, Mina. *Settlement Folk: Social Thought and the American Settlement Movement, 1885–1930.* Chicago: University of Chicago Press 1990

Cartier, Jacques. *Relations.* Édition critique par Michel Bideaux. Montreal: Presses de l'Université de Montréal 1986

Catalogue and Circular of the College for Women. Columbia, S.C.: Bryan Printing [1893]

Census of Newfoundland. St John's 1858, 1884, 1891, 1901, 1914

A Century of Butler Hospital, 1844–1944. Providence 1944

Chimmo, William. *Journal of a Voyage to the N.E. Coast of Labrador during the Year 1867.* Edited by William Kirwin. St John's 1989

Cohen, Norm. *Long Steel Rail: The Railroad in American Folksong.* Urbana: University of Illinois Press 1981

Cook, Frederick Albert. *My Attainment of the Pole.* New York and London: Mitchell Kennerley 1913

Crellin, John K. *Home Medicine: The Newfoundland Experience.* Montreal: McGill-Queen's University Press 1994

Dana, Ethel C. "Occupational Therapy in Labrador." In *U.S. Veterans' Bureau Medical Bulletin* (Washington), (1929): 392–3

Directory of Registered Occupational Therapists. New York: American Occupational Therapy Association 1937

Eaton, Allen H. *Handicrafts of New England.* New York: Harper 1949

Elliott, Noel Montgomery. *The Atlantic Canadians, 1600–1900*. 3 vols. Toronto: Genealogical Research Library 1994

Falk, Peter Hastings, ed. *Who Was Who in American Art*. Madison, Conn.: Sound View Press 1985

[Feild, Edward] *Journal of a Voyage of Visitation ... in the Year 1849*. London: Society for the Propagation of Christian Knowledge 1850

Fleming, Archibald Lang. *Archibald the Arctic*. New York: Appleton-Century-Crofts 1956

Fogarty, Robert S. *All Things New: American Communes and Utopian Movements, 1860–1914*. Chicago: University of Chicago Press 1990

Gilbert, W.S. *Plays*. Edited by George Rowell. Cambridge: University Press 1982

Gobineau, Joseph Arthur de. *Voyage à Terre-Neuve*. Paris: Hachette 1861

– *A Gentleman in the Outports: Gobineau and Newfoundland*. Edited and translated by Michael Wilkshire. Ottawa: Carleton University Press 1993

Grant, Douglas R. *Quaternary Geology of St Anthony–Blanc Sablon Area, Newfoundland and Quebec*. Geological Survey of Canada Memoir 427. Ottawa: Minister of Supply and Services Canada 1992

Grenfell, Lady [Anne]. "The Industrial Effort." *ADSF* (October 1930): 99–105

Grenfell, W.T. *What the Church Means to Me*. Boston: Pilgrim Press 1911

– "A New Industry for Labrador Women." *Toilers* 28 (1913): 153

– "The St Anthony Mat Industry." *Toilers* 31 (1916): 124–5

– *A Labrador Doctor*. London: Hodder and Stoughton [1919]

Grenfell, W.T., et al. *Labrador: The Country and the People*. New York: Macmillan 1909

Hahn, Hans-Christoph, and Hellmut Reichel. *Zinzendorf und die Herrnhuter Brüder: Quellen zur Geschichte der Brüder-Unität von 1722 bis 1760*. Hamburg: Friedrich Wittig 1977

Hall, Herbert J. "The Systematic Use of Work as a Remedy in Neurasthenia and Allied Conditions." *Boston Medical and Surgical Journal* 152 (1905): 29–32

– "Work as a Remedy for Nervous Exhaustion." *Good Housekeeping* (October 1905): 351–5.

– "Marblehead Pottery." *Keramic Studio* 10 (June 1908): 30–1

– *Bedside and Wheel-Chair Occupations*. New York: Red Cross Institute for Crippled and Disabled Men 1919

– *O.T.: A New Profession*. Concord: Rumford Press 1923

Hall, Herbert J., and Mertice M.C. Buck. *The Work of Our Hands: A Study of Occupations for Invalids*. New York: Moffat, Yard 1915

– *Handicrafts for the Handicapped*. New York: Moffat, Yard 1916

Halpert, Herbert, and G.M. Story, eds. *Christmas Mumming in Newfoundland: Essays in Anthropology, Folklore, and History*. Toronto: University of Toronto Press 1969

Hamer, Frank, and Janet Hamer. *The Potter's Dictionary of Materials and Techniques*. 3rd edn. London: A. & C. Black; Philadelphia: University of Pennsylvania Press 1991

Hare, F. Kenneth. "The Climate of the Island of New foundland: A Geographical Analysis." *Geographical Bulletin* 2 (1952): 36–88

Harrisse, Henry. *Découverte et évolution cartographique de Terre-Neuve et des pays circonvoisins, 1497–1501–1769.* London: Stevens & Stiles; Paris: Welter 1900

Hiller, James K. "The 1904 Anglo-French Newfoundland Fisheries Convention: Another Look." *Acadiensis* 25, no. 1 (1995): 82–98

– "The Newfoundland Fisheries Issue in Anglo-French Treaties, 1713–1904." *Journal of Imperial and Commonwealth History* 24 (1996): 1–23

Himmelfarb, Gertrude. *Poverty and Compassion: The Moral Imagination of the Late Victorians.* New York: Knopf 1991

Hofstadter, Richard. *The Age of Reform from Brian to F.D.R.* New York: Knopf 1955

Holland, Clive. *Arctic Exploration and Development, c. 500 B.C. to 1915: An Encyclopedia.* New York: Garland 1994

Hoppin, Laura Brackett, ed. *History of the World War Reconstruction Aides: Being an Account of the Activities and Whereabouts of Physiotheraphy and Occupational Therapy Aides Who Served in U.S. Army Hospitals in the United States and in France during the World War.* Millbrook, N.Y.: William Tyldsley 1933

Horowitz, Helen Lefkowitz. *Culture and the City: Cultural Philanthropy in Chicago from the 1880s to 1917.* Lexington: University Press of Kentucky 1976

Hull-House Bulletin. Vols. 1–7. 1896–1906

"The Hull-House Labor Museum." *Chautauquan* 38 (1903): 60–1

Jacobs, Jane, ed. *A Schoolteacher in Old Alaska: The Story of Hannah Breece.* Toronto: Random House 1955

Jarvis, J.R. "Laplanders for Labrador." *Toilers* 23 (1908): 255–6

Journal of the House of Assembly of Newfoundland. St John's: House of Assembly 1833–1933.

Kaplan, Wendy. *"The Art That Is Life": The Arts and Crafts Movement in America, 1875–1920.* Boston: Museum of Fine Arts 1987

Karp, Ivan, and Steven D. Lavine, eds. *Exhibiting Cultures: The Poetics and Politics of Museum Display.* Washington: Smithsonian Institution Press 1991

[King, W.L. Mackenzie] "Co-operative Stores among Atlantic Fishermen." *Labour Gazette* 3 (1903): 680–1

Kirwin, William J. "Selected French and English Fisheries Synonyms in Newfoundland." *Regional Language Studies: Newfoundland* 9 (1980): 10–21

Laframboise, Lisa. "Travellers in Skirts." PHD dissertation, University of Alberta 1997

Laut, Agnes C. "Cruising on the French Treaty Shore of Newfoundland." *Westminster Review* 151 (1899): 381–7

Lavaud, Charles. "Instructions pour naviguer sur la côte orientale de l'île de Terre-Neuve, depuis le cap de Bonavista jusqu'au cap Normand, à l'entrée du détroit de Belle-Isle." *Annales maritimes et coloniales* 2 (1837): 1231–86

Laverty, Paula. "Hooked Mats of the Grenfell Mission." *Piecework* 4, no. 6 (1996): 56–9

Lawrence, Karen R. *Penelope Voyages: Women and Travel in the British Literary Tradition.* Ithaca: Cornell University Press 1994

Lears, T.J. Jackson. *No Place of Grace: Antimodernism and the Transformation of American Culture, 1880–1920.* Chicago: University of Chicago Press 1983

Lewis, Arthur. *The Life and Work of E.J. Peck among the Eskimos.* London: Armstrong 1904

Little, Mrs John M. [Ruth Keese]. "Jessie Luther" [obituary]. *ADSF* (January 1953): 114–15

Lovell's Province of Newfoundland Directory for 1871. Montreal 1871

Luther, Jessie. "The Labor Museum at Hull House." *The Commons: A Monthly Record Devoted to Aspects of Life and Labor from the Settlement Point of View* 7 (May 1902): 1–13

– "Hand-weaving." *House Beautiful*: I Materials Required (May 1906): 39–41; II Small Looms and Simple Patterns (June 1906): 31–4; III Large Looms (July 1906): 26–8

– "Development of the Industrial Work in Dr Grenfell's Mission at St Anthony." *ADSF* (January 1907): 11–14

– "Grenfell's Reindeer Land." *Boston Evening Transcript,* 25 January 1908. Reprinted as "The Landing of the Reindeer." *ADSF* (April 1908): 19–22

– "The Story of Dr Grenfell's Rescue [as told by George Andrews]." *ADSF* (January 1909): 19–20

– "Industrial Development at St Anthony." *ADSF* (April 1909): 27–33

– "Industrial Work." *ADSF* (January 1914): 21–7

– "The Industrial Work." *ADSF* (April 1915): 3–11

– "After Ten Years." *ADSF* (April 1916): 4–9

– "Hooked Mats: How a Native Handicraft of the Women of Newfoundland and Labrador Was Placed on a Paying Basis." *House Beautiful* 14 (1916): 78–9, 106

– "Dr Little at Work." *ADSF* (January 1917): 128–32

– "Occupational Treatment in Nervous Disorders." *Modern Hospital* (July 1918): 11–15

– "In Retrospect." *ADSF* (October 1930): 112–24

– "Early Days." *ADSF* (April 1942): 4–5

[Lynch, Colleen] *Helping Ourselves: Crafts of the Grenfell Mission.* St John's: Newfoundland Museum 1985

McAlpine's St John's Directory. Halifax: McAlpine 1906–10

McGrath, P.T. "France in Newfoundland." *Nineteenth Century* 45 (1899): 46–55

McLeod, Ellen. *In Good Hands: The Women of the Canadian Handicrafts Guild.* Montreal: McGill-Queen's University Press 1999

Mannion, John J., ed. *The Peopling of Newfoundland: Essays in Historical Geography.* St John's: Institute of Social and Historical Research 1977

The Medical Directory 1912. 68th edn. London: Churchill 1912

Messet, Georges. "Les rochelais à Terre-Neuve, particulièrement de 1523 à 1550." *Bulletin de la Société de géographie commerciale de Paris* 14 (1891–92): 137–45

Mikkelson, Ejnar. *Conquering the Arctic Ice.* London: Heinemann 1909

Miller, Timothy. *American Communes, 1860–1960: A Bibliography.* New York: Garland 1990

Mills, Sara. *Discourses of Difference: An Analysis of Women's Travel Writing and Colonialism.* London: Routledge 1991

Moore, Maude. "History of the College for Women, Columbia, South Carolina." MA dissertation, University of South Carolina 1932

Morandière, Charles de la. *Histoire de la pêche française de la morue dans l'Amérique septentrionale.* 3 vols. Paris: G.-P. Maisonneuve et Larose 1962–66

Murray, Alexander, and James P. Howley. *Reports of Geological Surveys of Newfoundland from 1881 to 1909.* St John's: Robinson 1918

National Directory of Qualified Occupational Therapists Enrolled in 1931 in the National Register. New York: American Occupational Therapy Association 1932

Neary, Peter. "The French and American Shore Questions as Factors in Newfoundland History." In James Hiller and Peter Neary, eds., *Newfoundland in the Nineteenth and Twentieth Centuries: Essays in Interpretation,* 95–122. Toronto: University of Toronto Press 1980

Northend, Mary H. "Where Women Work and Rest." *Craftsman* 8 (1905): 341–3

Opie, Iona, and Peter Opie, eds. *The Oxford Dictionary of Nursery Rhymes.* Oxford: University Press 1997

Peary, Robert. *The North Pole.* London: Hodder and Stoughton 1909

Peck, M.A. *Sketch of the Activities of the Canadian Handicraft Guild and of the Dawn of the Handicraft Movement in the Dominion.* Montreal: Canadian Handicraft Guild 1929

[Pierre, G.] *Plan du havre de Saint Antoine (côte nord est de Terre Neuve).* Levé en 1857 par Mr. G. Pierre, Lieutenant de vaisseau commandant la goëlette la "Fauvette." Paris 1860

Pocius, Gerald L. "Hooked Rugs in Newfoundland: The Representation of Social Structure in Design" *Journal of American Folklore* 92 (1979): 273–84

Pratt, Jane. "From Merton Abbey to Old Deerfield." *Craftsman* 5 (1903): 183–91

Providence City Directory. Providence and Boston 1824–

Providence Journal. Providence 1829–

Prowse, D.W., ed. *The Newfoundland Guide Book, 1905, Including Labrador and St Pierre.* London: Bardbury, Agnew [1905]

Quiroga, Virginia A.M. *Occupational Therapy: The First Thirty Years: 1900 to 1930.* Bethesda, Md: American Occupational Therapy Association 1995

Reynolds, Minnie J. "A Revival of Feminine Handicrafts." *Everybody's Magazine* (July 1902): 32–6

Rhode Island School of Design. Yearbooks. Providence, R.I. 1878–

Robertson, A.W., F.C. Pollett, and O.A. Olsen. *Peatland Flora of Newfoundland.* Information Report N-X-93. St John's: Newfoundland Forest Research Centre 1973

Rompkey, Ronald. "Elements of Spiritual Autobiography in Sir Wilfred Grenfell's *A Labrador Doctor.*" *Newfoundland Studies* 1 (1985): 17–28

– "Heroic Biography and the Life of Sir Wilfred Grenfell." *Prose Studies* 12 (1989): 159–73
– *Grenfell of Labrador: A Biography*. Toronto: University of Toronto Press 1991
– ed. *Labrador Odyssey: The Journal and Photographs of Eliot Curwen on the Second Voyage of Wilfred Grenfell, 1893*. Montreal and Kingston: McGill-Queen's University Press 1996
Rowland, John T. *North to Adventure*. New York: W.W. Norton 1963
Sailing Directions for the First Part of the North American Pilot. London 1794
Sargent, Irene. "Chinese Pots and Modern Faïence." *Craftsman* 4 (1903): 415–25
Sayre, Francis Bowes. *Glad Adventure*. New York: Macmillan 1957
– "The Grenfell Work: Fifty Years Ago and Now." *ADSF* (January 1960): 101–4
Schriber, Mary Suzanne. *Writing Home: American Women Abroad, 1830–1920*. Charlottesville: University Press of Virginia 1997
Seary, E.R. *Place Names of the Northern Peninsula*. Ed. Robert Hollett and William J. Kirwin. St John's: ISER Books 2000
South, G. Robin. *Biogeography and Ecology of the Island of Newfoundland*. The Hague–Boston–London: Dr W. Junk 1983
Spencer, Arnold Hale. *The Spencers from June 1842*. Blenheim, N.Z.: Privately printed 1988
Story, G.M., W.J. Kirwin, and J.D.A. Widdowson. *Dictionary of Newfoundland English*. Toronto: University of Toronto Press 1982
Thompson, Frederic F. *The French Shore Problem in Newfoundland: An Imperial Study*. Toronto: University of Toronto Press 1961
Thornton, Patricia. "Dynamic Equilibrium: Settlement, Population and Ecology in the Strait of Belle Isle, Newfoundland, 1840–1940." 2 vols. PHD dissertation, University of Aberdeen 1979
Thoulet, J. *Un voyage à Terre-Neuve*. Paris: Berger-Levrault 1891
Thuen, Trond. *Quest for Equity: Norway and the Saami Challenge*. St John's: ISER Books 1995
Tiemann, Isabel C., and Florence Murphy. "Occupational Therapy at Butler Hospital: A Point of View." *American Journal of Occupational Therapy* 1 (1947): 373–5
Toilers of the Deep: A Monthly Record of Work amongst Them. London: Royal National Mission to Deep Sea Fishermen. Vols. 1–107 1886–1994
Townsend, Charles Wendell. *Along the Labrador Coast*. Boston: Dana Estes 1907
Turgeon, Laurier G. "Colbert et la pêche française à Terre-Neuve." In *Un nouveau Colbert*, 255–73. Paris: Éditions CDU et Sedes réunis 1985
Úlehla, Karen Evans, ed. *The Society of Arts and Crafts, Boston: Exhibition Record, 1897–1927*. Boston: Boston Public Library 1981
Valverde, Mariana. *The Age of Light, Soap, and Water*. Toronto: McClelland & Stewart 1991
Wallace, Dillon. *The Lure of the Labrador Wild*. New York: Fleming H. Revell 1905

Washburne, Marion Foster. "A Labor Museum." *Craftsman* 6 (1904): 570–80

Watt, J.S.C. "Labrador Year." *Beaver* (June 1937): 20–9

Waugh, Elizabeth, and Edith Foley. *Collecting Hooked Rugs.* New York: Century 1927

Whisnant, David E. *All That Is Native and Fine: The Politics of Culture in an American Region.* Chapel Hill: University of North Carolina Press 1983

White, Hayden. *Tropics of Discourse: Essays in Cultural Criticism.* Baltimore: Johns Hopkins University Press, 1978

Who's Who in and from Newfoundland, 1927. St John's: R. Hibbs 1927

Williams, Blanche E. Wheeler. *Mary C. Wheeler: Leader in Art and Education.* Boston: Marshall Jones 1934

Wingate, Isabel R., ed. *Fairchild's Dictionary of Textiles.* New York: Fairchild 1967

Wood, Francis H. "Reindeer, Dr Grenfell, and Labrador." *Toilers* 22 (1907): 176–8

– "Reindeer and Labrador." *Toilers* 23 (1908): 41–3

Worcester, Elwood. *The Issues of Life.* New York: Moffat, Yard 1915

Wynne, Madeleine Yale. "The Exhibition of the Chicago Arts and Crafts Society." *House Beautiful* (January 1902): 125–30

Year Book and Almanac of Newfoundland. St John's: Queen's Printer 1906–10

Index